CHAPMAN & HALL/CRC
Monographs and Surveys in
Pure and Applied Mathematics 121

NONLINEAR HYPERBOLIC WAVES IN MULTI-DIMENSIONS

CHAPMAN & HALL/CRC
Monographs and Surveys in
Pure and Applied Mathematics 121

NONLINEAR HYPERBOLIC WAVES IN MULTI-DIMENSIONS

PHOOLAN PRASAD

CHAPMAN & HALL/CRC

Boca Raton London New York Washington, D.C.

Library of Congress Cataloging-in-Publication Data

Prasad, Phoolan.
Nonlinear hyperbolic waves in multi-dimensions / by Phoolan Prasad.
p. cm.— (Chapman & Hall/CRC monographs and surveys in pure and applied mathematics ; 121)
Includes bibliographical references and index.
ISBN 1-58488-072-4
1. Nonlinear wave equations. 2. Differential equations, Hyperbolic. I. Title. II. Series.

QA927 .P73 2001
531′.1133′01515355—dc21 2001017331

Direct all inquiries to CRC Press LLC, 2000 N.W. Corporate Blvd., Boca Raton, Florida 33431.

International Standard Book Number 1-58488-072-4
Library of Congress Card Number 2001017331
Printed in the United States of America 1 2 3 4 5 6 7 8 9 0
Printed on acid-free paper

About this book

The book introduces necessary mathematical ideas and develops a weakly nonlinear ray theory and a shock ray theory to determine the complete history of curved nonlinear wavefronts and shock fronts as they evolve. The book includes extensive numerical computations which traces these fronts in a compressible medium and shows that geometrical features of nonlinear wavefronts and shock fronts are topologically thc same.

Many new mathematical ideas and results of the theory of hyperbolic conservation laws are presented: bicharacteristic lemma in an extended form, kinematical conservation laws and the kink phenomenon, Huygens' method in general, Fermat's principle for a nonstationary medium, the stability of steady transonic flows, the resolution of a caustic and the corrugation stability of a front are some of the subjects discussed.

The first three chapters of the book, though they contain some new results, are intended to be an introduction to nonlinear waves in multi-dimensions. The remainder of the book contains recent research. The book will be useful to final year undergraduates in applied mathematics and researchers in applied mathematics, gas dynamics and wave propagation.

Preface

Propagation of a curved nonlinear wavefront is influenced simultaneously by two important physical processes: (i) Different points of the front travel with different speeds depending on the local amplitude leading to a longitudinal stretching of rays and (ii) A lateral deviation of rays is produced due to non-uniform distribution of the amplitude on the front. This book introduces all necessary mathematical concepts in the first three chapters, which also have some new results, and then develops mathematical models and methods to calculate the complete history of not only a curved nonlinear wavefront as it evolves in time but also of a shock front. The evolution of a shock front becomes far more complex due to an additional effect: The nonlinear waves ahead of the shock and behind it interact with the shock and modify its evolution.

Some exact solutions and extensive numerical computation with the model equations of the weakly nonlinear ray theory (WNLRT) and shock ray theory (SRT) show that the geometrical features of a nonlinear wavefront and a shock front are topologically the same but with one difference – a nonlinear wavefront is self-propagating whereas a shock front is not.

The material contained in this book is an outcome of collaborative work carried out mainly at the Indian Institute of Science, Bangalore. The author happily acknowledges his association with many contributors. Their work made it possible to achieve a self-contained development of the subject. The names of these collaborators appear in many places throughout the text and in references. The author sincerely thanks his student S. Baskar for his help in the preparation of the figures. Mrs. S. Jayashree typed and retyped hundreds of manuscript pages, drew many figures and transformed a hardly legible draft into camera ready pages. I sincerely thank her for this painstaking work.

Bangalore
January 2001

Phoolan Prasad

About the author

Phoolan Prasad has been at the Indian Institute of Science, Bangalore since 1965. He held the distinguished chair of Mysore Sales International Limited Professorship from 1993 to 1996 and was Chairman of the Department of Mathematics from 1996 to 2000. Prasad won many medals for proficiency in mathematics from Calcutta University from 1962 to 1964 while he was a bachelor of science (Honors) and master of science student. He was awarded the S. S. Bhatnagar Prize in 1983 by the Government of India for his research contribution in nonlinear waves. He is a Fellow of all three science academies in India: the National Academy of Sciences, the Indian Academy of Sciences and the Indian National Science Academy. He has played a very important role in the development of quality education in mathematics from school education to research. He organized training camps for the International Mathematical Olympiads and worked for Indian participation in the IMO which became a reality in 1989. Prasad edited *Nonlinear Waves in One-dimensional Dispersive Systems*, written by his teacher P. L. Bhatnagar in the Oxford Mathematical Monograph Series (1979) and co-authored with Professor (Mrs.) Renuka Ravindran *Partial Differential Equations* in 1985. Most of the material in his third book, *Propagation of a Curved Shock and Nonlinear Ray Theory* published by Longman Higher Education (1993), is based on the research work of his group. The author was a member of the National Board for Higher Mathematics, Department of Atomic Energy, Government of India, from 1986 to 1996. Presently, he is Vice-President of the Ramanujan Mathematical Society. He has written more than 65 research papers on mathematical theory of nonlinear waves and 20 articles on mathematics education.

Nonlinear Hyperbolic Waves in Multi-dimensions

Phoolan Prasad

Contents

Chapter 1

1	**An introduction to nonlinear hyperbolic waves**	1
1.1	A wave equation with genuine nonlinearity	1
1.2	Breakdown of a genuine solution	4
1.3	Conservation law and jump condition	7
1.4	Stability consideration, entropy condition and shocks	10
1.5	Some examples	16
1.6	Shock structure, dissipation and entropy condition	26
1.7	The persistence of a shock	33
1.8	Nonlinear wavefront and shock front	36
1.9	Hopf's result on the general solution	39
1.10	Equal area rule for shock fitting	41

Chapter 2

2	**Hyperbolic system - some basic results**	47
2.1	Hyperbolic system of first order equations in two independent variables	47
2.1.1	Definition of a hyperbolic system	47
2.1.2	A canonical form of a system of linear and semilinear equations	50
2.2	The wave equation in $m(>1)$ space dimensions	53
2.2.1	Space-like surface and time-like direction	54
2.2.2	Bicharacteristics and rays	56
2.2.3	Compatibility condition on a characteristic surface	58

2.2.4	Propagation of discontinuities in second order derivatives along rays	60
2.3	Hyperbolic system in more than two independent variables	62
2.3.1	Space-like surface and time-like direction	63
2.3.2	Explicit definition of a hyperbolic system	67
2.4	Bicharacteristic curves, rays and compatibility condition	69
2.5	Propagation of discontinuities of first order derivatives along rays	74

Chapter 3

3	**Simple wave, high frequency approximation and ray theory**	77
3.1	Simple wave	77
3.1.1	Example of a simple wave in gas dynamics	78
3.1.2	Simple wave in one space dimension	84
3.1.3	Simple wave in multi-dimensions	90
3.1.4	An initial value problem leading to a kth simple wave	94
3.2	High-frequency approximation, wavefront, Huygens' method and Fermat's principle	96
3.2.1	Definition of a wavefront	96
3.2.2	Huygens' method of wavefront construction	98
3.2.3	Huygens' method and ray theory	102
3.2.4	Fermat's principle	104
3.2.5	Fermat's principle in a stationary medium	106
3.2.6	Fermat's principle in a nonstationary medium	108
3.2.7	Weakly nonlinear ray theory (WNLRT) in an isotropic medium using Fermat's principle	110
3.3	Kinematics of a propagating curve	112
3.3.1	Caustic, wavefront folding and some other general properties	112
3.3.2	Ray coordinate system and kinematical conservation laws	117
3.3.3	Two types of singularities and jump conditions across a kink	121

3.3.4 Kinematical compatibility conditions on a surface of discontinuity in multi-dimensions 128
3.4 Breakdown of the continuity of a solution of a quasilinear system 128
3.4.1 Combined effect of genuine nonlinearity and geometrical divergence 129
3.4.2 Transport equation for discontinuities in derivatives for a system in multi-dimensions 133
3.5 Jump conditions on a curved shock 137

Chapter 4

4 **Weakly nonlinear ray theory (WNLRT): derivation** 143
4.1 A historical account 143
4.2 Derivation of CPW theory 146
4.3 A geometric derivation of WNLRT 149
4.3.1 WNLRT for a hyperbolic system 150
4.3.2 Upstream propagating waves in a steady flow of a polytropic gas 156
4.4 An asymptotic derivation of WNLRT 158
4.4.1 Derivation of eikonal and transport equations 158
4.4.2 Ray formulation of the asymptotic equations 163
4.4.3 Comparison with other theories 168

Chapter 5

5 **Stability of solutions near a singularity of sonic type** 173
5.1 Introduction 173
5.2 One-dimensional weakly nonlinear wave propagation 178
5.2.1 BKPS theory 180
5.2.2 Sonic type of singularity in self-similar solutions 188
5.3 Waves in a multi-dimensional steady transonic flow 195

Chapter 6

6	**WNLRT in a polytropic gas**	205
6.1	Basic equations	205
6.1.1	Non-dimensional form of equations of WNLRT in two-space-dimensions	207
6.1.2	A simple wave solution	211
6.2	Geometrical features of a nonlinear wavefront	213
6.2.1	Elementary wave solutions and their interpretation as elementary shapes	214
6.2.2	Solution of the Riemann problem and interpretation	217
6.2.3	Interaction of elementary shapes	220
6.3	Exact solution of an initial value problem	222
6.4	Conclusion and validity of WNLRT	226

Chapter 7

7	**Compatibility conditions on a shock: single conservation law**	229
7.1	Derivation of the infinite system of compatibility conditions	229
7.2	Existence and uniqueness of the solution of the infinite system	233
7.3	A new theory of shock dynamics: analytic considerations	236
7.4	A new theory of shock dynamics: comparison of numerical results with the exact solution	239
7.5	Conclusion	245

Chapter 8

8	**One-dimensional piston problem: an application of NTSD**	247
8.1	Formulation of the problem	248
8.2	Dynamical compatibility conditions	249
8.3	Initial conditions for the piston problem	255
8.4	Results and discussions	257

Chapter 9

9	**Compatibility conditions on a shock manifold in multi-dimensions**	265
9.1	Shock rays	265
9.2	Shock manifold equation for a weak shock	266
9.3	Geometrical and kinematical compatibility conditions	268
9.3.1	Preliminary geometrical ideas for a moving curve in two-space-dimensions	268
9.3.2	Geometrical compatibility conditions	270
9.3.3	Some results in a ray coordinate system	272
9.3.4	Kinematical compatibility conditions	273
9.4	Dynamical compatibility conditions	273
9.4.1	The first set of dynamical compatibility conditions	275
9.4.2	The second set of dynamical compatibility conditions	278
9.4.3	First and second set of equations in the shock ray theory	284
9.5	A weak shock ray theory	286

Chapter 10

10	**Propagation of a curved weak shock**	289
10.1	Governing equations of the NTSD	289
10.2	Conservation form of the equations for a two-dimensional shock propagation	293
10.3	Initial conditions, results and discussion	297
10.3.1	Propagation of a shock front initially parabolic in shape	298
10.3.2	Propagation of a shock front with initially sinusoidal shape and periodic amplitude distribution	305
10.3.3	Propagation of shock front with initially asymmetric but piecewise parabolic shape in each period	308

10.3.4 Propagation of a shock front with initially periodic but arbitrary shape in each period 310
10.3.5 When the initial shock front has a single smooth dent or bulge 311
10.4 Comparison with other theories 315
10.4.1 Qualitative verification of the shape of the front obtained by DNS to support the kink theory 315
10.4.2 Comparison with earlier theories 315
10.4.3 Comparison with weakly nonlinear ray theory 318
10.4.4 Comparison with Whitham's theory 319
10.5 Corrugational stability and persistence of a kink 322

References 325
Index 337

Chapter 1

An introduction to nonlinear hyperbolic waves

1.1 A wave equation with genuine nonlinearity

We are concerned in this monograph with hyperbolic waves having nonlinearity of a special type i.e., waves governed by quasilinear hyperbolic partial differential equations. We introduce the terms and concepts in this chapter with the help of a model equation and examples of its solutions.

The simplest example of a linear hyperbolic partial differential equation in two independent variables x and t is

$$u_t + cu_x = 0, \quad c = \text{ real constant }, \ (x,t) \in I\!R^2 \tag{1.1.1}$$

Its solution $u = u_0(x - ct)$, where $u_0 : I\!R \to I\!R$ is an arbitrary real function with continuous first derivatives, represents a wave. Every point of its profile propagates with the same constant velocity c. An extension of (1.1.1) is another linear equation

$$u_t + cu_x = \alpha u \ , \alpha = \text{ real constant} \tag{1.1.2}$$

which also represents a wave $e^{\alpha t}u_0(x - ct)$ in which the amplitude either decays to zero ($\alpha < 0$) or tends to infinity ($\alpha > 0$) but these limiting values are reached asymptotically in infinite time.

Consider now a nonlinear equation

$$u_t + cu_x = u^2 \tag{1.1.3}$$

The solution of this equation with the initial condition

$$u(x,0) = u_0(x) \ , \ \ u_0 \in C^1(I\!R) \tag{1.1.4}$$

is

$$u(x,t) = \frac{u_0(x-ct)}{1 - tu_0(x-ct)} \tag{1.1.5}$$

which also represents a wave propagating with the constant velocity c. However, the wave profile now deforms in such a way that a negative amplitude at a point on the initial pulse decays to zero as t tends to infinity but a positive amplitude of a point propagating with the velocity c tends to infinity in finite time. The topic of discussion in this book is nonlinear waves which appear as solutions of equations having a different type of nonlinearity. The simplest and yet the most beautiful example of an equation having this type of nonlinearity, called *genuine nonlinearity*, is

$$u_t + uu_x = 0 \tag{1.1.6}$$

whose solutions represent waves in which the velocity of propagation of a point on the pulse is equal to the amplitude at that point. This equation is called Burgers' equation.

The term genuine nonlinearity was first defined by P. D. Lax in 1957. It refers to a special property of wave propagation, namely dependence of the propagation velocity on the amplitude. It may be present in certain modes but not in others of the system. In a small amplitude unimodal* genuinely nonlinear wave in a homogeneous system, the propagation velocity of a point on the pulse exceeds a constant velocity c by a quantity whose leading term is proportional to the amplitude u of the wave. Denoting the distance in the frame of reference moving with velocity c also by x, rescaling the amplitude, and assuming that the wave is free from dispersion and diffusion, we get the approximate partial differential equation (1.1.6) for the unimodal waves (see section 5.2, also Jeffrey (1976),

*By unimodal wave in a general system, we mean a wave in which the amplitudes of waves in all other modes is small compared to that in one mode.

Bhatnagar (1979)). This equation also appears as a natural model for a large class of physical processes governed by a single conservation law in which the flux function depends approximately on the density alone (Whitham (1974)). Therefore, we note that Burgers' equation models a large class of important physical phenomena. The solution set of this equation has a very rich variety of properties which form the subject matter of discussion in this chapter.

Consider the solution of the equation (1.1.6) satisfying the initial condition

$$u(x,0) = e^{-x^2} \quad , \quad x \in \mathbb{R} \tag{1.1.7}$$

Properties of the solution will be discussed in subsequent sections. We take up here a simple geometrical construction of the successive shapes of the initially single humped pulse given by (1.1.7). The graph of the solution at any time t (i.e., the pulse at time t) is obtained by translating a point P on the pulse (1.1.7) by a distance in positive x-direction, the magnitude of the translation being equal to t times the amplitude of the pulse at the point P. Fig.1.1.1 shows

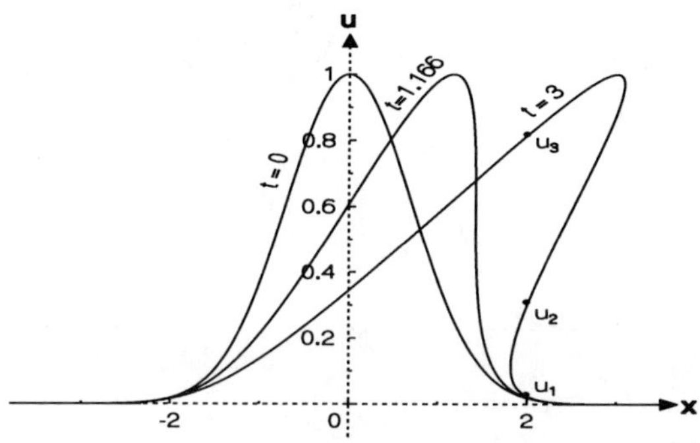

Fig. 1.1.1: As t increases, the pulse of the nonlinear wave deforms.

the pulse at times $t = 0,\ 1.166,\ 2$. We note that

(a) since different points of the pulse move with different velocity, the pulse now deforms;
(b) at a critical time t_c (for the initial pulse (1.1.7), $t_c = \sqrt{e/2} \cong 1.166$ (see section 1.2)) the pulse has a vertical tangent for the first time at some point on it; and
(c) after $t > t_c$, the pulse ceases to represent the graph of a function

(for example, at $x = 2$ it has three values u_1, u_2 and u_3) and the physical interpretation fails (for example, if u represents pressure in a fluid there can not exist three values of pressure at $x = 2$).

It has been observed in nature that a moving discontinuity appears in the quantity u immediately after the time t_c. This discontinuity at a point $x = X(t)$ is called a *shock*, which we shall define formally in section 1.4. When a shock appears in the solution, it fits into the multi-valued part of the solution in such a way that it cuts off lobes of areas on two sides of it in a certain ratio from the graph of the solution at any time $t > t_c$ and makes the solution single valued. The ratio in which the lobes on the two sides are cut off depends on a more primitive property (conservation of an appropriate density) of the physical phenomena represented by the equation (1.1.6) (see section 1.3 for details). When the primitive property is a conservation of the density $\rho(u) = u$, the shock cuts off lobes of an equal area on the two sides of it (for a proof see section 1.10).

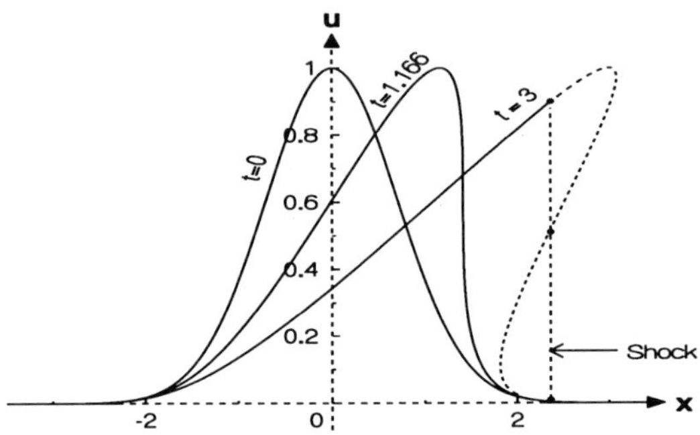

Fig. 1.1.2: The shock (shown by broken vertical line) fits into the multi-valued part of the curve at $t = 2$ assuming that the shock cuts off lobes of equal areas on two sides of it.

1.2 Breakdown of a genuine solution

A large number of physical phenomena are modelled by partial differential equations on the assumption that the variables which describe the state of a phenomenon are sufficiently smooth. In many realistic situations the state variables of these phenomena are smooth.

However, there exist other situations when they are not. This is reflected by the fact that for certain initial-boundary value problems associated with the equations either smooth solutions (also called *genuine solutions*) do not exist even locally or the solutions cease to be smooth after some critical time t_c even if the initial and boundary values are smooth. Burgers' equation (1.1.6) is an example for which a solution with a certain type of initial data, however smooth, always develops a singularity at a finite time.

An initial value problem or a Cauchy problem for the Burgers' equation† consists in finding a solution of

$$u_t + uu_x = 0, \ (x,t) \in \mathbf{R} \times \mathbf{R}_+ \tag{1.2.1}$$

satisfying initial data

$$u(x,0) = u_0(x), \ x \in \mathbf{R} \tag{1.2.2}$$

where $I\!R$ is the set of real numbers and $I\!R_+ = (0, \infty)$.

Definition A *genuine* (or *classical*) solution of the partial differential equation (1.2.1) in a domain D in (x,t)-plane is a function $u(x,t) \in C^1(D)$ which satisfies (1.2.1).

A sufficient condition for the existence of a local genuine solution (i.e., a solution valid for $0 < t < t_c$ with some $t_c < \infty$) of the initial value problem (1.2.1) and (1.2.2) is that $u_0(x) \in C^1(\mathbf{R})$ (John (1982), Prasad and Ravindran (1985)). A genuine solution can be obtained by solving the compatibility condition

$$\frac{du}{dt} = 0 \tag{1.2.3}$$

along the characteristic curves

$$\frac{dx}{dt} = u \tag{1.2.4}$$

†This equation is called the Burgers' equation by mathematicians but it was actually obtained by Airy (1845) in order to explain the solitary wave observed by Scott-Russell in (1844) in the Edinburgh-Glasgow canal. It is not only an injustice to Airy but also a distortion of the history of the subject because Burgers' paper appeared only in 1948.

These equations imply that u is constant along the characteristics $\xi \equiv x - ut =$ constant. Hence, the solution of (1.2.1) and (1.2.2) is given by

$$u = u_0(x - ut) \tag{1.2.5}$$

From the implicit function theorem, it follows that the relation (1.2.5) defines a C^1 function $u(x,t)$ as long as

$$1 + tu_0'(\xi) \neq 0 \tag{1.2.6}$$

which is satisfied for $\mid t \mid$ small. The x-derivative of the solution is given by

$$u_x = u_0'(\xi)/\{1 + tu_0'(\xi)\} \tag{1.2.7}$$

If the initial data is such that $u_0' < 0$ is on some interval of the x-axis, there exists a time $t_c > 0$ such that as $t \to t_c - 0$, the derivative $u_x(x,t)$ of the solution tends to $-\infty$ for some value of x and thus the genuine solution can not be continued beyond at $t = t_c$. The critical time t_c is given by

$$t_c = -\frac{1}{\min\limits_{\xi \in \mathbf{R}}\{u_0'(\xi)\}} > 0 \tag{1.2.8}$$

If $u_0'(x) > 0$ for all $x \in \mathbf{R}$; the relation (1.2.5) gives a genuine solution of (1.2.1) and (1.2.2) for all $t > 0$. Breakdown of the genuine solution at $t = t_c$ can be explained not only graphically as in the previous section but also from the geometry of the characteristic curves of the equation (1.2.1). Let I be an interval on the x-axis such that $u_0'(x) < 0$ for $x \in I$. The characteristics in the (x,t)-plane starting from the various points of the interval I converge and, in general, envelop a cusp starting from the time t_c. Consider the domain in the (x,t)-plane which is bounded by the two branches of the cusp. Three characteristics starting from some three points of I pass through any point of this domain. Since characteristics carry different constant values of the solution, u is not defined uniquely at interior points of this domain. Difficulty in continuation of a genuine solution beyond a finite time t_c is quite common for a hyperbolic system‡ of *quasilinear*

‡The meaning of a hyperbolic system of partial differential equations will be explained in sections 2.1.1 and 2.3.2. Equation (1.2.1) is a simple example of a quasilinear hyperbolic equation.

partial differential equations. A smooth solution of an initial value problem for the semilinear equation (1.1.3) also breaks down after a finite time due to an unbounded increase in the amplitude of the solution. However, the breakdown of the solution of (1.1.6) is due to a different reason: due to an unbounded increase in the absolute value of the first derivatives, the solution itself remains finite. Here, we are concerned with the breakdown of the type exhibited by a solution of (1.1.6). For a hyperbolic system of quasilinear equations, such a breakdown is due to a very special property of a characteristic velocity (or an eigenvalue), namely, the velocity of propagation (which is the same as the characteristic velocity) of the waves that depend essentially on the amplitude of the waves, that is, the characteristic field is "genuinely nonlinear" (see definition 3.2.1).

It is easy to solve the problem of determining the critical time when the amplitude of a discontinuity in the first derivatives of a solution tends to infinity (leading to the appearance of a discontinuity in a solution) on a suitably defined leading characteristic curve of a general hyperbolic system in two independent variables (Jeffrey (1976)) and the method can be easily extended to systems in more than two independent variables (see section 3.4).

1.3 Conservation law and jump condition

We have seen in the last section that a genuine solution of the initial value problem (1.2.1) and (1.2.2) ceases to be valid after a critical time t_c. However, equation (1.2.1) models quite a few physical phenomena, where the function u becomes discontinuous after the time t_c and the discontinuous states of the phenomena persist for all time $t > t_c$. Hence we must generalize the notion of a solution to permit u, which are not necessarily C^1. In order to do that we write (1.2.1) in a divergence form

$$\frac{\partial}{\partial t}(u) + \frac{\partial}{\partial x}(\frac{1}{2}u^2) = 0, (x,t) \in \mathbf{R} \times \mathbf{R}_+ \tag{1.3.1}$$

Definition A conservation law is an equation in a divergence form.

(1.3.1) is just one of an infinity of conservation laws,

$$\frac{\partial}{\partial t}(u^n) + \frac{\partial}{\partial x}(\frac{n}{n+1}u^{n+1}) = 0, \ n = \text{constant} \tag{1.3.2}$$

which can be derived from (1.2.1). Both these conservation forms are particular cases of a general form

$$\frac{\partial H(u)}{\partial t} + \frac{\partial F(u)}{\partial x} = 0 \tag{1.3.3a}$$

where the *density* H and the *flux* F are smooth functions of the state variable u. Equation (1.2.1) can be derived from (1.3.3a) if

$$H'(u) \equiv \frac{dH}{du} \neq 0 \text{ and } F'(u)/H'(u) = u \tag{1.3.3b}$$

In physics, a balance equation representing the conservation of a quantity such as mass, momentum or energy of a physical system is not expressed in the form of a differential equation (1.3.3a). The original balance equation is stated in terms of integrals, rather than in the form of (1.3.3a), as

$$\int_{x_1}^{x_2} H(u(\xi, t_2))d\xi - \int_{x_1}^{x_2} H(u(\xi, t_1))d\xi = \int_{t_1}^{t_2} \{F(u(x_1, t)) - F(u(x_2, t))\}dt \tag{1.3.4}$$

which holds for every fixed space interval (x_1, x_2) and for every time interval (t_1, t_2). This equation is meaningful even for a discontinuous function $u(x, t)$. We can now define a *weak solution* of the conservation law (1.3.3) to be a bounded measurable function $u(x, t)$ which satisfies the integral form (1.3.4). We shall give in the section 3.5 a definition of a weak solution which is mathematically more satisfactory (see equation (3.5.2)). We shall not use the balance equation in the general form such as (1.3.4) but in a more restricted form

$$\frac{d}{dt}\int_{x_1}^{x_2} H(u(\xi, t))d\xi = F(u(x_1, t)) - F(u(x_2, t)), \; x_1, x_2 \text{ fixed} \tag{1.3.5}$$

which we shall assume to be valid almost everywhere for $(x_1, x_2, t) \in \mathbb{R}^2 \times \mathbb{R}_+$. For smooth solutions, the equations (1.3.3), (1.3.4) and (1.3.5) are equivalent. We state this result:

Theorem 1.3.1 Every weak solution which is C^1 is a genuine solution of the partial differential equation $H'(u)u_t + F'(u)u_x = 0$.

Consider now a solution $u(x,t)$ of (1.3.5) such that $u(x,t)$ and its partial derivatives suffer discontinuities across a smooth isolated curve $\Omega : x = X(t)$ in the (x,t)-plane and is continuously differentiable elsewhere. It is further assumed that the limiting values of u and its derivatives as we approach Ω from either side exist. The function $u(x,t)$ is a genuine solution of (1.2.1) in the left and right subdomains of the curve of discontinuity Ω. Let the fixed points x_1 and x_2 be so chosen that $x_1 < X(t) < x_2$ for $t \in$ an open interval. Writing $\int_{x_1}^{x_2} H(u(\xi,t))d\xi = \int_{x_1}^{X(t)} H(u(\xi,t))d\xi + \int_{X(t)}^{x_2} H(u(\xi,t))d\xi$ in (1.3.5) and then taking its time derivative, we get

$$\int_{x_1}^{X(t)} H'u_t(\xi,t)d\xi + \int_{X(t)}^{x_2} H'u_t(\xi,t)d\xi + \dot{X}(t)\{H(u(X(t)-0,t)) \\ -H(u(X(t)+0,t))\} = \{F(u(x_1,t)) - F(u(x_2,t))\}$$

The first two terms tend to zero as $x_1 \to X(t)-$ and $x_2 \to X(t)+$. Hence taking the point x_1 on the left of $X(t)$ very close to it and the point x_2 on the right of $X(t)$ also very close to it, we get in the limit

$$\dot{X}(t)(H(u_\ell(t)) - H(u_r(t))) = F(u_l) - F(u_r) \tag{1.3.6}$$

where

$$u_\ell(t) = \lim_{x\to X(t)-0} u(x,t) \quad \text{and} \quad u_r(t) = \lim_{x\to X(t)+0} u(x,t) \tag{1.3.7}$$

(1.3.6) gives the following expression for the velocity of propagation $\dot{X}$ of the discontinuity

$$\dot{X}(t) = [F]/[H] \tag{1.3.8}$$

where the symbol [ ] is defined by

$$[f] = f(u_r) - f(u_l) \tag{1.3.9}$$

The relation (1.3.8), connecting the speed of propagation $\dot{X}(t)$ of a discontinuity and the limiting values u_ℓ and u_r on the two sides of the discontinuity, is called the *jump condition*. Such jump relations derived from the conservation laws of gas dynamics are called

Rankine–Hugoniot (RH) conditions. The usual initial and boundary conditions of the gas dynamics equations combined with the three RH conditions and an entropy condition (to be discussed in the next section) are sufficient to solve many problems of great practical importance (Courant and Friedrichs (1948)) and are also sufficient to prove many theorems on the existence and uniqueness of solutions (see Smoller (1983), Chapters 17 and 18). This mathematical completeness in the theory of discontinuous solutions was probably the main reason that mathematicians did not take up the question of further jump relations on the derivatives of the function $u(x,t)$. Derivation of the transport equations for the jumps in the first and higher order spatial derivatives of u and application of these relations in developing numerical methods for solving a weak solution having a single discontinuity forms the subject matter of discussion in Chapters 7 to 10 of this monograph.

For a discontinuity, the jump is non-zero i.e., $u_l \neq u_r$. Hence, for the conservation laws (1.3.1) and (1.3.2), the jump conditions (1.3.8) become

$$\dot{X}(t) = \frac{1}{2}(u_r + u_l) \tag{1.3.10}$$

and

$$\dot{X}(t) = \frac{n}{n+1}\left(\sum_{i=0}^{n} u_r^{n-i}\, u_l^{i}\right)\Big/\left(\sum_{i=0}^{n-1} u_r^{n-1-i}\, u_l^{i}\right) \tag{1.3.11}$$

1.4 Stability consideration, entropy condition and shocks

Let us discuss solutions of a number of initial value problems of the balance equation (1.3.5) with $H(u) = u, F(u) = \frac{1}{2}u^2$ i.e., weak solutions of (1.3.1). The solutions need not be smooth now. First we consider a smooth initial data

$$u(x,0) \;=\; 0 \quad (\equiv \varphi_1(x) \text{ say}) \tag{1.4.1}$$

One obvious solution with this initial data is

$$u(x,t) \;= 0, \quad (x,t) \in \mathbf{R} \times \mathbf{R}_+ \tag{1.4.2}$$

which is a genuine solution of (1.2.1). Consider now the function

$$u(x,t) = \begin{cases} 0\,, & x \le -\frac{1}{2}t, \\ -1\,, & -\frac{1}{2}t < x \le 0, \\ 1\,, & 0 < x \le \frac{1}{2}t, \\ 0\,, & \frac{1}{2}t < x \end{cases} \tag{1.4.3}$$

The characteristic curves of (1.2.1) are shown by broken lines in Fig. 1.4.1. The function (1.4.3) is constant and hence a genuine solution in each of the subdomains of the upper half of the (x,t)-plane in which it is divided by the straight lines $x = 0, x = -\frac{1}{2}t$ and $x = \frac{1}{2}t$. The function satisfies the jump relation (1.3.10) along the last two lines. Hence, (1.4.3) is another weak solution (now a discontinuous solution) of the equation (1.3.1) with the initial condition (1.4.1). This is a spectacular result of a non-zero solution coming out of a zero initial condition and is due to the fact that the discontinuity along the line $x = 0$ is not admissible which we shall explain later.

Note that even if the value of a function which satisfies (1.3.4) is changed at a set of moving points $x_i(t)$ finite in number (or of measure zero) in (x,y)-plane, the new function will still satisfy (1.3.4). Therefore, it is immaterial whether we take the weak solution of (1.3.1) to be continuous to the left (as in the case (1.4.3)) or to the right of the point of discontinuity.

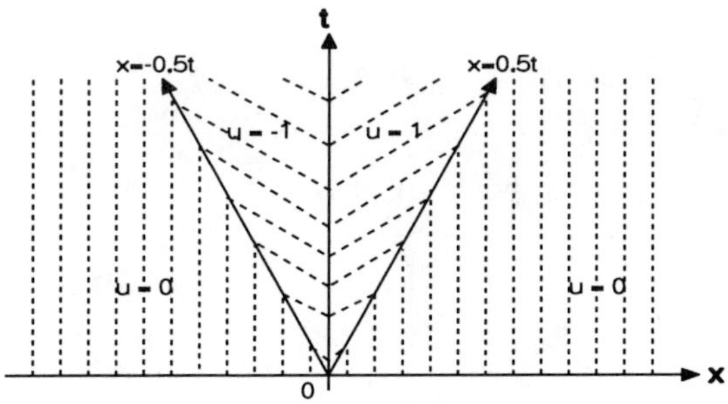

Fig. 1.4.1: A discontinuous solution of (1.3.1) with initial data given by $u(x,0) = 0$. The characteristic curves of (1.2.1) are shown by broken lines.

For a class of discontinuous solutions, we can consider not only smooth initial data but initial data which could be discontinuous. Let us consider a discontinuous initial data of the form

$$u(x,0) = \varphi_2(x) \equiv \begin{cases} 0\ , & x \le 0 \\ 1\ , & 0 < x \end{cases} \tag{1.4.4}$$

The equation (1.3.1) has an infinity of discontinuous weak solutions for this initial data:

$$u(x,t) = \begin{cases} 0 & , \quad x \le 0 \\ x/t & , \quad 0 < x \le \alpha t \\ \alpha & , \quad \alpha t < x \le \frac{1}{2}(1+\alpha)t \\ 1 & , \quad \frac{1}{2}(1+\alpha)t < x \end{cases} \tag{1.4.5}$$

It depends on a parameter α satisfying $0 \le \alpha \le 1$. This solution has been shown in Fig. 1.4.2.

The characteristic curves have been shown by broken lines. For $\alpha = 1$, the solution (1.4.5) becomes continuous but it is not a genuine solution of (1.2.1) since it is not continuously differentiable. We note that the non-uniqueness in the solution arises because of the possibility of fitting a line of discontinuity of an arbitrary slope $\frac{1}{2}(1+\alpha)$ joining a constant value $(u = \alpha)$ of the *centered wave* $u = x/t$ on the left and the constant state $u = 1$ on the right. Another simple solution with initial condition (1.4.4) is three constant states 0, $\frac{1}{2}$ and 1 separated by lines of discontinuities $x = \frac{1}{4}t$ and $x = \frac{3}{4}t$.

Instead of the initial data $\varphi_2(x)$, if we take

$$u(x,0) = \varphi_3(x) \equiv \begin{cases} 1 & , \quad x \le 0 \\ 0 & , \quad x > 0 \end{cases} \tag{1.4.6}$$

we get a discontinuous weak solution of (1.3.1):

$$u(x,t) = \begin{cases} 1 & , \quad x - \frac{1}{2}t \le 0 \\ \\ 0 & , \quad x - \frac{1}{2}t > 0 \end{cases} \tag{1.4.7}$$

In this case, it is not possible to have a centered wave with the center at the origin and hence (1.4.7) is the only weak solution. It has a curve of discontinuity through $(0,0)$.

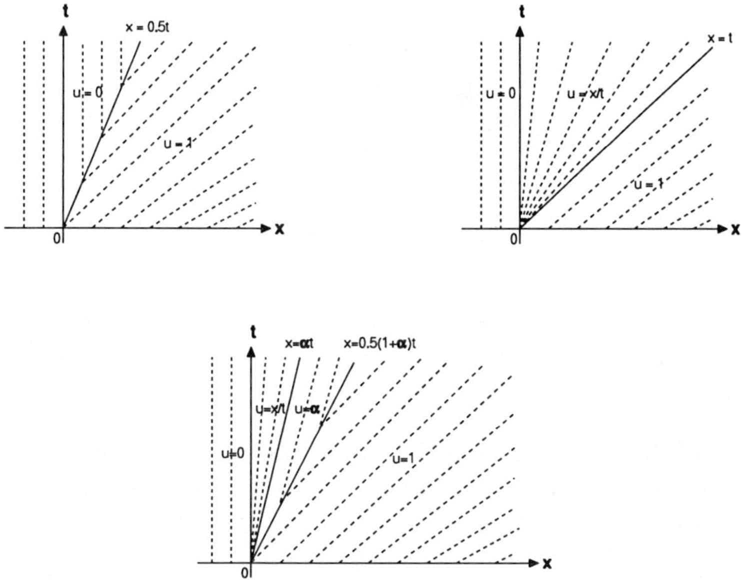

Fig. 1.4.2: Solution (1.4.5) for $\alpha = 0$
Solution (1.4.5) for $\alpha = 1$
Solution (1.4.5) for $0 < \alpha < 1$.
The characteristic curves have been shown by broken lines.

The above example shows that, in general, a discontinuous solution of an initial value problem for (1.3.5), i.e., a weak solution of the conservation law (1.3.1), is not unique. What is needed now is a mathematical principle characterizing a class of permissible solutions in which every initial value problem for the conservation law has a unique solution. We can deduce such a principle from the following consideration. A genuine solution satisfying smooth initial data is unique and this is true even for a solution with piecewise continuous derivatives such as (1.4.5) with $\alpha = 1$. In this case, given a point $(x, t) \in I\!R \times I\!R_+$, there exists a unique characteristic originating from a point on the initial line which passes through the point (x, t). However, for a discontinuous solution (1.4.5), the initial data (1.4.4) is unable to control the solution in the domain $\alpha < \frac{x}{t} < 1$. The failure

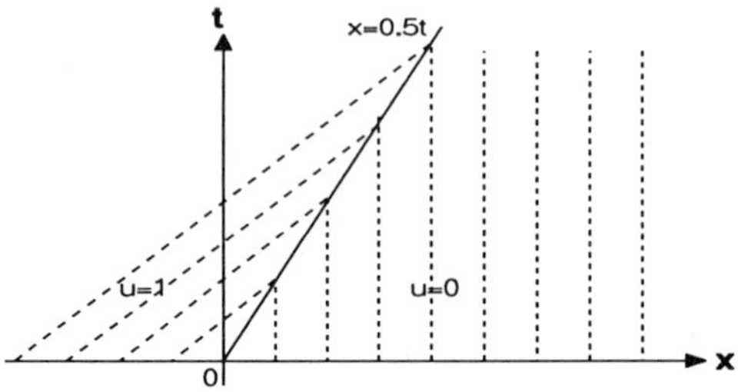

Fig. 1.4.3: Solution (1.4.7) with characteristic curves shown by broken lines.

of the initial data to control the solution in this domain leads to nonuniqueness of the curve discontinuity of the solution. From a curve of discontinuity in the solution (1.4.5) for $0 \le \alpha < 1$, the characteristics starting from a point $(\frac{1}{2}(1+\alpha)t, t)$ diverge (into the domains on the two sides of it) as t increases, so that discontinuity could have been replaced by a continuous centered wave from this point onward. The situation is different when the initial data is (1.4.6), which gives rise to a situation in which characteristics starting from the points on the two sides of the point of discontinuity converge and start intersecting as t increases. In this and all other such situations, a discontinuity must necessarily appear to prevent multivaluedness in the solution. Therefore, a discontinuity is permissible only if it prevents the intersection of the characteristics coming from the points of the initial line on the two sides of it, i.e.,

$$u_r(t) < \dot{X}(t) < u_\ell(t) \tag{1.4.8}$$

If we accept this as a principle, Fig. 1.4.1 and Fig. 1.4.2 show that the only admissible solution with initial data φ_1 is the zero solution, and that with the initial data φ_2 is the continuous solution obtained for $\alpha = 1$. The solution (1.4.7) is also the only admissible solution with the initial data φ_3.

The mathematical criterion (1.4.8), known as *Lax's entropy condition*, for admissible discontinuities can be derived from the following stability consideration (first shown by Gel'fand in 1962 and stated in this form by Jeffrey (1976)).

"A discontinuity is admissible if when small amplitude waves are incident upon the discontinuity, the resulting perturbations in the velocity of the discontinuity and the resulting waves moving away from the discontinuity are uniquely determined and remain small."

The derivation is trivial for the single conservation law.

Definition An admissible discontinuity satisfying the entropy condition (1.4.8) is called a *shock*.

Using (1.3.10), we find that for the conservation law (1.3.1) the stability condition (1.4.8) is equivalent to an easily verifiable condition

$$u_r(t) < u_\ell(t) \tag{1.4.9}$$

The term "shock" was first used for a compression discontinuity in gas dynamics, where an expansion discontinuity is ruled out by the second law of thermodynamics which implies that the specific entropy of the fluid particles must increase after crossing the discontinuity. Hence, the stability condition (1.4.8) is also called the *entropy condition*. We note two important results regarding solutions with shocks of nonlinear problems:

(1) In contrast to the results for linear equations, not only a discontinuity may appear in the solution of nonlinear equations with continuous data, but also a discontinuity in the initial data may be immediately resolved in the solution. This is shown by the continuous solution (1.4.5) with $\alpha = 1$ for the discontinuous initial data $\varphi_2(x)$.

(2) Physical processes described by continuous solutions of a hyperbolic system of quasilinear equations are reversible in time, i.e., if we know the solution at some time, we can use the differential equation to get the solution uniquely in the past as well as in future. However, if a process is described by a discontinuous solution (where discontinuities are shocks) of a system of balance equations, then it is irreversible. We shall illustrate this mathematically by means of an example. The weak solution of the conservation law (1.3.1) satisfying

$$u(x,0) = \begin{cases} 2 \ , & x \le \frac{1}{4} \\ 0 \ , & x > \frac{1}{4} \end{cases} \tag{1.4.10}$$

is

$$u(x,t) = \begin{cases} 2 \ , & x \le t + \frac{1}{4} \\ & \\ 0 \ , & t + \frac{1}{4} < x \end{cases} \qquad \text{for } t > 0 \tag{1.4.11}$$

and then satisfying

$$u(x,0) = \begin{cases} 2 \ , & x \le 0 \\ & \\ 1 \ , & 0 < x \le \frac{1}{2} \\ & \\ 0 \ , & \frac{1}{2} < x \end{cases} \tag{1.4.12}$$

is

$$\left.\begin{array}{l} u(x,t) = \begin{cases} 2 \ , & x \le \frac{3}{2}t \\ 1 \ , & \frac{3}{2}t < x \le \frac{1}{2}t + \frac{1}{2} \ , \quad \text{for } 0 < t < \frac{1}{2} \\ 0 \ , & \frac{1}{2}t + \frac{1}{2} < x \end{cases} \\ \\ u(x,t) = \begin{cases} 2 \ , & x \le t + \frac{1}{4} \\ & \qquad\qquad , \quad \text{for } t > \frac{1}{2} \\ 0 \ , & t + \frac{1}{4} < x \end{cases} \end{array}\right\} \tag{1.4.13}$$

The solution of (1.4.12), for $0 < t < \frac{1}{2}$, has two shocks and it is interesting to draw them in the (x,t)-plane.

(1.4.11) and (1.4.13) are admissible and unique solutions with two different initial values. However, both represent the same function for $t \ge \frac{1}{2}$. Thus, the same state given by the two solutions at time $t \ge \frac{1}{2}$ corresponds to two initial states. This shows irreversibility – the past can not be uniquely determined by the future.

1.5 Some examples

Even though the conservation form (1.3.1) of the Burgers' equation (1.2.1) looks innocently simple, an explicit solution (see section 1.9) of an initial value problem with arbitrary initial data (1.2.2) is so involved that it requires a lot of mathematical analysis to deduce even some simple properties of the solution. In this section we present a number of exact solutions and asymptotic forms of some other solutions which show that genuine nonlinearity significantly modifies the linear solution.

Example 1.5.1 A solution of an initial value problem for (1.3.1), with continuous initial data

$$u_0(x) = \begin{cases} 1 \ , & x \le -\frac{1}{2} \\ \frac{1}{2} - x \ , & -\frac{1}{2} < x \le \frac{1}{2} \\ 0 \ , & x > \frac{1}{2} \end{cases} \tag{1.5.1}$$

remains continuous for all t in the interval $0 \le t < 1$ and is given by

$$u(x,t) = \begin{cases} 1, \ , & x \le -\frac{1}{2} + t \\ \frac{(1/2)-x}{1-t} \ , & -\frac{1}{2} + t < x \le \frac{1}{2} \\ 0 \ , & x > \frac{1}{2} \end{cases} \tag{1.5.2}$$

For $t \ge 1$ the solution has a shock, which separates two constant states 1 and 0 and which moves along the path $x = X(t) \equiv \frac{1}{2}t$ as in the case of the solution satisfying the initial condition (1.4.6). The characteristic curves starting from the various points of the x-axis have been shown in Fig. 1.5.1.

Example 1.5.2 Consider an initial data

$$u_0(x) = \begin{cases} \frac{1}{2}A \ , & -1 < x \le 1 \\ 0 \ , & x \le -1 \text{ and } x > 1 \end{cases} \tag{1.5.3}$$

where we take $A > 0$. If the evolution of the initial data is described according to the conservation law (1.3.1) then the solution has two distinct representations in two different time intervals:

(i) $0 < t \le \frac{8}{A}$.

$$u(x,t) = \begin{cases} 0 \ , & x \le -1 \\ \frac{x+1}{t} \ , & -1 < x \le -1 + \frac{A}{2}t \\ \frac{1}{2}A \ , & -1 + \frac{A}{2}t < x \le 1 + \frac{A}{4}t \\ 0 \ , & 1 < x \end{cases} \tag{1.5.4}$$

which has been shown in the Fig. 1.5.2.

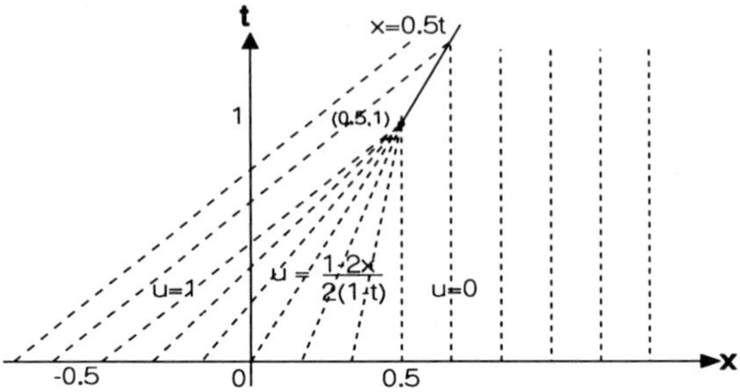

Fig. 1.5.1: All characteristic curves starting from the points of the initial data between $-\frac{1}{2} \leq x \leq \frac{1}{2}$ meet at the point $\left(\frac{1}{2}, 1\right)$.

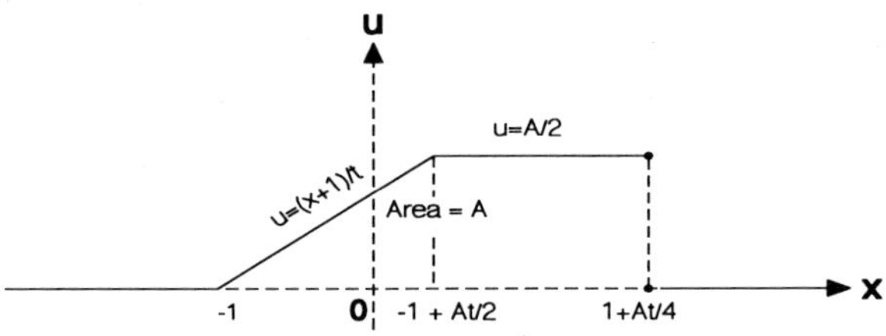

Fig. 1.5.2: Graph of the solution with initial value (1.5.3) valid in the time interval $0 < t < \frac{8}{A}$.

Note that there is a *centered wave* (see section 3.1.2 for definition) in the wedged shape region $-1 < x \leq -1 + \frac{A}{2}t$ and a shock along the curve $x = \frac{1}{2}At$ in the (x,t)-plane. At the time $t = \frac{8}{A}$ the leading front of the centered wave overtakes the shock at $x = 3$. After this time the shock interacts with the centered wave.

(ii) $t \geq \frac{8}{A}$.

The shock path $x = X(t)$ is obtained by solving

$$\frac{dX}{dt} = \frac{1}{2}(u_l(X(t)) + u_r(X(t)) = \frac{X+1}{2t}, \quad X\left(t = \frac{8}{A}\right) = 3$$

which gives

$$X(t) = -1 + \sqrt{2At} \tag{1.5.5}$$

At $x = X(t)$, the amplitude of the pulse i.e., the shock strength $u_l - u_r$ is given by

$$u = \frac{x+1}{t}\Big|_{x=X(t)} = \sqrt{\frac{2A}{t}} \tag{1.5.6}$$

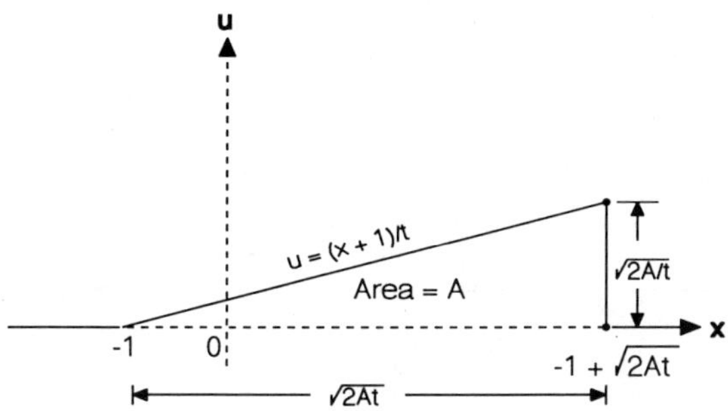

Fig. 1.5.3: Graph of the solution with initial condition (1.5.3) valid from $t > \frac{8}{A}$.

The pulse now takes a triangular shape whose base is spread over a distance $\sqrt{2At}$ and whose height is $\sqrt{\frac{2A}{t}}$ (Fig. 1.5.3). The total area of the pulse remains constant equal to A which is also the area of the initial pulse (1.5.3). This agrees with a general property of the conservation law (1.3.1) *"when the solution u vanishes outside a closed bounded interval of the x-axis, $\int_{-\infty}^{\infty} u(\xi, t)d\xi$ is independent of t."* Fig. 1.5.3 is the limiting form of the shape of the graph of any solution for which the initial data $u_0(x)$ is positive everywhere and is of compact support.

Example 1.5.3. Consider an initial data

$$u_0(x) = \begin{cases} 0 & , \quad -\infty < x < -1 \\ -x-1 & , \quad -1 < x \leq -\frac{1}{2} \\ x & , \quad -\frac{1}{2} < x \leq 1 \\ -x+2 & , \quad 1 < x \leq 2 \\ 0 & , \quad 2 < x < \infty \end{cases} \tag{1.5.7}$$

which has been shown in Fig. 1.5.4.

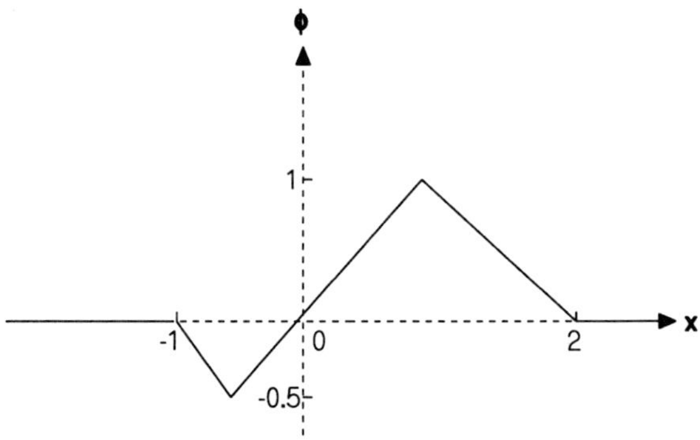

Fig. 1.5.4: Graph of the initial data (1.5.7).

The solution of the conservation law (1.3.1) with an initial condition (1.5.7) remains continuous for $t < 1$ and its expression can be easily written. During this initial stage, the interval in which the solution is non-zero remains fixed i.e., $(-1, 2)$. At $t = 1$, a pair of shocks appear at -1 and 2 and for $t > 1$ the solution is given by

$$u(x,t) = \begin{cases} 0 & , \quad -\infty < x \leq -\sqrt{\frac{1}{2}(1+t))} \\ \frac{x}{1+t} & , \quad -\sqrt{\frac{1}{2}(1+t)} < x < \sqrt{2(1+t)} \\ 0 & , \quad \sqrt{2(1+t)} < x < \infty \end{cases} \tag{1.5.8}$$

with shocks at the leading and the trailing ends. The values of u at these two ends are $\sqrt{2/(1+t)}$ and $-\sqrt{1/\{2(1+t)\}}$, respectively. The area of the positive pulse on the right side of $x = 0$ is

$$\frac{1}{2}\sqrt{2(1+t)}\sqrt{\frac{2}{1+t}} = 1$$

and that of the negative pulse on the left side of $x = 0$ is

$$\frac{1}{2}\sqrt{\frac{1+t}{2}}\,\frac{1}{\sqrt{2(1+t)}} = \frac{1}{4}$$

Thus, the areas on the two sides of the origin are conserved. This is because the flux function $\frac{1}{2}u^2$ in (1.3.1) vanishes at the origin.

The ultimate shape of the solution is depicted in Fig. 1.5.5 which is called an N wave.

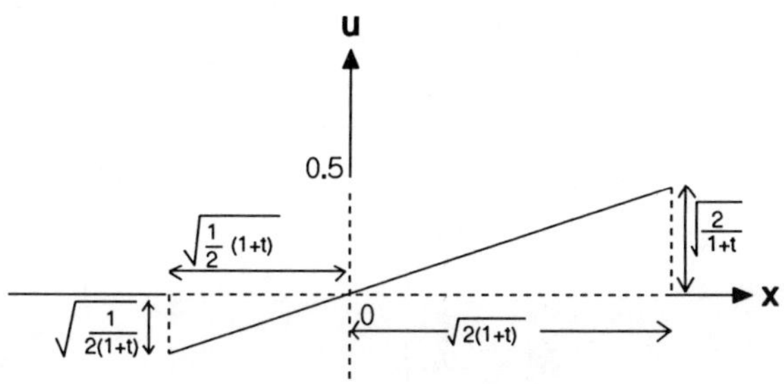

Fig. 1.5.5: An N wave solution (1.5.8).

The leading and trailing shocks are not of equal strength in this figure.

Example 1.5.4 Consider an initial data

$$u_0(x) = \begin{cases} 0 \quad , & -\infty < x \le -\lambda \\ 2a(x+\lambda)/\lambda \quad , & -\lambda < x \le -\frac{1}{2}\lambda \\ -2ax/\lambda \quad , & -\frac{1}{2}\lambda < x \le \frac{1}{2}\lambda \\ 2a(x-\lambda)/\lambda \quad , & \frac{1}{2}\lambda < x \le \lambda \\ 0 \quad , & \lambda < x < \infty \end{cases} \tag{1.5.9}$$

where $\lambda > 0$. This initial data has been shown graphically in Fig. 1.5.6.

The solution of the conservation law (1.3.1) with initial data (1.5.9) remains continuous for $0 < t < \frac{\lambda}{2a}$. At $t = \frac{\lambda}{2a}$, a shock appears at the origin with $u_l = a$ and $u_r = -a$. According to the jump relation, this shock does not move away from the origin but its amplitude decays. The solution in the interval $-\lambda < x \le -\frac{1}{2}\lambda + at$ for $t \le \frac{\lambda}{2a}$ and in $-\lambda < x < 0$ for $t > \frac{\lambda}{2a}$ is given by

$$u(x,t) = \frac{x+\lambda}{t+\frac{\lambda}{2a}} \tag{1.5.10}$$

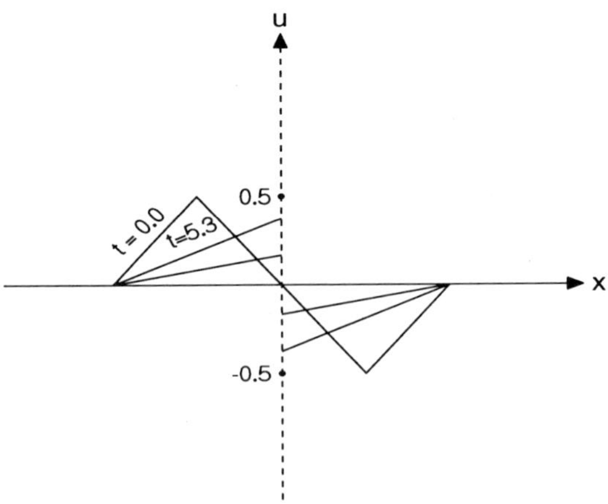

Fig. 1.5.6: The initial pulse is given by (1.5.9) with $\lambda = 1, a = \frac{1}{2}$. The solution develops a shock at the origin which decays to zero with time as $0(\frac{1}{t})$.

Since $u_r(t) = -u_l(t)$, (1.5.10) shows that the shock strength at the origin is

$$u_l - u_r = 2\lambda/(t + \frac{\lambda}{2a}) \tag{1.5.11}$$

showing that, unlike all previous examples where the shock strength decays as $0\left(\frac{1}{\sqrt{t}}\right)$, in this case, it decays as $0\left(\frac{1}{t}\right)$. The asymptotic solution as $t \to \infty$, retaining only the first term, is

$$u(x,t) = \begin{cases} 0 \ , & -\infty < x \leq -\lambda \\ (x+\lambda)/t \ , & -\lambda < x \leq 0 \\ (x-\lambda)/t \ , & 0 < x \leq \lambda \\ 0 \ , & \lambda < x < \infty \end{cases} \tag{1.5.12}$$

which has a shock of strength $2\lambda/t$. It is interesting to note that the asymptotic solution is the same whatever may be the initial amplitude but it remembers (apart from the initial total area of the pulse) the length 2λ where it is non-zero.

As another interesting example, the reader is asked to find the solution of (1.3.1) with initial data

$$u_0(x) = \begin{cases} 0 \ , & -\infty < x \leq -2 \\ x+2 \ , & -2 < x \leq -1 \\ -x \ , & -1 < x \leq \frac{1}{2} \\ x-1 \ , & \frac{1}{2} < x \leq 1 \\ 0 \ , & 1 < x < \infty \end{cases} \tag{1.5.13}$$

Example 1.5.5 We now consider an important example of a solution of (1.3.1) with periodic initial condition:

$$u_0(x) = -a \sin \frac{\pi x}{\lambda} \ , \ \lambda > 0 \tag{1.5.14}$$

The asymptotic form of the solution as $t \to \infty$ can be deduced with the help of the Lax-Oleinik formula (see section 1.9). Here we deduce the form by noting that at any time t, the solution must be periodic of the period 2λ and assuming that in the period $-\lambda < x < \lambda$, the solution is given by a pair of centered waves

$$u(x,t) = \begin{cases} (x+\lambda)/t\,, & -\lambda < x \leq 0 \\ (x-\lambda)/t\,, & 0 < x \leq \lambda \end{cases} \tag{1.5.15}$$

According to the jump condition, the shocks do not move away from the points $0, \pm 2\lambda, \pm 4\lambda, \ldots$; their strengths decay as $\frac{2\lambda}{t}$, as was the case in the previous example. The asymptotic periodic solution *forgets* the amplitude a of the initial data but *remembers* its period 2λ. The asymptotic solution in the form of a saw-tooth has been shown in Fig. 1.5.7.

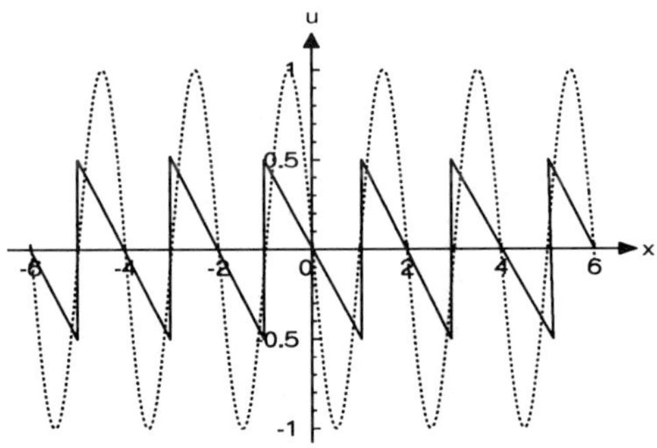

Fig. 1.5.7: The saw-tooth solution arising from a periodic initial data shown by a dotted line.

Example 1.5.6 An equation, governing the propagation of small perturbations trapped at a point on the sonic surface of a steady gas flow, is given by (Prasad (1973), see also (5.3.41 - 42)).

$$u_t + (u - Kx)u_x = Ku \tag{1.5.16}$$

The dependent and independent variables have been properly scaled. The constant K is proportional to the deceleration of the fluid elements at the sonic point in the steady flow. When the fluid is passing from a supersonic state to a subsonic state, $K > 0$.

Had the genuine nonlinearity not been present, the approximate equation would have been

$$u_t - Kxu_x = Ku \tag{1.5.17}$$

The solution of this equation with initial data

$$u(x,0) = u_0(x) \tag{1.5.18}$$

is

$$u(x,t) = e^{Kt}u_0(xe^{Kt}) \tag{1.5.19}$$

This solution shows that for $K > 0$ the amplitude u would tend to infinity as $t \to \infty$. However, for an initial data u_0 which is non-zero only on a bounded interval on the x- axis, the solution will get concentrated near the point $x = 0$.

A conservation form of the genuinely nonlinear equation (1.5.16) is

$$u_t + (\frac{1}{2}u^2 - Kxu)_x = 0 \tag{1.5.20}$$

It is simple to show that the jump relation for a shock appearing in a weak solution of (1.5.20) is

$$\frac{dX(t)}{dt} = \frac{1}{2}(u_l + u_r) - KX(t) \tag{1.5.21}$$

When a shock appears in a solution of an initial value problem for (1.5.20), it cuts off the growing part of the pulse as shown in Fig. 1.5.8 for a special initial data.

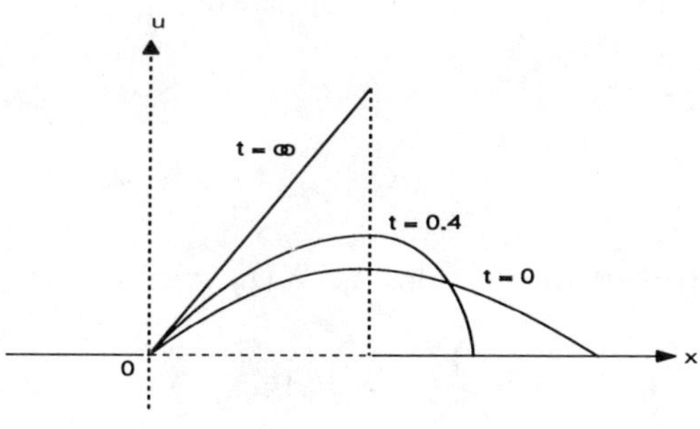

Fig. 1.5.8: Solution of (1.5.20), with an initial data which is positive and non-zero only on a closed bounded interval, gets trapped on the right hand side of the origin and ultimately attains a triangular shape.

1.6 Shock structure, dissipation and entropy condition

Shock front, a surface of discontinuity, is a mathematical idealization of a physical phenomenon in which a physical variable, say pressure, density or particle velocity vary continuously but rapidly in a narrow zone containing the surface. This idealization, in which the model conservation law (1.3.3) is valid, breaks down across a shock. In (1.3.3) the *flux* F has been taken to be a function of the *density* u alone whereas in certain physical phenomena F depends not only on u but also on its gradient u_x:

$$F = Q(u) - \nu u_x \, , \quad \nu > 0 \tag{1.6.1}$$

where ν is a small constant. In an interval where u_x is of order one, the term νu_x is negligible compared to $Q(u)$. But, in an interval where the gradient u_x is large and of the order of $1/\nu$, the two terms are comparable and the model equation (1.3.3) needs to be modified. Choosing $H = u, Q(u) = \frac{1}{2}u^2$ as in (1.3.1) and assuming that the function u is smooth, we get a modification of the equation (1.1.6) in the form (Lighthill (1954))

$$u_t + uu_x = \nu u_{xx}, \ \nu > 0 \tag{1.6.2}$$

In order to study how a shock front in a solution of (1.3.1) is replaced by a continuous solution of (1.6.2) with a narrow zone of rapid variation in u, we consider the solution

$$u(x,t) = \begin{cases} u_l \quad , \quad x \le St \\ \\ u_r \quad , \quad x > St \end{cases} \tag{1.6.3}$$

of (1.3.1) where the shock velocity $\dot{X}(t) = S$ satisfies

$$S = \frac{1}{2}(u_l + u_r) \, , \ u_l > u_r \tag{1.6.4}$$

This solution becomes steady in a frame of reference moving with the velocity S and hence we look for a solution of (1.6.2) in the form

$$u(x,t) = u(\xi), \ \ \xi = x - St \tag{1.6.5}$$

i.e., we solve the two point boundary value problem for the equation

$$-Su_\xi + uu_\xi - \nu u_{\xi\xi} = 0 \tag{1.6.6}$$

satisfying

$$\lim_{\xi\to-\infty} u(\xi) = u_l, \quad \lim_{\xi\to\infty} u(\xi) = u_r \tag{1.6.7}$$

We further assume that this solution tends to the two limiting values u_l and u_r smoothly i.e.,

$$\lim_{\xi\to\pm\infty} u_\xi(\xi) = 0 \tag{1.6.8}$$

Integrating (1.6.6) and using (1.6.7) we get $-Su + \frac{1}{2}u^2 - \nu u_\xi =$ a constant $= A$, say. In order that the solution tends to u_l and u_r as x tends infinity on the two sides, the constant A should be such that we can write this equation in the form

$$u_\xi = -\frac{1}{2\nu}(u_l - u)(u - u_r) \tag{1.6.9}$$

with

$$u_l = S + \sqrt{S^2 + 2A}\ , \quad u_r = S - \sqrt{S^2 + 2A} \tag{1.6.10}$$

Since u_l and u_r are real we must have $S^2+2A > 0$. Integrating (1.6.9) we get

$$u(x-St) \equiv u(\xi) = \frac{1}{2}\left[(u_l + u_r) - (u_l - u_r)\tanh\left\{\frac{u_l - u_r}{4\nu}(x - St)\right\}\right] \tag{1.6.11}$$

where we have chosen the constant of integration such that $u = \frac{1}{2}(u_l + u_r)$ at $\xi = 0$. We note that $u \to u_r$ as $x \to \infty$ and $u \to u_l$ as $x \to -\infty$.

(1.6.11) is a continuous solution of (1.6.2) joining the two *states*: u_l at $-\infty$ and u_r at $+\infty$ and thus represents the structure of a shock wave separating u_l on its left and u_r on its right. The shock speed $S = \frac{1}{2}(u_l + u_r)$ is the speed of translation of the whole continuous profile in the structure of the shock.

Since the transition from u_l to u_r takes place over an infinite distance, we measure the shock thickness by a distance over which transition equal to a given fraction $1-2\alpha$, where α is a small positive number, of the shock strength is observed. Let us denote by ξ_- the value of ξ where $u(\xi_-) = u_l - \alpha(u_l - u_r)$ and by ξ_+ that where $u(\xi_+) = u_r + \alpha(u_l - u_r)$. Here $0 < \alpha < \frac{1}{2}$. Then from (1.6.11), we obtain $\xi_+ + \xi_- = 0$ and the following expression for the shock thickness

$$\xi_+ - \xi_- = \frac{4\nu}{u_l - u_r} ln \frac{1-\alpha}{\alpha} \tag{1.6.12}$$

The shock thickness is inversely proportional to the shock strength $u_l - u_r$, showing that the transition through the shock from u_l to u_r takes place over a large distance for a weak shock and over a small distance for a strong shock. For a shock of moderate strength, the shock thickness is of the order of the *diffusion* coefficient ν. Since ν is small, the shock thickness is generally very small compared to the length scales that we normally consider in day-to-day life and, for all practical purposes, it can be treated as a surface of discontinuity as dealt with in the previous sections.

In the limit as $\nu \to 0+$, the travelling wave solution (1.6.11) satisfying (1.6.7) becomes a solution

$$u(x,t) = \begin{cases} u_l \quad , & x - \frac{1}{2}(u_l + u_r)t \le 0 \\ \\ u_r \quad , & x - \frac{1}{2}(u_l + u_r)t > 0 \end{cases} \tag{1.6.13}$$

of (1.3.1) with a single shock. We also note that an expansion shock solution (1.6.13) with $u_l < u_r$ can not be obtained as a limit of a solution of the *viscosity equation* (1.6.2) because the expression on the right hand side of (1.6.9) is negative for $u_l < u < u_r$ and hence, in any solution of (1.6.9), u can not increase from u_l to u_r as ξ increases. This is a general property of the conservation law (1.3.1): every weak solution of this equation with only discontinuities which are shocks can be obtained as a limit almost everywhere of a solution of the viscosity equation (1.6.2) as $\nu \to 0+$. The converse is also true. All solutions of (1.6.2) are smooth. The limit of any solution of this equation is a weak solution of (1.3.1) with discontinuities which, if any, are shocks. Thus, an entropy solution i.e., a stable solution of (1.3.1), can be defined as a limit of a viscosity equation like (1.6.2). We briefly discuss now how to define an entropy function and what role it plays.

Entropy function

We have seen in the section 1.3 that the scalar equation (1.2.1) has an infinity of conservation forms. This need not be true for a system of equations – there may not exist even one. In this section we restrict to the scalar equation and consider weak solutions of the conservation law (1.3.1). Let us take up one more conservation law (1.3.3), namely

$$(E(u))_t + (G(u))_x = 0,\ E'(u) \neq 0,\ G'(u) = uE'(u) \tag{1.6.14}$$

with an additional condition that E is a strictly convex function i.e.,

$$E''(u) > c > 0 \tag{1.6.15}$$

where c is a constant. Since such a function plays the role of entropy in gas dynamics, we call E also to be an *entropy function*. We also assume that $E \geq 0$. However, it turns out that unlike the physical entropy, E defined here decreases across a shock.

Consider now a solution u^ν of the viscosity equation (1.6.2) satisfying $u^\nu(x,0) = u(x,0)$. Multiplying (1.6.2) by E', we write it in the form

$$E_t + G_x = (\nu E' u_x^\nu)_x - \nu (u_x^\nu)^2 E'' \tag{1.6.16}$$

Assume that the solution u^ν tends to zero sufficiently rapidly as $x \to \pm\infty$ and the functions $E(u^\nu)$ and $G(u^\nu)$ also tend to zero sufficiently rapidly there. Integrating the equation (1.6.16) on the strip $S_T \equiv (-\infty < x < \infty,\ 0 < t < T)$ and using Green's theorem we get

$$\int_{-\infty}^{\infty} E(u^\nu(x,T))dx - \int_{-\infty}^{\infty} E(u^\nu(x,0))dx = -\nu \iint_{S_T} (u_x^\nu)^2 E''\, dxdt \tag{1.6.17}$$

Since $E''(u) > 0$, the right hand side of this result is negative for a non-constant solution and hence we get an important result in the limit

$$\int_{-\infty}^{\infty} E(u(x,T))dx \leq \int_{-\infty}^{\infty} E(u(x,0))dx \tag{1.6.18}$$

This result shows that in a solution of the conservation law (1.3.1) the integral of the quantity E at any time $T > 0$ is a nonincreasing function of T, showing that E has the property of the negative of the physical entropy dictated by the second law of thermodynamics. This law says that the physical entropy of a system can not decrease with time. Note that convexity condition $E^{''}u) > 0$ of E plays an important role in this result.

Using (1.6.15) in (1.6.17), we get an important result

$$\int_{-\infty}^{\infty} E(u^\nu(x,T))dx + \nu c \iint_{S_T} (u_x^\nu)^2 dxdt \leq \int_{-\infty}^{\infty} E(u^\nu(x,0))dx$$
$$= \int_{-\infty}^{\infty} E(u(x,0))dx \quad (1.6.19)$$

which implies that

$$\nu \iint_{S_T} (u_x^\nu)^2 dxdt \leq a \text{ constant independent of } \nu \qquad (1.6.20)$$

where the constant depends on T, c and the initial entropy. (1.6.20) remains true in the limit $\nu \longrightarrow 0+$. Thus, taking limit of (1.6.17) as $\nu \longrightarrow 0$, we get the following result

Let u be a weak solution of the conservation law (1.3.1) which decays to zero at infinity sufficiently rapidly and which is obtained as a limit of a solution of the viscosity equation (1.6.2), then for $E^{''} =$ constant,

$$\int_{-\infty}^{\infty} E(u(x,T))dx - \int_{-\infty}^{\infty} E(u(x,0))dx = C(T) \qquad (1.6.21)$$

where $C \leq 0$. Note that $C = 0$ when the weak solution is smooth.

It is interesting to calculate the value of C in a weak solution of (1.3.1) when the solution contains a single shock joining a state u_l on the left to a state u_r on the right. An example of such a solution is (1.5.4). In such a solution u_x (and u_x^ν for small ν) remains bounded in the shock free region of u and hence the only contribution to C is from a small neighbourhood of the shock. Therefore, the constant C can be calculated from the shock structure solution (1.6.11). In this case

$$u_x^\nu = u_\xi = -\frac{(u_l - u_r)^2}{8\nu} sech^2 \left(\frac{u_l - u_r}{4\nu} \xi \right) \qquad (1.6.22)$$

so that if we take $E(u) = \frac{1}{2}u^2$ i.e., $E''(u) = 1$, then

$$C = -\lim_{\nu \to 0} \nu \iint_{S_T} (u_x^\nu)^2 dxdt = -\frac{1}{12} \int_0^T (u_l - u_r)^3 dt \tag{1.6.23}$$

Thus we have proved the following theorem for $\frac{dC}{dT}$.

Theorem 1.6.1 The time rate of decrease of the mathematical entropy $E = \frac{1}{2}u^2$ due to a single shock in a weak solution of (1.3.1) is $\frac{1}{12}(u_l - u_r)^3$.

As we have pointed out, there is an infinity of strictly convex functions E for a single conservation law (1.3.1). It will be shown in Chapter 6 (see expression for $\mathbf{q}$ in (6.1.1)) that $E = \frac{1}{2}u^2$ represents the kinetic energy density in a weakly nonlinear high frequency wave running into a polytropic gas at rest. Therefore, if no fresh kinetic energy is supplied, the theorem gives an expression for the rate of decay or *dissipation* (due to a single shock) of the kinetic energy. Since the total energy remains conserved, the kinetic energy lost gets converted into the internal energy in the form of heat.

We have already given an argument (see soon after (1.6.13)) that a limit solution of the viscosity equation (1.6.2) can not contain an expansion shock (i.e., $u_l < u_r$). We present here a more general proof using the entropy function. This proof is valid also for a system of conservation law.

Let x_1, x_2 be two fixed points and an integral form of the equation (1.6.16) is

$$\frac{d}{dt}\int_{x_1}^{x_2} E(u^\nu(x,t))dx = \{G(u^\nu(x_1,t)) - G(u^\nu(x_2,t))\}$$

$$-\nu\{E'(x_1,t)u_x^\nu(x_1,t) - E'(x_2,t)u_x^\nu(x_2,t)\} - \nu\int_{x_1}^{x_2} (u_x^\nu)^2 E''(u^\nu)dx \tag{1.6.24}$$

Consider now a solution of (1.6.2) with small ν such that it has a sharp gradient in a region which tends to a point of discontinuity of a solution u of (1.3.1) at $x = X(t) \equiv St$ in the interval (x_1, x_2) as $\nu \to 0+$. The derivative u_x^ν remains finite at x_1 and x_2 as $\nu \to 0$ so that the second term on the right hand side of (1.6.24) tends to zero as $\nu \rightarrow 0$. According to (1.6.20) the third term on the right hand side remains bounded and negative as $\nu \rightarrow 0$. Therefore, taking the limit of (1.6.24) as $\nu \rightarrow 0$, we get

$$\frac{d}{dt}\int_{x_1}^{x_2} E(u(x,t))dx < G(u(x_1,t)) - G(u(x_2,t)) \tag{1.6.25}$$

Following the procedure in the derivation of (1.3.5) from (1.3.4) we get from (1.6.25) an important result

$$\dot{X}(t)\{E(u_l(t)) - E(u_r(t))\} < \{G(u_l(t)) - G(u_r(t))\} \tag{1.6.26}$$

for an entropy function E. Now we have proved a theorem.

Theorem 1.6.2 Let u be a piecewise continuous solution of (1.3.3) obtained as a limit of the viscosity equation (1.6.2) as $\nu \rightarrow 0$ and let E be an entropy function satisfying (1.6.15) and $G(u) = \int^u uE'(u)du$, then along each discontinuity u satisfies (1.6.26), where $\dot{X}$ is given by (1.3.8).

The condition (1.6.26) is called an *entropy condition.* We shall now show that for the entropy function $E = \frac{1}{2}u^2$ the condition (1.6.26) i.e.,

$$\frac{1}{2}\dot{X}(t)\{u_l^2 - u_r^2\} < \frac{1}{3}\{u_l^3 - u_r^3\} \tag{1.6.27}$$

with

$$\dot{X}(t) = \frac{1}{2}(u_l + u_r) \tag{1.6.28}$$

is equivalent to (1.4.8) i.e.,

$$u_r(t) < \dot{X} < u_l(t) \tag{1.6.29}$$

The proof is quite simple. The relation (1.6.27) i.e.,

$$(u_l - u_r)\{\frac{3}{2}\dot{X}(u_l + u_r) - (u_l^2 + u_l u_r + u_r^2)\} < 0$$

with (1.6.28) can be written as

$$(u_l - u_r)\{\dot{X}^2 + \dot{X}(u_l + u_r) - (u_l^2 + u_l u_r + u_r^2)\} < 0$$

i.e., $(u_l - u_r)\{\dot{X}^2 - \dot{X}(u_l + u_r) + 2\dot{X}(u_l + u_r) - (u_l^2 + u_l u_r + u_r^2)\} < 0$

which again with the help of (1.6.28), becomes

$$(u_l - u_r)\{\dot{X}^2 - \dot{X}(u_l + u_r) + u_l u_r\} < 0$$

$$\text{i.e.,} \ -(u_l - u_r)(\dot{X} - u_r)(u_l - \dot{X}) < 0 \tag{1.6.30}$$

Since $\dot{X}$ is the mean of u_l and u_r, the product of the last two factors is positive. Hence, (1.6.29) is true if $u_l > u_r$ which is equivalent to (1.6.28).

1.7 The persistence of a shock

In section 1.2 we showed that if in an initial value problem for (1.2.1), the initial value $u_0(x)$ has a negative gradient (i.e., $u_0'(x) < 0$), a shock necessarily appears in the solution at a finite time. Thus, a shock may originate in the (x, t)-plane from a point at a finite distance from the origin. In this section we shall prove another result which implies that a shock can not disappear in a solution at a finite time, i.e., a shock once formed will persist for all time. In order to prove this important result, we need another result which is given by the equation (1.7.5) below.

We assume a weak solution of (1.3.1) with a single curve of discontinuity $\Omega : x = X(t)$, which is a shock and which meets the initial line $t = 0$ at $x = X(0) = X_0$, say. In the left subdomain $x < X(t)$, let us draw the characteristic curve C_ℓ which starts from a point $x = \xi_\ell(t)$ and meets the shock at the point $(X(t), t)$. Similarly, we draw the characteristic curve C_r in the right subdomain $x > X(t)$ starting from a point $x = \xi_r(t)$ and meeting Ω at the same point $(X(t), t)$. C_ℓ and C_r are straight line segments as shown in Fig. 1.7.1. In the domain $\mathcal{D}_\ell$ (or $\mathcal{D}_r$) bounded by C_ℓ (or C_r), a part I_ℓ (or I_r) of the initial line $t = 0$ between $\xi_\ell(t)$ (or $\xi_r(t)$) and X_0, and the section Ω^t of Ω between $(X_0, 0)$ and $(X(t), t)$, the solution is smooth.

Integrating (1.3.1) over $\mathcal{D}_\ell$ we get

$$0 = \int_{\mathcal{D}_\ell} (u_t + (\frac{1}{2}u^2)_x) dx dt = \int_{I_\ell + \Omega^t + C_\ell} (u n_t + \frac{1}{2}u^2 n_x) ds \tag{1.7.1}$$

where (n_t, n_x) are components of the unit normal and ds is an element of arc length. Now

$$\int_{I_\ell} (un_t + \frac{1}{2}u^2 n_x)ds = -\int_{\xi_\ell(t)}^{X_0} u_0(\xi)d\xi$$
$$\int_{\Omega^t} (un_t + \frac{1}{2}u^2 n_x)ds = \int_{\Omega^t} (\frac{1}{2}u_\ell^2 - \dot{X}u_\ell)dt$$

and

$$\int_{C_\ell} (un_t + \frac{1}{2}u^2 n_x)ds = \frac{1}{2}u_\ell^2 t$$

since C_ℓ is a straight line along which $\frac{dx}{dt} = u_\ell$, u_ℓ being constant.

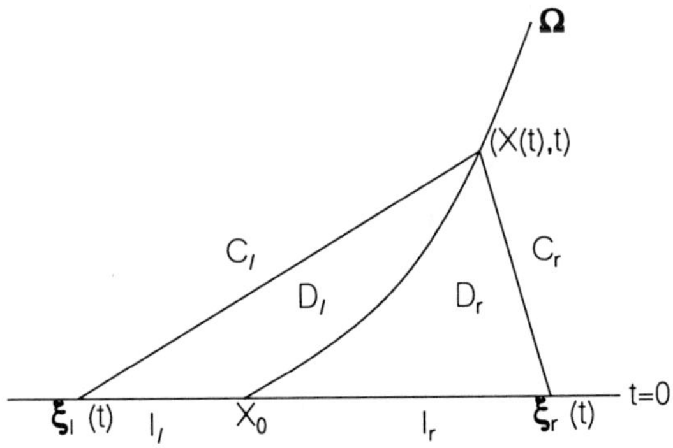

Fig. 1.7.1: $\Omega : x = X(t)$ is the only curve of discontinuity in an otherwise smooth solution.

Substituting in (1.7.1) we get

$$0 = -\int_{\xi_\ell(t)}^{X_0} u_0(\xi)d\xi + \int_{\Omega^t} (\frac{1}{2}u_\ell^2 - \dot{X}u_\ell)dt + \frac{1}{2}u_\ell^2 t \qquad (1.7.2)$$

Similarly integrating (1.3.1) over $\mathcal{D}_r$ we get

$$
\begin{aligned}
0 &= \int\limits_{\mathcal{D}_r} (u_t + (\frac{1}{2}u^2)_x)dxdt = \int\limits_{I_r+\Omega^t+C_\ell} (un_t + \frac{1}{2}u^2n_x)ds \\
&= -\int\limits_{X_0}^{\xi_r(t)} u_0(\xi)d\xi - \int\limits_{\Omega^t} (\frac{1}{2}u_r^2 - \dot{X}u_r)dt - \frac{1}{2}u_r^2 t
\end{aligned} \tag{1.7.3}
$$

Adding (1.7.2) and (1.7.3); and using the jump condition in the form (1.3.10)

$$
\frac{1}{2}(u_\ell^2(t) - u_r^2(t))t = \int\limits_{\xi_\ell(t)}^{\xi_r(t)} u_0(\xi)d\xi \tag{1.7.4}
$$

Since $u_r(t) = u_0(\xi_r(t))$ and $u_\ell(t) = u_0(\xi_\ell(t))$ we can write this relation purely in terms of the initial data:

$$
t = \frac{2}{u_0^2(\xi_\ell(t)) - u_0^2(\xi_r(t))}\left[\int\limits_{\xi_\ell(t)}^{\xi_r(t)} u_0(\xi)ds\right] \tag{1.7.5}
$$

Now we are ready to prove the main result of the persistence of the shock. We state it more precisely in the form of a theorem.

Theorem 1.7.1 A shock path can not terminate at a point at a finite distance from the origin in the (x,t)-plane. It will either extend up to infinity or will join another shock to form a third shock.

Proof If a shock path terminates at a point (x^*, t^*), then all points (x,t) on the shock would satisfy $t < t^*$. If $u_0(\xi_\ell(t^*)) \neq u_0(\xi_r(t^*))$, the shock path extends further as a non-characteristic curve according to (1.3.10). Therefore,

$$
\lim_{t\to t^*-0} \left\{u_0(\xi_\ell(t)) - u_0(\xi_r(t))\right\} = 0 \tag{1.7.6}
$$

The quantity in the square bracket on the right hand side of (1.7.5) represents the area $A(t)$ between the x-axis and the curve representing the initial data $u_0(x)$ between $x = \xi_\ell(t)$ and $x = \xi_r(t)$. Now we consider two cases:

Case (a):

$$A(t^*) = \int_{\xi_\ell(t^*)}^{\xi_r(t^*)} u_0(x)dx \neq 0 \tag{1.7.7}$$

From (1.7.6) and (1.7.7) it follows that the right hand side of (1.7.5) tends to infinity as $t \to t^* - 0$. Hence, (1.7.6) implies that t^* can not be finite.

Case (b): $A(t^*) = 0.$

We note that the equation $u_t + uu_x = 0$ and the jump condition (1.3.10) are invariant under the transformation

$$t' = t, \quad x' = x - at, \quad u' = u - a \tag{1.7.8}$$

Therefore, by choosing a to be smaller than a lower bound of $u_0(x)$ on $\mathbf{R}$, we can always make the above change of variables such that the initial data $u_0'(x)$ for u' is greater than 0 for all $x \in \mathbf{R}$. Then $A' > 0$. Therefore, this case can also be reduced to a problem in which $A \neq 0$ and the arguments of the case (a) applies.

This completes the proof of the theorem.

The theorem is also true for a shock appearing in the genuinely nonlinear characteristic field of an arbitrary system of conservation laws. This is because when a shock path terminates at a point (x^*, t^*), the shock strength must vanish at this point. Therefore, due to continuity, shock strength remains arbitrarily small just before the time t approaches t^*. We shall show (see arguments leading to the equation (5.2.15)) that the propagation of a weak shock (in a genuinely nonlinear characteristic field) is governed approximately by the conservation law (1.3.1). Hence, the result of (1.3.1) regarding the persistence of a shock is valid for a general system of hyperbolic conservation laws.

1.8 Nonlinear wavefront and shock front

Solution $u = u_0(\xi)$, $\xi = x - ct$, of the partial differential equation (1.1.1) satisfying the initial condition $u(x, 0) = u_0(x)$ represents a pulse which is characterized by one parameter family of wavefronts represented by $x = ct + \xi$ where $\xi \in I\!R$ is the parameter. u_0 need not be $C^1(I\!R)$ in which case, u is a suitably defined generalized or weak solution. If u_0 is continuous, we get a continuous pulse and the amplitude of the wave varies continuously across all wavefronts, each one of which moves with the same velocity c. If u_0 has a discontinuity at a point x_0, we get a wavefront $x = ct + x_0$ across which the amplitude is discontinuous. For a linear equation like (1.1.1), kinematic properties of a wavefront across which the amplitude is continuous and another for which it is discontinuous are the same. Both types of fronts move with the same velocity c.

For a nonlinear equation (1.1.6) with a suitable conservation law, say (1.3.1), a distinction has to be made between a one parameter family of wavefronts $x = tu_0(\xi) + \xi$ forming a continuous pulse given by $u = u_0(x - ut)$ and a discontinuous front i.e., a shock front. A shock is also a wavefront, if we interpret the meaning of the word as used commonly but a wavefront (across which the amplitude varies continuously) and a shock front have different kinematic properties as explained below.

In this monograph, we shall make a clear distinction between the use of the words *nonlinear wavefront* and *shock front.* Consider a pulse in a genuinely nonlinear characteristic field. *When the amplitude of the pulse varies continuously across a front, it will be called a nonlinear wavefront and when the amplitude is discontinuous across it, it will be called a shock front.* This is not a new definition of a shock front but only a reiteration of the definition in section 1.4.

A nonlinear wavefront in a pulse governed by (1.1.6) is determined by solving the equations (1.2.3 - 4) with initial conditions $u = u(x_0)$, $x = x_0$ at $t = 0$. Thus, to calculate the successive positions of a nonlinear wavefront and amplitude on it, we need only its initial position and initial amplitude. Thus, a nonlinear wavefront is *self-propagating* in the sense that its evolution is determined only by the information on itself and is not influenced by wavefronts which follow or precede it. Kinematically, a shock front is not self-propagating. We shall show in Chapter 7 that its evolution is governed by solving the equation (1.3.10) and an infinite system of equations involving u_l and some other quantities for which we require initial position X_0 of the shock and initial value $u_0(x)$ for x on an interval of the real line and not just at X_0. We have already noted in examples in section 1.5 that the shock interacts with the nonlinear waves ahead of it and behind it and *swallows* them gradually so that information on an interval of increasing length gets lost. We have also noticed, in an example in section 1.4, that a physical process which has a shock is irreversible. This irreversibility and a shock being not self-propagating are intimately related. A nonlinear wavefront gets lost after it interacts with a shock front. Though we may integrate the equations (1.2.3 - 4) with respect to t even beyond the time the nonlinear wavefront interacts with the shock, the values of $x(t)$ and $u(t)$ obtained as a result of integration have no physical significance after the interaction.

Distinction between a linear wavefront, a nonlinear wavefront and a shock front is very easily understood in the case of one -dimensional or plane fronts. This becomes very complex for waves in multi-dimensions due to the convergence or divergence of rays and due to nonlinear waves which propagate on the fronts themselves. Some of these questions are raised in section 3.3.1. To understand the answers to these questions, the reader needs to read the entire monograph.

1.9 Hopf's result on the general solution

Examples in section 1.5 may give a false impression that given an initial value problem for the conservation law (1.3.1), it is simple to obtain the weak solution with shocks. The initial values there were chosen carefully so that not only their solutions could be easily derived but they would highlight the most important properties of solutions. For a general initial value $u_0(x)$, even the smooth solution is to be obtained by solving the implicit relation (1.2.5) which itself is a difficult problem. An explicit expression for the solution of an initial value problem with an arbitrary initial data for the equation (1.3.1) was first given by Hopf in 1950. We mention here Hopf's result in the form of a theorem:

Theorem 1.9.1 A weak solution of (1.3.1) with initial value (1.2.2) is given by

$$u(x, t) = \frac{x - y(x, t)}{t} \tag{1.9.1}$$

where $y(x, t)$ minimizes

$$G(x, y, t) \equiv U_0(y) + \frac{(x - y)^2}{2t} \tag{1.9.2}$$

with

$$U_0(y) = \int_{-\infty}^{y} u_0(x)dx \tag{1.9.3}$$

It is assumed that the initial value u_0 is integrable on $\mathbb{R}$ so that U_0 given by (1.9.3) is defined.

Proof We first prove the theorem when u is a genuine solution of (1.3.1). Assuming $u_0(x)$ and $u(x,t)$ to tend to zero as $x \to -\infty$ sufficiently rapidly, we define U_0 by (1.9.3) and $U(x,t)$ by

$$U(x,t) = \int_{-\infty}^{x} u(\xi,t)d\xi \tag{1.9.4}$$

Then

$$u = U_x \tag{1.9.5}$$

The integral form (1.3.5) of (1.3.1) ($H = u, F = \frac{1}{2}u^2$) with $x_1 \to -\infty$ and $x_2 = x$ gives

$$U_t + \frac{1}{2}(U_x)^2 = 0 \tag{1.9.6}$$

Using the inequality

$$\frac{1}{2}u^2 \geq -\frac{1}{2}v^2 + uv \ , \quad \forall\, v \in \mathbb{R} \tag{1.9.7}$$

in the above with $u = U_x$, we get

$$U_t + vU_x \leq \frac{1}{2}v^2 \ , \quad \forall\, v \in \mathbb{R} \tag{1.9.8}$$

Let us denote now by y a point where the line with constant slope $\frac{1}{v}$ through (x,t) interacts the x-axis, then

$$v = \frac{x-y}{t} \tag{1.9.9}$$

Integrating (1.9.8) along this line we get

$$U(x,t) \leq U_0(y) + \frac{1}{2}tv^2 = U_0(y) + \frac{(x-y)^2}{2t} \ , \qquad \forall\, y \in \mathbb{R} \tag{1.9.10}$$

The equality in (1.9.7) and hence in (1.9.10) is valid only when $v = u$, hence,

$$U(x,t) = \min_{y \in \mathbb{R}} G(x,y,t) \tag{1.9.11}$$

The minimizing value $y(x,t)$ gives the solution $u(x,t)$ in (1.9.9). This completes the proof of the theorem when u is a genuine solution.

Consider now a weak solution of (1.3.1). The relation (1.9.5) and equation (1.3.5) are valid almost everywhere. As above, the inequality (1.9.7) leads to (1.9.8). If all discontinuities of the generalized solution $u(x,t)$ are shocks, then every point $(x,t), t>0$ can be connected to a point y on the initial line by a backward characteristic. For that value of y, the sign of equality holds in (1.9.10). Therefore the theorem 1.9.1 is true also for a generalized solution whose discontinuities are shocks. The converse of the theorem is also true.

Theorem 1.9.2 For an arbitrary *integrable* initial value $u_0(x)$, the formula (1.9.1) defines a unique function $u(x,t)$, possibly discontinuous. The function u so defined is a weak solution of (1.3.1), the discontinuities of u are shocks and $u_0(x)$ is a *weak limit* of $u(x,t)$ as $t \to 0+$.

With mathematical tools used so far, we can not prove this theorem – in fact we have not explained the meaning of *integrable* function and *weak limit*. We mention an important result which we derive while proving the theorem. At any fixed time $t>0$, the solution $u(x,t)$ can be shown to be of bounded variation on any finite interval on the x-axis. Thus, even though the initial data is merely integrable (an example of such a function is given below), the solution becomes fairly regular. For example, a function having locally bounded total variation has, at most, a countable number of jump discontinuities and is differentiable almost everywhere. This regularity is brought about by the genuine nonlinearity present in the equation. As an example of even more dramatic regularity produced by the genuine nonlinearity, we consider initial value problems with two initial data.

$$u_{01}(x) = e^{-x^2} p(x) \ , \ p(x) = \begin{cases} 1 \ , & \text{when } x \text{ is irrational} \\ 0 \ , & \text{when } x \text{ is rational} \end{cases} \tag{1.9.12}$$

and

$$u_{02}(x) = e^{-x^2} \tag{1.9.13}$$

The two initial values differ only on rational points, which though dense in $\mathbb{R}$, form a set of measure zero. The function $U(y)$ in (1.9.3) for both functions u_{0_1} and u_{0_2} is the same. The weak solution corresponding to both initial data is also the same, the solution being C^∞ in a strip: $x \in I\!R,\ 0 < t < t_c \simeq 1.166$ as shown in Fig. 1.1.1.

In the first half of the 1950s, Hopf's result was extended independently by Lax and Oleinik to the single conservation law

$$u_t + (F(u))_x = 0 \tag{1.9.14}$$

where the flux function F is convex and smooth but otherwise arbitrary. This extension gives an expression for $u(x,t)$ in terms of x, t and $y(x,t)$, where $y(x,t)$ minimizes a suitably defined function of y with x and t as parameters. This beautiful result is known as the Lax-Oleinik formula. A detailed discussion is available in a recent book by Evans (1998).

The explicit formula for the weak solution, though very elegant and also useful for deriving many interesting properties including asymptotic form of the solution as $t \to \infty$, are of limited use for pointwise numerical evaluation of the solution in the (x,t)-plane. Such an explicit formula is not available for a system of two or more conservation laws. Hence, it is of great importance to devise numerical or semi-analytical methods for the pointwise evaluation of the solution.

1.10 Equal area rule for shock fitting

A simple geometrical method for shock fitting in a solution of a single conservation law (1.3.1) is the famous equal area rule. This method is applicable when there is only one shock (after the critical time t_c, (see section 1.2)) in a solution arising out of an initially continuous data.

The solution given by (1.2.5) is obtained with the help of the equations (1.2.3 - 4) and can be parametrically represented as

$$u(\xi, t) = u_0(\xi) \tag{1.10.1}$$

$$x(\xi, t) = \xi + t u_0(\xi) \tag{1.10.2}$$

(1.10.2) represents the equation of a characteristic which starts from a point ξ on the x-axis and which carries a constant value $u_0(\xi)$. In what follows, we assume u_0 to be smooth but we also assume that when we solve (1.3.1) with this initial condition, a shock appears at

$t = t_c$ in the solution and, for $t > t_c$, this solution has no other shock. The function $u(x,t)$, given by (1.10.1 - 2), is not defined for $t > t_c$. However, for $t > t_c$ we denote by $\bar{u}(x,t)$ the *multi-valued function* (it is not a function in the strict sense) obtained by eliminating ξ from these equations.

The multi-valuedness in $\bar{u}$ appears because the partial derivative x_ξ of $x(\xi,t)$ obtained by differentiating (1.10.2) vanishes at points $\xi_0(t)$ for $t > t_c$. In general, there are two points ξ_{0-} and ξ_{0+} for each value of t and $\xi_{0-}(t_c) = \xi_{0+}(t_c)$. The corresponding points $x_{0-}(t) = x(\xi_{0-}(t),t)$ and $x_{0+}(t) = x(\xi_{0+}(t),t)$ trace the two branches of a cusp (starting at t_c) which bound the domain in the (x,t)-plane where the solution is multi-valued (Fig. 1.10.1). We denote by $u_-(t)$ and $u_+(t)$, the values

$$u_-(t) = u_0(\xi_{0-}(t)), \quad u_+(t) = u_0(\xi_{0+}(t)) \tag{1.10.3}$$

Let the characteristics starting from points $\xi_l(t)$ and $\xi_r(t)$ on the x-axis (initial line) meet the shock at $x = X(t)$ at a time $t > t_c$.

Then the values u_l and u_r on the two sides of the shock are obtained from

$$u_l(t) = u_0(\xi_l(t)), \quad u_r(t) = u_0(\xi_r(t),t) \tag{1.10.4}$$

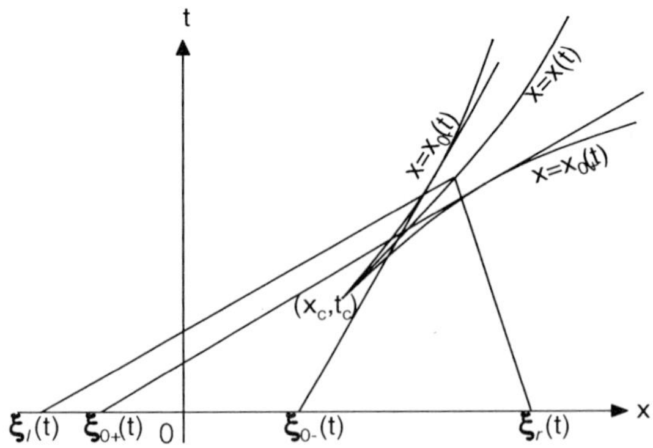

Fig. 1.10.1: Geometry of characteristics in (x,t)-plane.

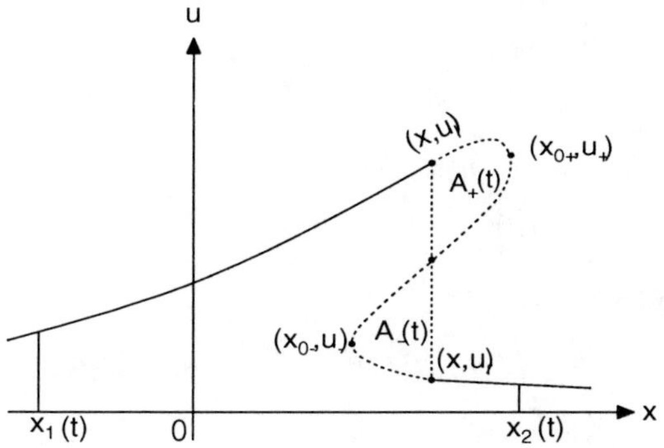

Fig. 1.10.2: The pulse at $t > t_c$ showing the values $u_-(t), u_+(t), u_l(t)$ and $u_r(t)$. The graph of the multi-valued function $\bar{u}(x,t)$ has been shown partly by a continuous line —— and partly by .... line. It also shows two points x_1 and x_2.

The function $u(x,t)$ is obtained from $\bar{u}(x,t)$ by cutting off a positive area $A_+(t)$ bounded by the line $x = X(t)$ and $\bar{u}(x,t)$ on the right of $X(t)$ and a negative area $A_-(t)$ with boundary $x = X(t)$ and $\bar{u}(x,t)$ on the left of $X(t)$ as shown in Fig. 1.10.2. We wish to prove that $A_+(t) + A_-(t) = 0$ for all $t > t_c$. Let ξ_1 and ξ_2 be two real numbers such that

$$x_1 = \xi_1 + tu_0(\xi_1) < X(t) < x_2 = \xi_2 + tu_0(\xi_2) \tag{1.10.5}$$

for t in a given time interval I_t. Then the areas A bounded by the graph of u between x_1 and x_2, and the x-axis is given by

$$A(t) = \int_{x_1}^{x_2} u(x,t)dx \tag{1.10.6}$$

Similarly, the area $\bar{A}$ bounded by the graph of $\bar{u}$ between x_1 and x_2 and the x-axis is given by

$$\bar{A}(t) = \int_{\xi_1}^{\xi_2} u_0(\xi)x_\xi d\xi \tag{1.10.7}$$

where x_ξ in (1.10.7) is obtained by differentiating (1.10.2). We can replace the integration with respect to x in (1.10.6) also by that with respect to ξ as follows

$$\begin{aligned} A(t) &= \int_{x_1(t)}^{X(t)} u(x,t)dx + \int_{X(t)}^{x_2(t)} u(x,t)dx \\ &= \int_{\xi_1}^{\xi_l(t)} u_0(\xi)x_\xi d\xi + \int_{\xi_r(t)}^{\xi_2} u_0(\xi)x_\xi d\xi \end{aligned} \tag{1.10.8}$$

From (1.10.6 - 8), it follows that the difference between the two areas is: $\Delta A = \bar{A} - A$ is given by

$$(\Delta A)(t) = \int_{\xi_l(t)}^{\xi_r(t)} u_0(\xi)x_\xi d\xi \tag{1.10.9}$$

Hence,

$$\frac{d(\Delta A)}{dt} = \int_{\xi_l(t)}^{\xi_r(t)} u_0(\xi)x_{\xi t}d\xi + u_0(\xi_r)(x_\xi)_r\dot{\xi}_r - u_0(\xi_l)(x_\xi)_l\dot{\xi}_l$$

From (1.10.2) we get $x_{\xi t} = u_0'(\xi)$ so that the first term on the right hand side becomes

$$\begin{aligned} \int_{\xi_l(t)}^{\xi_r(t)} u_0(\xi)\frac{du_0(\xi)}{d\xi}d\xi &= \tfrac{1}{2}\{u_0^2(\xi_r) - u_0^2(\xi_l)\} \\ &= -\tfrac{1}{2}\{u_0^2(\xi_r) - u_0^2(\xi_l)\} + \{u_0^2(\xi_r) - u_0^2(\xi_l)\} \end{aligned}$$

Therefore,

$$\begin{aligned} \tfrac{d}{dt}(\Delta A) = -\tfrac{1}{2}\{u_0^2(\xi_r) - u_0^2(\xi_l)\} &+ u_0(\xi_r)\{u_0(\xi_r) + (x_\xi)_r\dot{\xi}_r\} \\ &- u_0(\xi_l)\{u_0(\xi_l) + (x_\xi)_l\dot{\xi}_l\} \end{aligned} \tag{1.10.10}$$

From (1.10.2) $X(t) = tu_0(\xi_r(t)) + \xi_r(t)$, so that $\dot{X}(t) = u_0(\xi_r) + \{tu_0'(\xi_r) + 1\}\dot{\xi}_r = u_0(\xi_r) + (x_\xi)_r\,\dot{\xi}_r$, since again from (1.10.2)

$x_\xi = tu_0'(\xi) + 1$. Similar expressions are valid when the subscript r is replaced by l. The relation (1.10.10), after using $u_r = u_0(\xi_r)$ and $u_l = u_0(\xi_l)$, becomes

$$\frac{d}{dt}(\Delta A) = -\frac{1}{2}(u_r^2 - u_l^2) + (u_r - u_l)\dot{X} = 0 \tag{1.10.11}$$

The middle expression vanishes as written above due to the jump relation (1.3.10).

From (1.10.11), it follows that as t increases from t_c, ΔA remains constant and is equal to its value zero at $t = t_c$ i.e.,

$$\Delta A = \bar{A} - A = A_+ + A_- = 0, \quad \forall\ t > 0 \tag{1.10.12}$$

This is the equal area rule, which says that *the shock fits into the graph of the multi-valued function $\bar{u}$ in such a way that the positive area $A_+(t)$ is numerically equal to the negative area $A_-(t)$ for all $t > t_c$.*

The equal area rule is valid for solutions of more general equations in the form

$$u_t + (F(x, t, u))_x + c(x, t) = 0$$

where the flux function F is convex with respect to u (Anile, Hunter, Pantano and Russo (1993)) and c does not depend on u.

The equal area rule applied to the graph of $\bar{u}(x, t)$ can also be transferred to that of the initial value u_0. We first note that the graph of the initial data u_0 can be obtained from the graph of $\bar{u}$ by translating in parallel to x-direction each point of the latter by a distance $-\bar{u}t$. Let us translate similarly each point of the vertical line segment joining the points $P(X, u_l)$ and $Q(X, u_r)$ of Fig. 1.10.2. This translated line segment P_0Q_0 cuts off lobes of areas $A_{0+}(t)$ and $A_{0-}(t)$ from the graph of u_0. Note that the line segment P_0Q_0 joins two points (ξ_l, u_l) and (ξ_r, u_r) in (u, ξ)-plane. The mapping, which maps the graph of an initial data u_0 to a multi-valued solution $\bar{u}(x, t)$, also maps the line segment P_0Q_0 into the shock i.e., the line segment PQ. Under this mapping, the boundaries of the area $A_{0+}(t)$ and $A_{0-}(t)$ are mapped in those of $A_+(t)$ and $A_-(t)$. An argument, similar to that given above, implies that

$$\text{area of } A_{0+}(t) = \text{ area of } A_+(t), \qquad \text{area of } A_{0-}(t) = \text{ area of } A_-(t) \tag{1.10.13}$$

Therefore the pre-image P_0Q_0 *of the line segment* PQ *also cuts off lobes of equal area from the graph of the initial data.* As t increases, the line segment P_0Q_0 changes its position and the equal area property is being maintained all the time.

Chapter 2

Hyperbolic system - some basic results

Convention: In this chapter and in the rest of the book we shall use the summation convention that a repeated suffix in a term will represent the sum over the range of the suffix. Usually, the range of the suffix i, j or k will be $\{1, 2, \ldots, n\}$ and that of α, β or γ will be $\{1, 2, \ldots, m\}$.

Throughout the book all variables and constants are assumed to be real and all functions are real functions on appropriate domains.

2.1 Hyperbolic system of first order equations in two independent variables

2.1.1 Definition of a hyperbolic system

A first order system of n partial differential equations is of the form

$$A\frac{\partial \mathbf{u}}{\partial t} + B\frac{\partial \mathbf{u}}{\partial x} + \mathbf{C} = 0 \tag{2.1.1}$$

where $\mathbf{u} \in I\!R^n$, $\mathbf{C} \in I\!R^n$, $A \in I\!R^{n\times n}$, $B \in I\!R^{n\times n}$. When the coefficient matrices A and B and the vector $\mathbf{C}$ are functions of x and t only, the system is linear; for the system to be semilinear $\mathbf{C}$ depends also on $\mathbf{u}$. When A, B and $\mathbf{C}$ are functions of x, t and $\mathbf{u}$, the system is quasilinear. We assume that A is nonsingular. This system of equations is defined to be hyperbolic in an appropriate domain of the arguments of A and B if

(i) the matrix B has n real eigenvalues relative to the matrix A i.e., the n roots of the equation in λ

$$\det(B - \lambda A) = 0 \tag{2.1.2}$$

are real and

(ii) the dimension of the eigenspace of each eigenvalue is equal to its algebraic multiplicity.

We denote the eigenvalues by $c_1, c_2, \ldots, c_n$ and assume that

$$c_1 \leq c_2 \leq c_3 \leq \ldots \leq c_n \tag{2.1.3}$$

We also denote the left and right eigenvectors corresponding to c_i by $\mathbf{l}^{(i)}$ and $\mathbf{r}^{(i)}$, respectively:

$$\mathbf{l}^{(i)}(B - c_i A) = 0 \text{ and } (B - c_i A)\mathbf{r}^{(i)} = 0, \ i = 1, 2, \ldots, n \tag{2.1.4}$$

Note that $\mathbf{l}^{(i)}$ is a row vector and $\mathbf{r}^{(i)}$ is a column vector.

Characteristic curves corresponding to the eigenvalue c_i are curves in the (x, t)-plane given by the equation

$$\frac{dx}{dt} = c_i \tag{2.1.5}$$

For a linear or semilinear system $c_i = c_i(x, t)$ so that the characteristic curves are determined without any reference to a solution but for a quasilinear system they are determined for a particular solution $u(x, t)$ of (2.1.1) using $c_i = c_i(x, t, \mathbf{u}(x, t))$ in (2.1.5). Multiplying (2.1.1) by $\mathbf{l}^{(i)}$, we get the following *compatibility condition* along a characteristic curve of the *ith family*

$$\mathbf{l}^{(i)} A \left(\frac{\partial}{\partial t} + c_i \frac{\partial}{\partial x} \right) \mathbf{u} + \mathbf{l}^{(i)} \mathbf{C} = 0 \tag{2.1.6}$$

which means

$$\mathbf{l}^{(i)} A \frac{du}{dt} + \mathbf{l}^{(i)} \mathbf{C} = 0 \quad \text{along} \quad \frac{dx}{dt} = c_i \tag{2.1.7}$$

Note that if an eigenvalue c_i appears a number of times (say p) in (2.1.3) i.e., c_i is a multiple eigenvalue of multiplicity p, there exists p linearly independent eigenvectors corresponding to this eigenvalue and we get p independent compatibility conditions corresponding to it. The compatibility condition (2.1.6) shows that *in a Cauchy problem for the system (2.1.1) with a characteristic curve as a datum curve, the Cauchy data for* $\mathbf{u}$ *can not be arbitrarily prescribed on the characteristic curve.*

A system of *conservation laws* is a system of equations in divergence form

$$\mathbf{H}_t + \mathbf{F}_x = 0 \tag{2.1.8}$$

where $\mathbf{H} : I\!R^n \to I\!R^n$, $\mathbf{F} : I\!R^n \to I\!R^n$ are functions of the dependent variable $\mathbf{u} \in I\!R^n$. (2.1.8) is a *hyperbolic system of conservation laws* in a domain of $(u_1, \ldots, u_n)$-space if the system of partial differential equations derived from it i.e.,

$$A\mathbf{u}_t + B\mathbf{u}_x = 0 \tag{2.1.9}$$

where

$$A = \nabla_{\mathbf{u}}\mathbf{H}, \quad B = \nabla_{\mathbf{u}}\mathbf{F} \tag{2.1.10}$$

is hyperbolic in that domain.

One of the most important examples of a hyperbolic system of partial differential equations is the system of Euler's equations governing the motion of a polytropic gas for

$$\mathbf{u} = (\rho, q, p)^T \tag{2.1.11}$$

where ρ is the mass density, q the fluid velocity, p the pressure and γ is the ratio of specific heats assumed to be constant. These equations are

$$\rho_t + q\rho_x + \rho q_x = 0 \tag{2.1.12}$$

$$q_t + qq_x + \frac{1}{\rho}p_x = 0 \tag{2.1.13}$$

$$p_t + qp_x + \gamma p q_x = 0 \tag{2.1.14}$$

so that $A = I$, the identity matrix, $\mathbf{C} = 0$ and

$$B = \begin{bmatrix} q & \rho & 0 \\ 0 & q & \frac{1}{\rho} \\ 0 & \gamma p & q \end{bmatrix} \tag{2.1.15}$$

The eigenvalues are

$$c_1 = q - a\,, \quad c_2 = q\,, \quad c_3 = q + a \tag{2.1.16}$$

where a is the local sound velocity in the fluid relative to the fluid particles and is given by

$$a^2 = \gamma p/\rho \tag{2.1.17}$$

Since the eigenvalues are real and distinct, the eigenspace is complete and hence the system (2.1.12 - 14) is hyperbolic. The eigenvectors corresponding to the eigenvalue c_3 are

$$\mathbf{l}^{(3)} = \left[0,\ 1,\ \frac{1}{\rho a}\right]\ ,\ \mathbf{r}^{(3)} = \left[\frac{\rho}{a},\ 1,\ \rho a\right]^T \tag{2.1.18}$$

The compatibility condition along this family of characteristic curves is

$$\left(\frac{\partial}{\partial t} + (q+a)\frac{\partial}{\partial x}\right) q + \frac{1}{\rho a}\left(\frac{\partial}{\partial t} + (q+a)\frac{\partial}{\partial x}\right) p = 0 \tag{2.1.19}$$

The specific entropy is proportional to $p\rho^{-\gamma}$ (see also the relation (2.3.22)). We also state without proof that sound waves i.e., pressure waves propagate with the velocities $q \pm a$ and entropy or density waves propagate with the fluid velocity q.

The differential equations (2.1.12 - 14) can be derived from the three fundamental conservations laws representing conservation of mass, momentum and energy. In this case, the functions $\mathbf{H}$ and $\mathbf{F}$ in (2.1.8) are given by

$$\mathbf{H} = \begin{bmatrix} \rho \\ \rho q \\ \frac{1}{\gamma-1}p + \frac{1}{2}\rho q^2 \end{bmatrix}\ ,\ \mathbf{F} = \begin{bmatrix} \rho q \\ \rho q^2 + p \\ \frac{\gamma}{\gamma-1}qp + \frac{1}{2}\rho q^3 \end{bmatrix} \tag{2.1.20}$$

Note that specific internal energy density e is given by $e = \frac{1}{\gamma-1}\frac{p}{\rho}$ so that $\frac{1}{\gamma-1}\,p = \rho e$ represents the internal energy of the gas per unit volume. The quantity $\rho e + \frac{1}{2}\rho q^2$ is the total energy (= internal + kinetic) density.

2.1.2 A canonical form of a system of linear and semilinear equations

Consider a hyperbolic system (2.1.1) where the matrices A and B are functions of x and t only. The column vector $\mathbf{C}$ may depend nonlinearly on $\mathbf{u}$. The left and right eigenvectors for distinct eigenvalues

c_i, c_j(i.e., $c_i \neq c_j$) satisfy

$$\mathbf{l}^{(i)} A \mathbf{r}^{(j)} \begin{cases} = 0 & \text{for } i \neq j \\ \neq 0 & \text{for } i = j \end{cases} \tag{2.1.21}$$

Even for multiple eigenvalues, when $c_i = c_j$ we can choose $\mathbf{l}^{(i)}$ and $\mathbf{r}^{(j)}$ in such a way that the above relation is satisfied.

We make a change of dependent variables from $\mathbf{u} = (u_1, \ldots, u_n)^T$ to $\mathbf{w} = (w_1, \ldots, w_n)^T$ by

$$\mathbf{u} = \sum_{j=1}^{n} \mathbf{r}^{(j)} w_j \equiv R\,\mathbf{w} \tag{2.1.22}$$

where the column vectors of the matrix

$$R = (\mathbf{r}^{(1)}, \ldots, \mathbf{r}^{(n)}) \equiv (r_{ij} = r_i^{(j)}) \tag{2.1.23}$$

are the right eigenvectors. Substituting (2.1.22) in (2.1.1), pre - multiplying the result by $\mathbf{l}^{(i)}$ and dividing by $\mathbf{l}^{(i)} A \mathbf{r}^{(i)}$ we get

$$\frac{\partial w_i}{\partial t} + c_i \frac{\partial w_i}{\partial x} + \sum_{j=1}^{n} \Gamma_{ij} w_j + \Gamma_i = 0, \quad i = 1, 2, \ldots, n \tag{2.1.24}$$

where

$$\Gamma_{ij} = \mathbf{l}^{(i)} A \left(\frac{\partial \mathbf{r}^{(j)}}{\partial t} + c_i \frac{\partial \mathbf{r}^{(j)}}{\partial x} \right) / (\mathbf{l}^{(i)} A \mathbf{r}^{(i)}) \tag{2.1.25}$$

and

$$\Gamma_i = \mathbf{l}^i \mathbf{C} / (\mathbf{l}^{(i)} A \mathbf{r}^{(i)}) \tag{2.1.26}$$

Let Λ be the diagonal matrix with diagonal elements $c_1, c_2, \ldots, c_n$ and $\mathbf{F}$ be the column vector with elements $\sum_{j=1}^{n} \Gamma_{ij} w_j + \Gamma_i, \ i = 1, 2, \ldots, n$. The semilinear hyperbolic system (2.1.1) now reduces to the canonical form

$$\frac{\partial \mathbf{w}}{\partial t} + \Lambda \frac{\partial \mathbf{w}}{\partial x} + \mathbf{F} = 0 \tag{2.1.27}$$

In this process, we have proved the following theorem.

Theorem 2.1.1 Any semilinear hyperbolic system in two independent variables is equivalent to a symmetric hyperbolic system $I \frac{\partial \mathbf{w}}{\partial t} + \Lambda \frac{\partial \mathbf{w}}{\partial x} + \mathbf{F} = 0$ in which the matrices I and Λ are symmetric and one matrix I is positive definite. In the canonical form (2.1.27), I is an identity matrix and Λ is a diagonal matrix.

Definition 2.1.1 We call w_i *characteristic variable* of the *ith characteristic field.*

The importance of the canonical form (2.1.24) lies in the fact that the ith equation gives the time rate of change of the ith characteristic variable along the characteristic curves of the ith family. (2.1.22) expresses that any solution of the semilinear hyperbolic system consists of n parts $w_1, w_2, \ldots, w_n$. The part w_i propagates along the x-axis with a velocity equal to the ith eigenvalue c_i. The mutual interaction between these parts takes place only through the last term in (2.1.27), namely $F_i = \sum_{j=1}^{n} \Gamma_{ij} w_j + \Gamma_i$. Corresponding to a multiple characteristic c_q, say, of multiplicity p_q, there are p_q parts which move with the same velocity c_q, again the mutual interaction within these p_q parts and interaction with other parts take place only through the third term $\mathbf{F}$. For a linear homogeneous system (so that $\mathbf{C} = 0$) with constant coefficients, $\mathbf{F} = 0$ so that these parts propagate independently of each other, each part being governed by a simple equation of the form (1.1.1).

This monograph aims to study those approximate solutions of a hyperbolic system in which the characteristic variable w_i corresponding to a simple ith genuinely nonlinear characteristic field dominates over other characteristic variables. More precisely, all ratios of other characteristic variables to w_i is of the order of a small quantity. Under such an approximation, the solution of a quasilinear hyperbolic system can be approximated by a simpler equation which is a generalization of the Burgers' equation (1.1.6).

The approximations mentioned above are closely related to discontinuities in the solutions and their derivatives. The results in section 2.5, when written for a particular case of two independent variables, show that discontinuities in the derivatives of a solution of a first order system can exist only across a characteristic curve. If there are discontinuities in a solution itself (a suitably defined weak solution) of a first order system of linear equations, we can show that the discontinuities can exist only along a characteristic curve. Suppose such a discontinuity exists along a characteristic curve γ of ith family (it is assumed that the solution and its derivatives remain bounded as we approach the curve of discontinuity from both sides of it). We can then show that the corresponding characteristic variable i.e., w_i must be discontinuous across γ and $w_1, \ldots, w_{i-1}, w_{i+1}, \ldots,$ and w_n must be continuous across γ. In this case, it is simple to deduce the transport equation giving the variation along γ of the jump in w_i across γ (Prasad and Ravindran (1985), section 2.3). The transport equation states that if the jump in w_i is prescribed at one point of γ, it can be uniquely determined at all other points of γ. More general results of this type will be discussed later in this monograph.

2.2 The wave equation in $m(> 1)$ space dimensions

In this section, we shall view the wave equation as a particular case of a general hyperbolic equation and gradually but briefly introduce those properties which can be generalized to other hyperbolic equations in more than two independent variables. The wave equation in m-space dimension for the function $u(\mathbf{x}, t)$ is

$$u_{tt} - a_0^2(u_{x_1x_1} + \ldots + u_{x_mx_m}) = 0, \quad a_0 = \text{ constant } > 0 \tag{2.2.1}$$

The generalization of a characteristic curve discussed in the last section is a *characteristic surface* in $(\mathbf{x}, t)$-space and can be defined to be the surface across which discontinuities in the second and higher order derivatives of a solution of (2.2.1) can exist (see section 2.2.4). Every characteristic surface represented by $\psi(\mathbf{x}, t) = 0$ can be embedded into a one parameter family of characteristic surfaces given by $\phi(\mathbf{x}, t) = \alpha$, $\alpha =$ constant (Prasad (1993), section 4.2) such that ϕ satisfies the characteristic partial differential equation

$$Q(\nabla\phi, \phi_t) \equiv \phi_t^2 - a_0^2(\phi_{x_1}^2 + \ldots + \phi_{x_m}^2) = 0, \quad \nabla\phi = (\phi_{x_1}, \ldots, \phi_{x_m}) \tag{2.2.2}$$

The original characteristic surface $\psi = 0$ corresponds to a particular value of the parameter α, say $\alpha = \alpha_0$.

If we interpret $\Omega : \phi(x_\alpha, t) = \alpha$, $\alpha =$ constant, as the locus of a moving surface Ω_t in $(x_1, \ldots, x_m)$-space, unit normal $\mathbf{n}$ to the *wavefront* Ω_t is given by

$$\mathbf{n} = \frac{\nabla\phi}{|\nabla\phi|} \tag{2.2.3}$$

The wavefront Ω_t propagates with a *velocity* $-\phi_t/|\nabla\phi|$ which from (2.2.2) becomes $\pm a_0$.

The most important solution of the characteristic equation (2.2.2) represents a *characteristic conoid* in $(\mathbf{x}, t)$-space with its vertex at an arbitrary point $P_0(\mathbf{x}_0, t_0)$:

$$\phi \equiv (t - t_0) \pm \frac{1}{a_0}|\mathbf{x} - \mathbf{x}_0| = 0 \tag{2.2.4}$$

where + and − signs represent respectively the lower and upper branches of the right circular conoid in space-time with vertex at P_0. For simplicity of discussion we take $t_0 > 0$. Intersection of the conoid by the hyperplane $t = 0$ is a sphere

$$S_0 : |\mathbf{x} - \mathbf{x}_0|^2 = a_0^2(t_0)^2 \tag{2.2.5}$$

in $(\mathbf{x})$-space.

An explicit form of the solution of the wave equation (see Courant and Hilbert (1962), Section 11, Chapter VI) shows that if the initial data is changed only outside the sphere S_0, the solution remains unchanged in the closed domain bounded by the lower part of the conoid (2.2.4) and the sphere S_0. More precisely, the solution shows that the *domain of dependence* on the plane $t = 0$ of the point P_0 is either the set of points on the sphere S_0 (when m is odd except $m = 1$) or the set of all points on and inside the sphere S_0 when m is even. It also follows that the value of u at the point P_0 in space-time influences the solution at all points of the upper part of the characteristic conoid (2.2.5) when m is odd (except $m = 1$) and for all other values of m (including $m = 1$) it influences the solution everywhere on the conoid and also in its interior. These point sets constitute the *domain of influence* of P_0.

2.2.1 Space-like surface and time-like direction

Consider any solution u of the wave equation (2.2.1). An m- dimensional surface R in space-time is said to be *space-like* if the value of u at any point P on R does not influence the solution u at other points of R.

An example of a space-like surface for the wave equation (2.2.1) is a hyperplane $t =$ constant. Any other plane

$$v(t-t_0) - \langle \mathbf{n}, (\mathbf{x}-\mathbf{x}_0) \rangle = 0, \quad (v, \mathbf{n} = \text{ constant } \neq 0) \tag{2.2.6}$$

through $P_0(\mathbf{x}_0, t_0)$ such that it intersects the characteristics conoid (2.2.4) through P_0 only at P_0, is also an example of a space-like surface. It is simple to show that the condition on the coefficients v and $\mathbf{n}_0$ for (2.2.6) to be space-like is (Prasad and Ravindran (1985))

$$\frac{v^2}{|\mathbf{n}|^2} > a_0^2 \tag{2.2.7}$$

If we choose $|\mathbf{n}| = 1$, then v is the normal velocity of the moving plane in $(\mathbf{x})$-space represented by (2.2.6). Thus (2.2.7) implies that a space-like plane (2.2.6) represents a locus in $(\mathbf{x})$-space of points starting from $\mathbf{x}_0$ at t_0 and moving with a speed greater than the speed a_0.

A direction in space-time, which lies in a space like plane is called a *space-like direction.*

Time-like direction and curve Consider a straight line in space-time passing through a point P_0 $(\mathbf{x}_0, t_0)$. If the straight line lies in the interior of the characteristic conoid through the point P_0, then the direction of the straight line is said to be a *time-like* direction for the wave equation. A curve

$$\mathbf{x} = \mathbf{x}(\sigma) \ , \ t = t(\sigma) \tag{2.2.8}$$

in space-time is said to be a *time-like curve* if its tangent direction is always a time-like direction. This implies

$$\left(\frac{dt}{d\sigma}\right)^2 - \frac{1}{a_0^2}\left|\frac{d\mathbf{x}}{dt}\right|^2 > 0 \tag{2.2.9}$$

A generator of a characteristic conoid is neither space-like nor time-like.

It is a matter of very deep physical importance that all time-like directions have the same significance for the wave equation. This is seen easily from the special theory of relativity, which is based on two axioms:

(i) the propagation of light in an inertial frame is governed by the wave equation (2.2.1) and

(ii) the velocity of light a_0 appearing in (2.2.1) is the same constant in all inertial frames.

Transformations of spatial coordinates and time, complying with the above postulates (excluding the two trivial ones: translation of the origin and similarity transformation), are called the *Lorenz transformations* (Petrovsky (1954)). They express the transformation from an inertial frame of reference K to another frame K' which travels with respect to K with a velocity v, say for simplicity along the positive x_1-axis:

$$x_1' = \frac{x_1 - vt}{\sqrt{1 - v^2/a_0^2}}, \; x_2' = x_2, \ldots, x_m' = x_m, t' = \frac{t - (v/a_0^2)x_1}{\sqrt{1 - v^2/a_0^2}} \tag{2.2.10}$$

It is simple to verify that the wave equation (2.2.1) and the characteristic conoid at the origin represented by $|x^2| - a_0^2 t^2 = 0$ are invariant under a Lorenz transformation, which maps the t axis into a time like line and $x_1, \ldots, x_m$ axes into space-like lines.

2.2.2 Bicharacteristics and rays

Consider one parameter family of characteristic surfaces: $\varphi(\mathbf{x}, t) = \alpha$, of the wave equation (2.2.1) so that the function φ satisfies the characteristic partial differential equation (2.2.2) i.e., $Q(\nabla\varphi, \varphi_t) = 0$. The characteristic curves of the first order equation (2.2.2) in space-time are defined to be the *bicharacteristic curves* of the wave equation. These are curves in space-time and are given by a system of ordinary differential equations for t, x, φ_t and $\nabla\varphi = (\varphi_{x_1}, \varphi_{x_2}, \ldots, \phi_{x_m})$:

$$\frac{dt}{d\sigma} = \frac{1}{2}Q_{\varphi_t} = \varphi_t, \quad \frac{dx_\alpha}{d\sigma} = \frac{1}{2}Q_{\varphi_\alpha} = -a_0^2\varphi_{x_\alpha}, \tag{2.2.11}$$

$$\frac{d\varphi_t}{d\sigma} = -\frac{1}{2}Q_t = 0, \quad \frac{d\varphi_{x_\alpha}}{d\sigma} = -\frac{1}{2}Q_{x_\alpha} = 0 \tag{2.2.12}$$

From the theory of first-order nonlinear partial differential equations, it follows that a characteristic surface of (2.2.1) i.e., an integral surface $\varphi = \alpha$ of (2.2.2) is generated by a family of bicharacteristic curves of (2.2.1). Successive positions Ω_t of a wavefront of (2.2.1) form a one-parameter family of surfaces in $(x_1, \ldots, x_m)$-space given by $\varphi(x_\alpha, t) = \alpha$ for a fixed t and α as the parameter.

A *ray* associated with a wavefront Ω_t is a curve in the $(\mathbf{x})$-space traced by a moving point $\mathbf{x}(\sigma)$ according to the equations (2.2.11 - 12) and starting from a point of the wavefront at a particular time, say $t = 0$. Thus a ray is the projection of a bicharacteristic curve on $(\mathbf{x})$-space. Similarly, the wavefront Ω_t is the projection on the $(\mathbf{x})$-space of the section of the characteristic surface Ω by a $t =$ constant plane. Thus the moving point on a ray remains on the wavefront Ω_t at any time t. The equations of a point $\mathbf{x}$ on a *ray* of the wave equation (2.2.1) and the unit normal $\mathbf{n}$ of the wavefront at that point, can be derived from (2.2.11 - 12) using $n_\alpha = \varphi_{x_\alpha}/|\operatorname{grad}\varphi|$ and $\varphi_t = -a_0|\operatorname{grad}\varphi|$ (for a forward facing wavefront):

$$\frac{dx_\alpha}{dt} = n_\alpha a_0, \quad \frac{dn_\alpha}{dt} = 0 \tag{2.2.13}$$

Therefore the rays of the wave equation starting from a point $\mathbf{x}_0$ at time $t = t_0$ are straight lines normal to the successive positions of the wavefront :

$$\mathbf{x} = \mathbf{x}_0 + \mathbf{n}a_0(t - t_0) \tag{2.2.14}$$

This means that the wavefront at any time is the locus of the tips of the normals of length $a_0(t - t_0)$ drawn from the various points of the wavefront at $t = t_0$. This is equivalent to the Huyghens' wavefront construction stated later in section 3.2.2, which, in essence, contains a very fundamental fact that a linear wavefront is *self-propagating.* This means that a moving wavefront is completely determined by the information only on the wavefront at any fixed time and is not influenced by the wavefronts which follow or precede it. For a linear wave, the information required is simply the position and the geometry of the wavefront and not the amplitude of the wavefront. One of the aims of this monograph is to deduce results which can be interpreted as extensions of the Huygens' wavefront construction to a nonlinear wavefront and a shock front.

The wave equation represents a wave phenomenon in a medium which is a particular case of wave propagation in an *isotropic medium* where the rays are orthogonal to the successive positions of the wavefront. In a general isotropic medium a_0 need not be constant and the right hand side of the second equation in (2.2.13) is no longer zero. If a_0 is a known function of $\mathbf{x}$ and t (and is not dependent on the amplitude), the geometry of the successive positions of the wavefront and the associated rays (which are not necessarily straight lines) can be uniquely determined from the ray equations i.e., the equations for $\mathbf{x}$ and $\mathbf{n}$. For the wave equation with constant a_0, the situation is very simple since rays are straight lines. In this case, if there is a concave part of the wavefront, the rays starting from different points of the wavefront start intersecting after some time. This leads to the formation of a caustic (as shown in Fig. 3.3.1 for two space dimensions), which is an envelope of rays. In the region bounded by the caustic, the rays starting from different points of the concave part of the initial wavefront intersect.

2.2.3 Compatibility condition on a characteristic surface

We write the wave equation (2.2.1) in a new coordinate system $(\mathbf{x}', \varphi)$ (see equations (3.4.14 - 16)) where

$$\mathbf{x}' = \mathbf{x},\ \varphi = \varphi(\mathbf{x}, t) \tag{2.2.15}$$

Equation (2.2.1) transforms to

$$(\varphi_t^2 - a_0|\nabla\phi|^2)\frac{\partial^2 u}{\partial\varphi^2} + (\varphi_{tt} - a_0^2\nabla^2\varphi)\frac{\partial u}{\partial\phi} - 2a_0^2\varphi_{x_\alpha}\frac{\partial}{\partial x'_\alpha}\left(\frac{\partial u}{\partial\varphi}\right) - a_0^2\frac{\partial^2 u}{\partial x'_\alpha \partial x'_\alpha} = 0 \tag{2.2.16}$$

The operator $\frac{d}{d\sigma}$ for the directional derivative along a bicharacteristic curve becomes

$$\begin{aligned}\frac{d}{d\sigma} &= \frac{dt}{d\sigma}\frac{\partial}{\partial t} + \frac{dx_\alpha}{d\sigma}\frac{\partial}{\partial x_\alpha} = \phi_t\frac{\partial}{\partial t} - a_0^2\varphi_{x_\alpha}\frac{\partial}{\partial x_\alpha} \\ &= (\varphi_t^2 - a_0^2|\nabla\phi|^2)\frac{\partial}{\partial\varphi} - a_0^2\varphi_{x_\alpha}\frac{\partial}{\partial x'_\alpha} = -a_0^2\varphi_{x_\alpha}\frac{\partial}{\partial x'_\alpha}\end{aligned} \tag{2.2.17}$$

since φ satisfies (2.2.2). Setting

$$\frac{\partial u}{\partial\varphi} = v \tag{2.2.18}$$

and using (2.2.2) and (2.2.17) we deduce the following relation from (2.2.16)

$$2\frac{dv}{d\sigma} - a_0^2 \frac{\partial^2 u}{\partial x'_\alpha \partial x'_\alpha} + (\varphi_{tt} - a_0^2 \nabla^2 \varphi)v = 0 \tag{2.2.19}$$

Equation (2.2.19) is an important form of the wave equation. The derivative $\frac{d}{d\sigma}$ in the direction of a bicharacteristic curve in space-time is a tangential derivative on a characteristic surface $\varphi =$ constant. The derivative $\frac{\partial}{\partial x'_\alpha}$ is also a tangential derivative on $\varphi =$ constant. Thus, all derivatives of u and $v \left(= \frac{\partial u}{\partial \varphi}\right)$ appearing in (2.2.19) are tangential derivatives on a characteristic surface $\Omega : \varphi =$ constant. Therefore, the equation (2.2.19), restricted to a characteristic surface, represents a compatibility condition on it. This compatibility condition involves two quantities u and v which are required to be prescribed in a Cauchy problem on a m dimensional surface in space-time. The compatibility condition shows that u and v cannot be arbitrarily prescribed on a characteristic surface. The compatibility condition (2.2.19) is in a very special form in the sense that one of the interior derivatives is in a bicharacteristic direction.

Before we present another form of the compatibility condition, we define a quantity called *ray tube area*, say A, along a ray associated with the successive positions Ω_t of a wavefront. Let us take a surface element dS_0 on the wavefront Ω_0 at $t = 0$. Let us draw rays from the various points of the boundary of dS_0; then these rays form a tube, called a *ray tube*. The intersection of the ray tube with the wavefront at any time $t \neq 0$ gives a surface element dS on Ω_t. The ray tube area A is defined by

$$A = \lim_{dS_0 \to 0} \frac{dS}{dS_0} \tag{2.2.20}$$

where $dS_0 \to 0$ means that the largest diameter of dS_0 tends to zero.

Let $\frac{d}{dl}$ represent the space rate of change of a quantity as we move along a ray. Then it is simple to show that the divergence of the unit normal $\mathbf{n}$ of wavefront, i.e., $\langle \nabla, \mathbf{n} \rangle = \frac{\partial n_\alpha}{\partial x_\alpha}$, is related to the space rate of change of the ray tube area by

$$\frac{1}{A}\frac{dA}{dl} = \langle \nabla, \mathbf{n} \rangle \tag{2.2.21}$$

The quantity $-\frac{1}{2}\langle\nabla, \mathbf{n}\rangle$ is defined to be the *mean curvature* of the surface Ω_t and we denote it by Ω:

$$\Omega = -\frac{1}{2}\langle\nabla, \mathbf{n}\rangle \tag{2.2.22}$$

so that

$$\frac{1}{A}\frac{dA}{dl} = -2\Omega \tag{2.2.23}$$

Using the expression $n = \nabla\varphi/|\nabla\varphi|$ we deduce

$$\langle\nabla, \mathbf{n}\rangle = -(\varphi_{tt} - a_0^2\nabla^2\varphi)/\{a_0^2|\nabla\phi|\} \tag{2.2.24}$$

Also, from (2.2.11)

$$\frac{d}{dl} = \frac{1}{a_0}\frac{d}{dt} = \frac{1}{a_0\varphi_t}\frac{d}{d\sigma} = -\frac{1}{a_0^2|\nabla\varphi|}\frac{d}{d\sigma}$$

since $\varphi_t = -a_0|\nabla\varphi|$. Substituting these results in (2.2.19) we get

$$\frac{d}{dl}(v^2A) + \frac{Av}{|\nabla\varphi|}\frac{\partial^2 u}{\partial x'_\alpha \partial x'_\alpha} = 0 \tag{2.2.25}$$

The equation (2.2.25) is a very interesting form of the compatibility condition on the characteristic surface. We shall show below and in subsequent sections that the propagation of singularities in the solution, say of discontinuities in second derivatives, is governed by an equation in which only the first term of (2.2.25), with v replaced by the amplitude of the singularity, is equated to zero. Thus, for the wave equation, the amplitude of a singularity in the wave varies inversely as the square root of the ray tube area. As a point of a caustic is approached, the ray tube area tends to zero showing that the amplitude of a singularity tends to infinity at the caustic.

2.2.4 Propagation of discontinuities in second order derivatives along rays

Let us consider a generalized solution $u(\mathbf{x}, t)$ of (2.2.1) which is C^1 in a domain D of space-time but is C^2 in D except for jump discontinuities in the second order derivatives of u across an m-dimensional surface $\Omega : \phi(\mathbf{x}, t) = 0$ which divides D into two subdomains D_l and D_r. In terms of a new set of independent variables $(\mathbf{x}', \phi)$ introduced by (2.2.15), the wave equation reduces to the equation (2.2.16) which is valid separately in D_l and D_r. Since u and its first derivatives are continuous across Ω, the second order tangential derivatives $\frac{\partial^2 u}{\partial x'_\alpha \partial x'_\beta}$, and tangential derivatives of the first order transversal derivative, namely $\frac{\partial}{\partial x'_\alpha}\left(\frac{\partial u}{\partial \phi}\right)$, are continuous across Ω. Since not all second order derivatives can be continuous, $\frac{\partial^2 u}{\partial \phi^2}$ must be discontinuous across Ω (see also section 3.3.4). Writing (2.2.16) at a point P_1 in D_l, and at a point P_2 in D_r, taking the limit as P_1 and P_2 both tend to P on Ω, and subtracting the resultant equations we get

$$(\phi_t^2 - a_0^2|\nabla\phi|^2)\left[\frac{\partial^2 u}{\partial \phi^2}\right] = 0 \tag{2.2.26}$$

where $\left[\frac{\partial^2 u}{\partial \phi^2}\right]$ represent the jump of the quantity $\frac{\partial^2 u}{\partial \phi^2}$ across Ω.

Since $\left[\frac{\partial^2 u}{\partial \phi^2}\right] \neq 0$, it follows that $\phi_t^2 - a_0^2|\nabla\phi|^2 = 0$ on $\phi = 0$ showing that *the surface of discontinuity* Ω *of the second order derivatives must be a characteristic surface.* As mentioned earlier in this section, we can embed Ω into a family of one parameter characteristic surfaces so that ϕ satisfies $\phi_t^2 - a_0^2|\nabla\phi|^2 = 0$ not only on $\phi = 0$ but it satisfies (2.2.2) as a partial differential equation. The equation (2.2.16) now becomes

$$(\phi_{tt} - a_0^2\nabla^2\phi)\frac{\partial u}{\partial \phi} - 2a_0^2\phi_{x_\alpha}\frac{\partial}{\partial x'_\alpha}\left(\frac{\partial u}{\partial \phi}\right) - a_0^2\frac{\partial^2 u}{\partial x'_\alpha \partial x'_\alpha} = 0 \tag{2.2.27}$$

The quantities $\phi_{tt} - a_0^2\nabla^2\phi$ and ϕ_{x_α} appearing in the coefficients above can be expressed as function of ϕ and $\mathbf{x}'$. We write

$$\phi_{tt} - a_0^2\nabla^2\phi = M(\mathbf{x}', \phi), \quad \phi_{x_\alpha} = N^{(\alpha)}(\mathbf{x}', \phi) \tag{2.2.28}$$

Differentiating (2.2.27) with respect to ϕ and denoting $u_{\phi\phi}$ by v, we get

$$(\phi_{tt} - a_0^2\nabla^2\phi)v + M_\phi u_\phi \quad -2a_0^2\phi_{x_\alpha} v_{x'_\alpha}$$
$$-2a_0^2 N_\phi^{(\alpha)} u_{\phi x'_\alpha} - a_0^2 u_{\phi x'_\alpha x'_\alpha} = 0 \tag{2.2.29}$$

We note that the first order derivative u_ϕ and hence its tangential derivatives $u_{\phi x'_\alpha}$, $u_{\phi x'_\alpha x'_\alpha}$ are continuous across the characteristic surface $\phi = 0$. Also, all the coefficients in (2.2.29) are continuous across it. Writing the equation (2.2.29) on two sides of Ω and taking the difference we get

$$(\phi_{tt} - a_0^2 \phi_{x_\alpha x_\alpha})[w_2] - 2a_0^2 \left(\phi_{x_\alpha} \frac{\partial}{\partial x'_\alpha} \right) [w_2] = 0 \qquad (2.2.30)$$

where $w_2 = [v] = [u_{\phi\phi}]$. Using $\frac{d}{d\sigma}$ for the bicharacteristic derivatives as in (2.2.17) we finally write this equation in the form

$$2\frac{d[w_2]}{d\sigma} + (\phi_{tt} - a_0^2 \nabla^2 \phi)[w_2] = 0 \qquad (2.2.31)$$

or using (2.2.22 - 24) and the relation between $d\sigma$ and dl as given by the relation there, we get

$$\frac{d[w_2]}{dl} = \Omega(l)[w_2] \qquad (2.2.32)$$

We can also write this relation in terms of the ray tube area A (using (2.2.23)) so that on integration, we get

$$A^2[w_2] = \text{ constant } = A_0^2[w_2]_0 \qquad (2.2.33)$$

Any one of the last three equations gives the *transport* equation for the propagation of the discontinuity in the second derivatives of the solution. As mentioned earlier, (2.2.33) implies that the amplitude $[w]$ of the discontinuity is inversely proportional to the ray tube area. As we approach a point of a caustic, where neighbouring rays tend to meet, the ray tube area A tends to zero, the amplitude $[w]$ of the discontinuity tends to infinity.

An important consequence of a transport equation for the propagation of a singularity along a ray for a linear hyperbolic equation, such as the wave equation, is the fact that the magnitude of the singularity remains finite at any finite distance along the ray unless either a singularity of the rays (such as a caustic) is encountered or a singularity of the coefficients (in a general linear system the coefficients are functions of $\mathbf{x}$ and t) is encountered.

2.3 Hyperbolic system in more than two independent variables

To start with, we shall not distinguish between the spatial variables $x_\alpha, \alpha = 1, 2, \ldots, m$ and time t. Our aim is to suitably identify a time-like variable for a hyperbolic system and then use symbol t for it. Readers not interested in the detailed discussion of space-like surface and time-like direction may skip the subsection 2.3.1 and go to the subsection 2.3.2.

2.3.1 Space-like surface and time-like direction

Consider a system of n first order partial differential equations in the form

$$\sum_{p=1}^{m+1} B^{(p)} \frac{\partial \mathbf{u}}{\partial x_p} + C = 0 \tag{2.3.1}$$

for n dependent variables $u_i, i = 1, 2, \ldots, n$ forming the components of $\mathbf{u} \in \mathbf{R^n}$. Note that the range of the suffix p is $1, 2, \ldots, m, m+1$. The system may be linear, semilinear or quasilinear. In the last case, we shall first take a known solution $\mathbf{u}_0(x_p)$ and substitute it for the function $\mathbf{u}$ in the matrices $B^{(p)}$. However, we shall have to remember that our results are true only for the particular solution under consideration. The characteristic equation of (2.3.1) is a nonlinear first order partial differential equation

$$Q(x_p, \phi_{x_p}) \equiv \det \left[\sum_{p=1}^{m+1} B^{(p)} \phi_{x_p} \right] = 0 \tag{2.3.2}$$

where $\phi =$ constant is a one-parameter family of characteristic surfaces. We set

$$\phi_{x_p} = k_p, \quad p = 1, 2, \ldots, m+1. \tag{2.3.3}$$

We shall discuss the algebraic property of the characteristic polynomial as a function of ϕ_{x_p} at a fixed point (x_{p0}) of (x_p)-space. Therefore we shall denote the characteristic polynomial $Q(x_{p0}, \phi_{x_p})$ simply by $Q(\phi_{x_p})$. Using (2.3.3) we see that the characteristic equation

$$Q(k_p) \equiv \det \left[\sum_{p=1}^{m+1} B^{(p)} k_p \right] = 0 \tag{2.3.4}$$

is a homogeneous algebraic equation of degree n in $\mathbf{k}$ and represents the equation of a conoid (in $\mathbf{k}$ space) which is called *normal conoid* at the point (x_{p0}). The *characteristic conoid* at (x_{p0}) is obtained as the envelope of planes which are orthogonal * to the generators of the normal conoid i.e., planes

$$k_1(x_1 - x_{10}) + k_2(x_2 - x_{20}) + \ldots + k_m(x_{m+1} - x_{m+1,0}) = 0 \quad (2.3.5)$$

where $\mathbf{k}$ satisfies (2.3.4). For the wave equation with $m = 2$ and $x_{m+1} = t$, the normal and the characteristic conoids are circular cones with upper and lower sheets. In higher dimensions these are generalizations of circular cones. For other hyperbolic equations the two cones may have complicated geometrical features (see Courant and Hilbert, 1962). For example, the normal conoid of

$$\left\{\left(\frac{\partial}{\partial x_3} - \frac{\partial}{\partial x_2}\right)\left(\frac{\partial}{\partial x_3} + \frac{\partial}{\partial x_2}\right)^2 - \frac{\partial^2}{\partial x_1^2}\left(2\frac{\partial}{\partial x_3} + \frac{\partial}{\partial x_2}\right)\right\} u = 0$$

has branches extending to infinity and its characteristic conoid is not convex (see Prasad and Ravindran (1985) for details).

Definition of hyperbolicity: The first order system (2.3.1) of n equations is said to be hyperbolic in a domain of $(x_1, \ldots, x_{m+1})$ space if, at each point $P(x_p)$, there exist directions $\boldsymbol{\zeta} = (\zeta_1, \ldots, \zeta_m, \zeta_{m+1})$ such that all straight lines (except the one passing through the vertex) parallel to the vector $\boldsymbol{\zeta}$ intersect the normal conoid in exactly n distinct points.

We shall extend later in the next subsection the definition of hyperbolicity to include characteristics of higher multiplicity.

Algebraically, the above definition is equivalent to the following one. If $\boldsymbol{\theta} = (\theta_1, \theta_2, \ldots, \theta_{m+1})$ is a nonzero vector not parallel to $\boldsymbol{\zeta}$, then the equation

$$Q(\lambda\boldsymbol{\zeta} + \boldsymbol{\theta}) = 0 \qquad (2.3.6)$$

in λ must have n real and distinct roots.

Space-like surface If a vector $\boldsymbol{\zeta}$ satisfying the above condition exists, the plane element at P orthogonal to $\boldsymbol{\zeta}$ is called a space-like element. A surface in $m+1$ dimensional space is defined to be space-like if its surface elements are space-like.

The vectors of the type $\boldsymbol{\zeta}$, which are orthogonal to space-like elements at P, form the *inner core* of the normal conoid bounded

*Two vectors $\mathbf{x}$ and $\mathbf{y}$ in s-dimensional Euclidean space $\mathbb{R}^s$ are said to be orthogonal if $\langle \mathbf{x}, \mathbf{y} \rangle \equiv x_1 y_1 + \ldots + x_s y_s = 0$.

by the *inner sheet* of the conoid. It can be proved that this inner core of the normal conoid is convex. Geometrically then, the normal conoid may be visualized as one consisting of the closed inner sheet bounding the inner core into which normals to space-like surface elements point, and of further sheets which form subsequent shells around the core (Duff (1960)). The outer sheets may be closed or may extend up to infinity. We can also prove that the boundary of the cone supported by the planes orthogonal to the generators of the convex inner sheet of the normal conoid is the convex hull Γ of the local characteristic conoid; more specifically, it is the hull of the *outer shell* of the characteristic conoid.

Time-like direction and curve : Every direction from a point P into the convex hull Γ of the outer shell of the characteristic conoid at P is called time-like. A curve in the $(m+1)$ dimensional space is called time-like if its direction is everywhere time-like.

Now we notice that if a system (2.3.1) is hyperbolic at a point P, then there exist m dimensional space-like surface elements at P and time-like directions through P. We can choose a local coordinate system at P such that the direction of the x_{m+1} axis is time-like and the m-dimensional sub-space spanned by the unit vectors along x_α coordinate axes $(\alpha = 1, 2, \ldots, m)$ is space-like at P. So far we have presented a local description at a point P. We now move to a global discussion. We assume that we have a first order system of equations which is hyperbolic in a domain D of the $m+1$ dimensional $(x_1, x_2, \ldots, x_{m+1})$-space and it is possible to introduce a coordinate system such that the x_{m+1} coordinate axis is time-like and the other axes lie in a space-like surface S at every point of D. We now designate the time-like coordinate x_{m+1} by t, the matrix $B^{(m+1)}$ by A and write the system (2.3.1) in the form

$$A\frac{\partial \mathbf{u}}{\partial t} + B^{(\alpha)}\frac{\partial \mathbf{u}}{\partial x_\alpha} + C = 0 \tag{2.3.7}$$

We note that the spatial coordinate system $\mathbf{x} = (x_1, x_2, \ldots, x_m)$ in the space-like surface $t = 0$ at any point P may be chosen to be orthogonal but the coordinate system $(\mathbf{x}, t)$ at P is not necessarily orthogonal. We take a vector $\boldsymbol{\zeta}$ orthogonal to the space-like $t = 0$ at P and resolve it into two components one $\boldsymbol{\zeta}_t$ parallel to the t-axis another $\boldsymbol{\zeta}_x$ in the tangent plane to S at P. Similarly, we resolve a vector $\boldsymbol{\theta}$ into two components $\boldsymbol{\theta}_t$ and $\boldsymbol{\theta}_x$, and combine $\lambda\ \boldsymbol{\zeta}_x$ and $\boldsymbol{\theta}_x$ to form a new arbitrary vector $\bar{\boldsymbol{\theta}}$. Then

$$Q(\boldsymbol{\theta} + \lambda\boldsymbol{\zeta}) = Q(\bar{\boldsymbol{\theta}} + \lambda\boldsymbol{\zeta}_t + \boldsymbol{\theta}_t) \tag{2.3.8}$$

In the coordinate system $(\mathbf{x}, t)$, we note that the last (i.e., $(m+1)$th) component of $\bar{\theta}$ is zero and the first m components of $\lambda\boldsymbol{\zeta}_t + \boldsymbol{\theta}_t$ are zeroes. We denote

$$\bar{\boldsymbol{\theta}} = (n_1, \ldots, n_m, 0) \quad \text{and} \quad \lambda\boldsymbol{\zeta}_t + \boldsymbol{\theta}_t = (0, 0, \ldots, 0, -c) \tag{2.3.9}$$

Hyperbolicity of the system (2.3.7) implies that $Q(\boldsymbol{\theta} + \lambda\boldsymbol{\zeta}) = 0$ has n real and distinct roots for λ for an arbitrary $\boldsymbol{\theta}$ which implies n real distinct roots for c satisfying

$$Q(\bar{\boldsymbol{\theta}} + \lambda\boldsymbol{\zeta}_t + \boldsymbol{\theta}_t) \equiv Q(\mathbf{n}, -c) = 0 \tag{2.3.10}$$

where $\mathbf{n} = (n_1, n_2, \ldots, n_m)$. As $\boldsymbol{\theta}$ is arbitrary, $\bar{\boldsymbol{\theta}}$ is also arbitrary (even though it contains λ). Hence, hyperbolicity of (2.3.7) implies existence of n real and distinct values of c from the equation

$$Q \equiv \det\left[n_\alpha B^{(\alpha)} - cA\right] = 0 \tag{2.3.11}$$

with arbitrary values of $n_1, n_2, \ldots, n_m$.

Since the equation (2.3.7) is homogeneous of degree n in n_α and c, it is sufficient if we choose

$$|\mathbf{n}| = 1 \tag{2.3.12}$$

In this case, we denote the n values of c by

$$c_1, c_2, \ldots, c_n \tag{2.3.13}$$

which are called eigenvalues or characteristic velocities. Our assumption of l hyperbolicity implies that the real roots $c_i (i = 1, 2, \ldots, n)$ are finite in D. The necessary and sufficient condition for finiteness is that the matrix A is nonsingular in D, i.e.,

$$\det A \neq 0 \quad \text{in} \quad D \tag{2.3.14}$$

In all physical systems which evolve with time and which are governed by the hyperbolic equations, the time variable t is always time-like and the physical space containing the spatial coordinates x_α is always a space-like surface in space-time. However, there are examples of time-independent physical systems which are governed by hyperbolic equations and where the time-like directions and space-like manifolds are not immediately clear. An example of such a system is the three-dimensional steady supersonic flow of a compressible gas where the Mach cone at a point plays the role of the characteristic cone. In this case, the direction of the axis of the Mach cone is time-like.

2.3.2 Explicit definition of a hyperbolic system

Consider a first order system (2.3.7) of n equation in $m+1$ independent variables. We assume that the matrix A is nonsingular i.e., (2.3.14) is valid in a domain D of the space-time. *We define the system (2.3.7) as hyperbolic in D with t as time-like variable if given an arbitrary unit vector* $\mathbf{n}$, *the characteristic equation (2.3.11) has n distinct real roots (which we denote as $c_1, c_2, \ldots, c_n$) at each point of D.*

For the simple root c_i of the characteristic equation (2.3.11), the matrix $n_\alpha B^{(\alpha)} - cA$ has rank $n-1$ and there exist unique (except for a scalar multiplier) left and right eigenvectors $\mathbf{l}^{(i)}$ and $\mathbf{r}^{(i)}$ respectively satisfying

$$\mathbf{l}^{(i)}\ (n_\alpha B^{(\alpha)}) = c_i \mathbf{l}^{(i)} A, \quad (n_\alpha B^{(\alpha)})\mathbf{r}^{(i)} = c_i A \mathbf{r}^{(i)} \tag{2.3.15}$$

Unlike the case of the two independent variables, the eigenvalue c_i, the left eigenvector $\mathbf{l}^{(i)}$ and the right eigenvector $\mathbf{r}^{(i)}$ not only depend on the position (x_α, t) in space-time, but also on the m arbitrary numbers $n_1, n_2, \ldots, n_m$ satisfying (2.3.12).

Hyperbolic system with characteristics of uniform constant multiplicity > 1

For a system given in the form (2.3.7) satisfying (2.3.14), we can easily extend the definition of hyperbolicity even if the eigenvalues $c_1, c_2, \ldots, c_n$ are not distinct. *Suppose an eigenvalue c_i is repeated p_i times in the set (2.3.13), then the system (2.3.7) is defined to be hyperbolic if eigenspace of c_i is complete i.e., the number of linearly independent left eigenvectors (and hence also right eigenvectors) corresponding to c_i is p_i.*

Example The Euler's equations, governing the motion of a polytropic gas in two and three space variables, represent an elegant system of hyperbolic equations with multiple eigenvalues. We presented one space variable case in the section 2.1.1 where all eigenvalues were simple. In three-dimensions, the equations are

$$\rho_t + \langle \mathbf{q}, \nabla \rho \rangle + \rho \langle \nabla, \mathbf{q} \rangle = 0 \tag{2.3.16}$$

$$\mathbf{q}_t + \langle \mathbf{q}, \nabla \rangle \mathbf{q} + \frac{1}{\rho} \nabla p = 0 \tag{2.3.17}$$

$$p_t + \langle \mathbf{q}, \nabla \rangle p + \rho a^2 \langle \nabla, \mathbf{q} \rangle = 0 \tag{2.3.18}$$

where a is given by (2.1.17). It is a system of 5 first order equations. Taking $\mathbf{u} = (\rho, \mathbf{q} = (q_1, q_2, q_3), p)^T$, we find $A = I$ and the matrix

$$B^{(\alpha)} = \begin{bmatrix} q_\alpha & \rho\delta_{1\alpha} & \rho\delta_{2\alpha} & \rho\delta_{3\alpha} & 0 \\ 0 & q_\alpha & 0 & 0 & \rho^{-1}\delta_{1\alpha} \\ 0 & 0 & q_\alpha & 0 & \rho^{-1}\delta_{2\alpha} \\ 0 & 0 & 0 & q_\alpha & \rho^{-1}\delta_{3\alpha} \\ 0 & \rho a^2\delta_{1\alpha} & \rho a^2\delta_{2\alpha} & \rho a^2\delta_{3\alpha} & q_\alpha \end{bmatrix}$$

The five eigenvalues are

$$c_1 = \langle \mathbf{n}, q \rangle - a, \ \ c_2 = c_3 = c_4 = \langle \mathbf{n}, \mathbf{q} \rangle, \ \ c_5 = \langle \mathbf{n}, \mathbf{q} \rangle + a \tag{2.3.19}$$

We can easily check that there are three linearly independent left (or right) eigenvectors corresponding to $\langle \mathbf{n}, \mathbf{q} \rangle$ so that the system (2.3.16 - 18) is hyperbolic.

The system of conservation laws representing conservation of mass, the three components of momentum and energy of gas elements from which the equations (2.3.16 - 18) can be derived, are

$$\mathbf{H}_t + \langle \nabla, F \rangle = 0 \tag{2.3.20}$$

$$\mathbf{H} = \begin{bmatrix} \rho \\ \rho u_1 \\ \rho u_2 \\ \rho u_3 \\ \rho(e + \frac{1}{2}\mathbf{q}^2) \end{bmatrix}, \qquad \mathbf{F} = \begin{bmatrix} \rho\mathbf{q} \\ \rho(q_1^2 + p,\ q_1q_2,\ q_1q_3) \\ \rho(q_2q_1,\ q_2^2 + p,\ q_2q_3) \\ \rho(q_3q_1,\ q_3q_2,\ q_3^2 + p) \\ \rho\mathbf{q}(e + \frac{p}{\rho} + \frac{1}{2}\mathbf{q}^2) \end{bmatrix} \tag{2.3.21}$$

The specific internal energy (denoted by e) and the relation between pressure, density and specific entropy (denoted by σ) for a polytropic gas are given by

$$e = \frac{p}{(\gamma - 1)\rho} \quad \text{and} \quad p = A(\sigma)\rho^\gamma \tag{2.3.22}$$

where A is a function of σ and γ is a constant.

2.4 Bicharacteristic curves, rays and compatibility condition

Consider a system of first order equations (2.3.7). Its characteristic equation (2.3.11) written in terms of the derivatives of the function ϕ

$$Q(\mathbf{x}, t; \nabla\phi, \phi_t) \equiv \det(A\phi_t + B^{(\alpha)}\phi_{x_\alpha}) = 0 \tag{2.4.1}$$

is a first order nonlinear partial differential equation for the function ϕ. The characteristic curves of (2.4.1) are called *bicharacteristic curves* of (2.3.7). These are curves in space-time whose parametric representation is obtained after solving the ordinary differential equations

$$\frac{dt}{d\sigma} = \frac{1}{2}Q_q, \qquad \frac{dx_\alpha}{d\sigma} = \frac{1}{2}Q_{p_\alpha} \tag{2.4.2}$$

and

$$\frac{dq}{d\sigma} = -\frac{1}{2}Q_t, \qquad \frac{dp_\alpha}{d\sigma} = -\frac{1}{2}Q_{x_\alpha} \tag{2.4.3}$$

where

$$q = \phi_t, \qquad p_\alpha = \phi_{x_\alpha}, \quad \alpha = 1, 2, \ldots, m \tag{2.4.4}$$

The solution $\mathbf{p}(\sigma), q(\sigma), \mathbf{x}(\sigma), t(\sigma)$ of these equations along a bicharacteristic curve must satisfy the relation $Q(\mathbf{p}, q, \mathbf{x}, t) = 0$. From the theory of a first order partial differential equation, it follows that a bicharacteristic curve lies in a characteristic surface. Further, a characteristic surface $\phi = 0$ is generated by a $(m-1)$-parameter family of bicharacteristic curves.

As explained in section 2.2, rays are the projections of the bicharacteristic curves on the hyperplane $t = 0$. These are curves in $\mathbf{x}$-space.

If the coefficient matrices A and $B^{(\alpha)}$ are constant matrices, then $Q_t = 0$ and $Q_{x_\alpha} = 0$. This implies that $\mathbf{p}$ and q are constant along the bicharacteristic curves or rays. The equations (2.4.2) imply that bicharacteristic (or rays) are straight lines in space-time (or in $(\mathbf{x})$-space).

Lemma on bicharacteristics: Consider a characteristic surface given by $\phi(\mathbf{x}, t) =$ constant, of the first order system (2.3.7) corresponding to an eigenvalue c. For the validity of the lemma below, it is not necessary to assume that (2.3.7) is hyperbolic but only that c is a simple real eigenvalue. The statement of the lemma is

Lemma Along a bicharacteristic curve, variation of $\mathbf{x}$ and

$$\mathbf{n} = \frac{\nabla\phi}{|\nabla\phi|} = \frac{\mathbf{p}}{|\mathbf{p}|} \tag{2.4.5}$$

is given by

$$\frac{dx_\alpha}{dt} = \frac{\mathbf{l}B^{(\alpha)}\mathbf{r}}{\mathbf{l}A\mathbf{r}} = \chi_\alpha, \quad \text{say} \tag{2.4.6}$$

and

$$\frac{dn_\alpha}{dt} = -\frac{1}{\mathbf{l}A\mathbf{r}}\mathbf{l}\left\{n_\beta\left(-c\frac{\partial A}{\partial \eta^\alpha_\beta} + n_\gamma\frac{\partial B^{(\gamma)}}{\partial \eta^\alpha_\beta}\right)\right\}\mathbf{r} = \Psi_\alpha, \text{ say} \tag{2.4.7}$$

where

$$\frac{\partial}{\partial \eta^\alpha_\beta} = n_\beta\frac{\partial}{\partial x_\alpha} - n_\alpha\frac{\partial}{\partial x_\beta} \tag{2.4.8}$$

Note: The first part of the lemma i.e., the equation (2.4.6), is available in Courant and Hilbert (1962). When we choose $\mathbf{n}$ as in (2.4.5), the relation (2.3.12) is satisfied. The eigenvalue c is given by

$$c = -\phi_t/|\nabla\phi| \tag{2.4.9}$$

Proof :

Post-multiplication of the first relation in (2.3.15) by $\mathbf{r}$ (dropping the subscript i from c, $\mathbf{l}$ and $\mathbf{r}$) gives another relation satisfied by ϕ_t and ϕ_{x_α}

$$Q(\mathbf{x}, t, \nabla\phi, \phi_t) \equiv (\mathbf{l}A\mathbf{r})\phi_t + (\mathbf{l}B^{(\alpha)}\mathbf{r})\phi_{x_\alpha} = 0 \tag{2.4.10}$$

From (2.4.5), (2.4.9 - 10), it follows that the eigenvalue c, which is a root of (2.3.11), is given by

$$c = n_\alpha \frac{\mathbf{l}B^{(\alpha)}\mathbf{r}}{\mathbf{l}A\mathbf{r}} = n_\alpha \chi_\alpha \tag{2.4.11}$$

It follows that the factor of the left hand side of (2.4.1), which makes it vanish is the expression on the left hand side of (2.4.10). Therefore, in the equations (2.4.2 - 3), it is sufficient if we take the expression of Q as given in (2.4.10). The equations (2.4.2) immediately give the result (2.4.6). Equations (2.4.3) give

$$\frac{dp_\alpha}{dt} = -\frac{1}{\mathbf{l}A\mathbf{r}}\mathbf{l}(A_{x_\alpha}q + B^{(\beta)}_{x_\alpha}p_\beta)\mathbf{r} \tag{2.4.12}$$

From (2.4.5)

$$\frac{dn_\alpha}{dt} = \frac{1}{|\mathbf{p}|^3}\left\{|\mathbf{p}|^2\frac{dp_\alpha}{dt} - p_\alpha\left(p_\beta\frac{dp_\beta}{dt}\right)\right\}$$

which after some rearrangement of terms gives

$$\frac{dn_\alpha}{dt} = \frac{1}{|\mathbf{p}|^3}\left\{p_\beta\left(p_\beta\frac{dp_\alpha}{dt} - p_\alpha\frac{dp_\beta}{dt}\right)\right\} \tag{2.4.13}$$

Substituting (2.4.12) in (2.4.13) and using $\mathbf{n} = \mathbf{p}/|\mathbf{p}|$ we finally get the result (2.4.7).

Relations between bicharacteristic curves, rays, characteristic surfaces and wavefronts have been explained for the wave equation in the previous section. We note here that for a quasilinear system A and $B^{(\alpha)}$ are functions of $\mathbf{x}, t$ and $\mathbf{u}$. From (2.4.11), c is a function of $\mathbf{x}, t, \mathbf{u}$ and $\mathbf{n}$. Since the elements of the matrix $qA + p_\alpha B^{(\alpha)}$ are linear homogeneous expressions of q and p_α, it is possible to express the left and right null vectors $\mathbf{l}$ and $\mathbf{r}$ of this matrix only in terms of $\mathbf{n}$ in addition to their dependence on $\mathbf{x}, t$ and u. Thus, the right hand sides of (2.4.6 - 7) are functions of $\mathbf{x}, t, \mathbf{u}$ and $\mathbf{n}$. These *ray equations* do not form a closed system because of their dependence on $\mathbf{u}$. In this case the hyperbolic system (2.3.7) is linear, $\mathbf{u}$ is no longer present in these equations and the rays can be traced once for all without any reference to a solution $\mathbf{u}$.

Compatibility condition on a characteristic surface :

Multiplying the system (2.3.7) by $c\mathbf{l}$ and using $c\mathbf{l}A = \mathbf{l}\left(n_\alpha B^{(\alpha)}\right)$, we get

$$\mathbf{l}B^{(\alpha)}\left(n_\alpha \frac{\partial}{\partial t} + c\frac{\partial}{\partial x_\alpha}\right)\mathbf{u} + c\mathbf{l}\mathbf{C} = 0 \tag{2.4.14}$$

The unit normal $\mathbf{n}$ and the eigenvalue c are related to the function $\phi(\mathbf{x}, t)$ by (2.4.5) and (2.4.9) respectively. Therefore, equation (2.4.14) becomes

$$\mathbf{l}B^{(\alpha)}\left(\phi_{x_\alpha} \frac{\partial}{\partial t} - \phi_t\frac{\partial}{\partial x_\alpha}\right)\mathbf{u} + c|\nabla\phi|\mathbf{l}\mathbf{C} = 0 \tag{2.4.15}$$

For a given α, the expression $\phi_{x_\alpha}\frac{\partial}{\partial t} - \phi_t\frac{\partial}{\partial x_\alpha}$ represents a tangential differentiation in the characteristic surface. Hence in (2.4.15) and also in (2.4.14), only tangential derivatives with respect to a characteristic surface appear. Therefore, equation (2.4.14) or (2.4.15) represents a compatibility condition on a characteristic surface.

A canonical form of the compatibility condition:

Through any point on a characteristic surface, there exists a bicharacteristic direction tangential to the surface. Therefore, the tangential derivatives appearing in the compatibility condition can be written as linear combinations of a derivative along a bicharacteristic curve and other $m-1$ independent tangential derivatives. We shall derive such a form of the compatibility condition on a characteristic surface $\phi =$ constant.

The first part (2.4.6) of the lemma on bicharacteristics gives the direction of the bicharacteristic curves in space-time. The operator

$$\frac{d}{dt} = \frac{\partial}{\partial t} + \chi_\alpha \frac{\partial}{\partial x_\alpha} \tag{2.4.16}$$

appearing in the lemma represents the time-rate of change when one moves along a bicharacteristic with the ray velocity

$$\boldsymbol{\chi} = (\chi_1, \chi_2, \ldots, \chi_m), \quad \chi_\alpha = (\mathbf{l}B^{(\alpha)}\mathbf{r})/(\mathbf{l}A\mathbf{r}) \tag{2.4.17}$$

A linear combination of the scalar equations in (2.1), containing only tangential derivatives in $\phi =$ *constant* of the components $u_i (i = 1, 2, \ldots, n)$ of $\mathbf{u}$, is obtained by pre-multiplying by $\mathbf{l} : \mathbf{l}A\mathbf{u}_t + \mathbf{l}B^{(\alpha)}\mathbf{u}_{x_\alpha} + \mathbf{l}\mathbf{C} = 0$ which reduces to (2.4.14). Using (2.4.16), we can write it in the form

$$\mathbf{l}A\frac{d\mathbf{u}}{dt} + \mathbf{l}(B^{(\alpha)} - \chi_\alpha A)\frac{\partial \mathbf{u}}{\partial x_\alpha} + \mathbf{l}C = 0 \qquad (2.4.18)$$

The derivative $\tilde{\partial}_j = s_j^\alpha \frac{\partial}{\partial x_\alpha} \equiv l_i(B_{ij}^{(\alpha)} - \chi_\alpha A_{ij})\frac{\partial}{\partial x_\alpha}$ on u_j in the second term is a special tangential derivative with respect to the characteristic surface, it is a tangential derivative also with respect to the *wavefronts* Ω_t. This follows from

$$n_\alpha s_j^\alpha = l_i A_{ij}(c - n_\alpha \chi_\alpha) = 0, \text{ for each } j \qquad (2.4.19)$$

since $c = n_\alpha \chi_\alpha$.

The form (2.4.18) of the compatibility condition has a very special feature. The derivative $\frac{d}{dt}$ in the bicharacteristic direction is the only derivative in this equation which contains $\frac{\partial}{\partial t}$. The other n tangential derivatives $\tilde{\partial}_j (j = 1, 2, \ldots, n)$ contain only spatial derivatives and can be expressed in terms of any $m-1$ of the m tangential derivatives L_α appearing in (2.4.7), where

$$L_\alpha = n_\beta \frac{\partial}{\partial \eta_\beta^\alpha} \quad , \alpha = 1, 2, \ldots, m \qquad (2.4.20)$$

(2.4.18) is important in formulating numerical methods using bicharacteristic curves (Reddy, Tikekar and Prasad (1982); Lukacova-Medvidova, Morton and Warnecke (2000)).

We state all the results in this section in a form of a theorem.

Theorem: The bicharacteristic curves in a simple characteristic field of (2.3.7) satisfy (2.4.6 - 7). Along a bicharacteristic curve the compatibility condition (2.4.18) holds.

We note that the operator L_α can also be written in the form

$$\begin{aligned} L_\alpha &= n_\beta \left(n_\beta \frac{\partial}{\partial x_\alpha} - n_\alpha \frac{\partial}{\partial x_\beta} \right) \\ &= \frac{\partial}{\partial x_\alpha} - n_\alpha \left(n_\beta \frac{\partial}{\partial x_\beta} \right) \end{aligned}$$

so that

$$\mathbf{L} = (L_1, L_2, \ldots, L_m) = \nabla - \mathbf{n}\langle \mathbf{n}, \nabla \rangle \qquad (2.4.21)$$

i.e., $\mathbf{L}$ is the projection of the gradient ∇ on the tangent plane of the surface $\Omega_t : \phi(\mathbf{x}, t) = 0$, for each constant t.

Example: compatibility condition for the Euler's equations of a polytropic gas

We consider here the eigenvalue $c_5 = \langle \mathbf{n}, \mathbf{q} \rangle + a$ of the system of equations (2.3.16 - 18). Left and right eigenvectors $\mathbf{l}$ and $\mathbf{r}$ corresponding to c_5 can be chosen to be

$$\mathbf{l} = (0, n_1, n_2, n_3, \frac{1}{\rho a}), \quad \mathbf{r} = (\rho/a, n_1, n_2, n_3, \rho a) \tag{2.4.22}$$

The characteristic partial differential equation (2.4.10) corresponding to this eigenvalue is

$$Q \equiv \phi_t + \langle \mathbf{q}, \nabla\phi \rangle + a|\nabla\phi| = 0 \tag{2.4.23}$$

The bicharacteristic equations (2.4.6 - 7) become

$$\frac{d\mathbf{x}}{dt} = \mathbf{q} + \mathbf{n}a \tag{2.4.24}$$

and

$$\frac{d\mathbf{n}}{dt} = -\mathbf{L}a - n_\beta L q_\beta \tag{2.4.25}$$

Multiplying the equations in (2.3.16 - 18) by components of $\mathbf{l}$ and adding the results, we derive the compatibility condition on the characteristic surface as

$$a\frac{d\rho}{dt} + \rho\langle \mathbf{n}, \frac{dq}{dt} \rangle + \rho a \langle \mathbf{L}, q \rangle = 0 \tag{2.4.26}$$

This is the form of the compatibility condition (2.4.18) for the Euler's equations (2.3.16 - 18).

2.5 Propagation of discontinuities of first order derivatives along rays

In this section we shall extend the results of the section (2.2.4) on the propagation of discontinuities in a solution of the wave equation. Consider a linear hyperbolic system of n equations for $\mathbf{u} \in I\!R^n$

$$\mathcal{L}\mathbf{u} \equiv A(\mathbf{x}, t)\mathbf{u}_t + B^{(\alpha)}(\mathbf{x}, t)\mathbf{u}_{x_\alpha} + F(\mathbf{x}, t)\mathbf{u} = \mathbf{f}(\mathbf{x}, t) \tag{2.5.1}$$

where A, B, F are $n \times n$ matrices with sufficiently smooth elements and $f \in I\!R^n$ is also sufficiently smooth vector function (for the discussion in this section, they need to be only C^1 functions in a domain of space-time). We shall first show that *a surface of discontinuity in the first derivatives of a generalized solution of (2.5.1) is a characteristic surface* and then we shall derive a *transport equation* for an *amplitude* of the discontinuity. This transport equation will be given in terms of an ordinary differential equation along rays.

Let us take a generalized solution $\mathbf{u}(\mathbf{x}, t)$ of (2.5.1) which is C^1 in a domain D of $(\mathbf{x}, t)$ -space except for discontinuities in the first order derivatives of $\mathbf{u}$ across a surface $\Omega : \phi(\mathbf{x}, t) = 0$ which divides D into two subdomains D_l and D_r. In terms of the coordinate system (x', ϕ) introduced in (2.2.15) (see relations (3.4.14 - 16)), the equation (2.5.1) reduces to

$$(A\phi_t + B^{(\alpha)}\phi_{x_\alpha})\mathbf{u}_\phi + B^{(\alpha)}\mathbf{u}_{x'_\alpha} + F\mathbf{u} = \mathbf{f} \tag{2.5.2}$$

Since $\mathbf{u}$ itself is continuous in D, we can follow the arguments presented in the section 2.2.4 to show that u_ϕ must be discontinuous across S, but $u_{x'_\alpha}$ is continuous across it and further

$$(A\phi_t + B^{(\alpha)}\phi_{x_\alpha})[\mathbf{u}_\phi] = 0 \tag{2.5.3}$$

This implies that $\Omega : \phi(x_\alpha, t) = 0$ is a characteristic surface and the jump in the transversal derivative $\mathbf{u}_\phi$ is given by

$$[\mathbf{u}_\phi] = \mathbf{r}w_1 \tag{2.5.4}$$

where $\mathbf{r}$ is a right eigenvector corresponding to an eigenvalue $c(\equiv -\phi_t/|\text{ grad } \phi|)$ and w_1 is a scalar function defined on Ω. Multiplying (2.5.2) by the corresponding left eigenvector $\mathbf{l}$, using $\mathbf{l}(A\phi_t + B^{(\alpha)}\phi x_\alpha) = 0$ and differentiating the result with respect to ϕ we get

$$(\mathbf{l}B^{(\alpha)})_\phi \mathbf{u}_{x'_\alpha} + (\mathbf{l}B^{(\alpha)})\mathbf{u}_{\phi x'_\alpha} + (\mathbf{l}F)_\phi \mathbf{u} + (\mathbf{l}F)\mathbf{u}_\phi = (\mathbf{l}\mathbf{f})_\phi \tag{2.5.5}$$

In the equation (2.5.5), all quantities except $\mathbf{u}_\phi$ are continuous across Ω. Hence, taking the jump across Ω, we get (using Hadamard's lemma, see (3.3.39))

$$(\mathbf{l}B^{(\alpha)})[\mathbf{u}_\phi]_{x'_\alpha} + (\mathbf{l}F)[\mathbf{u}_\phi] = 0 \tag{2.5.6}$$

Substituting $[\mathbf{u}_\phi]$ from (2.5.4), we get

$$(\mathbf{l}B^{(\alpha)}\mathbf{r})w_{1x'_\alpha} + \left(\mathbf{l}B^{(\alpha)}\mathbf{r}_{x'_\alpha} + \mathbf{l}F\mathbf{r}\right)w_1 = 0 \tag{2.5.7}$$

Using the lemma on bicharacteristics i.e., the result (2.4.6) and changing from (x_α, t) coordinates to (x'_α, ϕ), we find that the operator giving the rate of change along the bicharacteristic curve is

$$\frac{d}{d\sigma} = (\mathbf{l}A\mathbf{r})\frac{\partial}{\partial t} + (\mathbf{l}B^{(\alpha)}\mathbf{r})\frac{\partial}{\partial x_\alpha} = \mathbf{l}B^{(\alpha)}\mathbf{r}\frac{\partial}{\partial x'_\alpha} \tag{2.5.8}$$

We also note that

$$\mathbf{l}\left(A\frac{\partial}{\partial t} + B^{(\alpha)}\frac{\partial}{\partial x_\alpha}\right) = \mathbf{l}B^{(\alpha)}\frac{\partial}{\partial x'_\alpha} \tag{2.5.9}$$

Therefore, the equation (2.5.7) can be written as

$$\frac{dw_1}{d\sigma} + \mathbf{l}(\mathcal{L}\mathbf{r})w_1 = 0 \tag{2.5.10}$$

where $\mathcal{L}$ represents the linear differential operator appearing on the left hand side of (2.5.1).

Along a bicharacteristic curve the function $\mathcal{L}\mathbf{r}$ can be expressed as a function of σ. Therefore, the equation (2.5.10) is a linear, homogeneous, ordinary differential equation for an amplitude w_1 of the discontinuity and gives the rate of change of w_1 along the bicharacteristic curves on the characteristic surface. It follows immediately that if there exists a discontinuity in a transversal derivative of $\mathbf{u}$ at some point of a characteristic surface, it persists (i.e., it remains non-zero) at all points on the bicharacteristic curve through that point. Interpreted in the language of wave propagation, *discontinuities propagate along rays.*

We have seen in this chapter that the propagation of discontinuities is a remarkable feature of hyperbolic equations. The analysis of propagation of discontinuities gives rise to the concept of generalized solutions which are the physically meaningful solutions. The structure of solutions of a hyperbolic equation is dominated by characteristic surfaces and rays. The main features of the solution can be analysed using the essential character of the differential operator along the characteristic manifolds.

Chapter 3

Simple wave, high - frequency approximation and ray theory

3.1 Simple wave

We shall discuss in this section a class of solutions, known as simple waves, which are the only known exact solutions (except for some solutions of particular equations) and which can be easily evaluated numerically. These are plane or one-dimensional wave solutions of what is known as a *reducible system of equations*, in which the coefficient matrices A and B's (either in (2.1.1) or (2.3.7)) are functions of $\mathbf{u}$ only and the nonhomogeneous term is absent i.e., $C = 0$. Simple waves form the building block of more general solutions. For example, for a reducible pair of equations, any solution in a characteristic quadrilateral type of domain bounded by two pairs of intersecting characteristic curves of different families can not only be obtained as a result of interaction of a pair of simple waves but can also give rise to such a pair. We shall also see later in this chapter that a large class of solutions which are obtained in high frequency limit are extensions or modifications of simple wave solutions. We start with an example of a simple wave.

3.1.1 Example of a simple wave in gas dynamics

Let us consider the motion produced by a moving piston in an initially undisturbed polytropic gas in a uniform state and contained in a semi-infinite tube bounded on the left by the piston. Equations of one-dimensional motion, written in the matrix form are

$$\begin{bmatrix} 1 & 0 \\ 0 & 1 \end{bmatrix} \frac{\partial}{\partial t} \begin{bmatrix} \rho \\ q \end{bmatrix} + \begin{bmatrix} q & \rho \\ \frac{a^2}{\rho} & q \end{bmatrix} \frac{\partial}{\partial x} \begin{bmatrix} \rho \\ q \end{bmatrix} = 0 \tag{3.1.1}$$

where ρ is the mass density, q the particle velocity and a the sound velocity satisfying

$$a^2 \equiv a^2(\rho) = k^2 \rho^{\gamma-1}, \quad k^2 = \text{ constant}, \quad \gamma = \text{ constant} > 1 \tag{3.1.2}$$

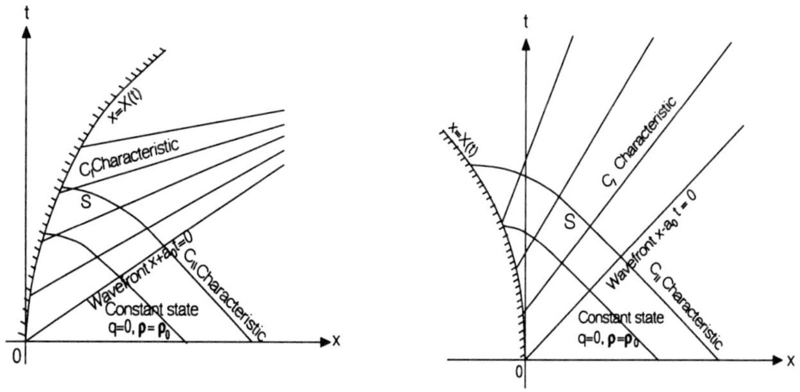

(a) Accelerating piston produces a compression wave (b) Decelerating piston produces an expansion wave

Fig. 3.1.1: Waves produced by a piston starting with zero velocity.

Equations in (3.1.1) form a reducible hyperbolic system with distinct characteristic velocities

$$c_1 = q + a, \quad c_2 = q - a \tag{3.1.3}$$

Let the characteristics of the first and second family be denoted by C_I and C_{II}, respectively.

Let the equation of the piston be given by

$$x = X_p(t), \quad X_p(0) = 0 \tag{3.1.4}$$

where we assume that initially the piston starts with zero velocity i.e.,

$$X_p'(0) = 0 \tag{3.1.5}$$

For an accelerating piston motion we have

$$X_p''(t) > 0 \tag{3.1.6}$$

and for a decelerating motion

$$X_p''(t) < 0 \tag{3.1.7}$$

The problem is to find the solution of the system (3.1.1) in the region on the right of the piston, i.e., in the domain D of the (x,t)-plane:

$$D : x > X_p(t), \quad t > 0 \tag{3.1.8}$$

Initially the gas in the tube is at rest with constant density, so the solution must satisfy the initial conditions

$$\rho(x,0) = \text{ constant } = \rho_0 \text{ (say)}, \quad q(x,0) = 0, \quad x \geq 0 \tag{3.1.9}$$

At the piston the fluid velocity is equal to the particle velocity, so the solution must satisfy the boundary condition

$$q(X_p(t),t) = X_p'(t), \qquad t \geq 0 \tag{3.1.10}$$

Since the initial and boundary values are continuous with a continuous value of q at (0,0) (see the condition (3.1.6)), the solution is also continuous at least in a small subdomain of the domain D near the origin (0,0).

The characteristic equations and the compatibility conditions are

$$\text{along } \frac{dx}{dt} = q+a, \quad dq+dl = 0 \text{ or } w \equiv \frac{1}{2}(q+l) = \text{ constant} \tag{3.1.11}$$

and

$$\text{along } \frac{dx}{dt} = q-a, \quad dq-dl = 0 \text{ or } \pi \equiv \frac{1}{2}(q-l) = \text{ constant} \tag{3.1.12}$$

where

$$l(\rho) \equiv \int \frac{a d\rho}{\rho} = \frac{2a}{\gamma-1} \tag{3.1.13}$$

Solving (3.1.11-12) for q and l in terms of w and π, we get

$$q = w + \pi \ , \quad l(\rho) = w - \pi \tag{3.1.14}$$

Since l is a monotic function of ρ, we can use (3.1.14) to solve ρ in terms of w and π.

An important property of a hyperbolic equation is the finiteness of the speed of propagation. We can observe it in this case from the following consideration as long as the solution is smooth. At every point of the x-axis i.e. initially, $q = 0$ and the sound velocity has a constant value a_0. This implies that C_I and C_{II} characteristics starting from the points on the x-axis carry constant values $\pm a_0/(\gamma - 1)$ of w and π, respectively. Consider now the characteristics C_{I0} of the first family (i.e., that given by (3.1.11)) and starting from the initial position of the piston i.e., the origin in the (x, t)-plane. The characteristics of the two families, which meet at a point (x, t) in a domain on the right of C_{I0}, are those which start from the points of the x-axis and hence have values $\pm a_0/(\gamma - 1)$ of w and π. This implies that in the domain on the right of the characteristic C_{I0}, $q = 0$ and $a = a_0$ and C_{I0} itself is a straight line. Thus, the effect of piston motion is not felt at those points in the (x, t)-plane which can be reached from the points of the x-axis by moving with a velocity greater than and equal to a_0 i.e., the signal propagation velocity is finite.

Further, the C_{II} characteristics from the points of the x-axis continue beyond the C_{I0} as t increases and carry the same constant value of $\pi = -\frac{1}{2}l_0 = -a_0/(\gamma - 1)$ which means that in the domain S bounded by the piston path and C_{I0}, the relation

$$q = l - l_0 = \frac{2}{\gamma - 1}(a(\rho) - a_0) \tag{3.1.15}$$

is valid or using (3.1.14) with $\pi = -\frac{1}{2}l_0$ we get

$$q = w - \frac{1}{2}l_0 \ , \quad l = w + \frac{1}{2}l_0 \ \text{ in } S \tag{3.1.16}$$

A nonconstant solution of the form (3.1.15 or 16), in which all members of the dependent variables of a reducible system are expressed in terms of a single function $w(x, t) \in C^1(S)$, is called a *simple* wave. In this particular case, we shall determine w and hence

q and l in S as functions of x and t from the piston motion. (3.1.11) shows that along C_I i.e., the first family of characteristics, the functions q and l in (3.1.16) satisfy one more relation $q + l =$ constant i.e., w constant. Thus, in S both q and a are constant along a C_I characteristic and since its slope $\frac{dx}{dt} = q + a$ also becomes constant, a C_I characteristic is a straight line given by

$$x = (q + a)t + \text{ constant} \tag{3.1.17}$$

We can evaluate the constant with the help of the boundary condition (3.1.10). Let the characteristic through the point (x, t) meet the piston path at a time η at the point $(\eta, X_p(\eta))$. Then on this characteristic $q = X'_p$ and from (3.1.15) $a = a_0 + \frac{\gamma-1}{2}q$ i.e.,

$$q(x,t) = X'_p(\eta) \ , \ a(x,t) = a_0 + \frac{\gamma - 1}{2}X'_p(\eta) \tag{3.1.18}$$

and its equation (3.1.17) becomes

$$x = X_p(\eta) + \{a_0 + \frac{\gamma + 1}{2}X'_p(\eta)\}(t - \eta) \tag{3.1.19}$$

Since the function $X_p(\eta)$ is given, (3.1.19) can be solved to give η as a function $\eta(x,t)$ of x and t. The simple wave solution in S is completely determined by substituting $\eta(x,t)$ in (3.1.18)

When the piston is accelerating i.e., $X''_p(\eta) > 0, a$ in S is greater than its value a_0 in the constant state in which the wave is running, and a increases (and hence ρ and p also increase) with t at a fixed x (a point with fixed x keeps on meeting characteristics of C_I family with larger values of $X'_p(\eta)$), therefore, the simple wave is a compression wave. Similarly, when the piston is decelerating i.e., $X''_p < 0$, it produces an expansion wave. For a detailed discussion, reference may be made to part B, chapter III of Courant and Friedrichs (1948).

The main aim of discussion in this section is to give an example of a simple wave that is governed by a *single* first order quasilinear partial differential equation. The compatibility condition (3.1.11) was obtained from the differential form of it

$$\left\{\frac{\partial}{\partial t} + (q + a)\frac{\partial}{\partial x}\right\}\left\{\frac{q}{2} + \frac{a}{\gamma - 1}\right\} = 0 \tag{3.1.20}$$

Eliminating a from this equation with the help of (3.1.15) we get

$$\frac{\partial q}{\partial t} + (a_0 + \frac{\gamma + 1}{2}q)\frac{\partial q}{\partial x} = 0 \tag{3.1.21}$$

which is nothing but the Burgers' equation (1.1.6), for $u = \frac{\gamma+1}{2}q$ in a coordinate system moving with the constant velocity a_0. For a simple wave in a *polytropic gas, Burgers' equation appears as an exact equation.*

Consistent with the definition 2.1.1, we call a variable w which remains constant on the first family of characteristic curves in a simple wave of the first family to be a *characteristic variable* * *of the first characteristic field.* The variable π is called the *Riemann invariant of the first characteristic field.* In the particular case of a hyperbolic system of two equations, π is a characteristic variable and w a Riemann invariant of the second characteristic field. A definition of these variables for a general hyperbolic system of quasilinear equations will be given in the next subsection.

Let us consider a particular case of the decelerating piston. Given $q_l > 0$ and $v > 0$, we take

$$X_p(t) = \begin{cases} -\frac{1}{2}q_l v t^2, & 0 \le t \le \frac{1}{v} \\ -q_l t + \frac{q_l}{2v}, & t > \frac{1}{v} \end{cases} \tag{3.1.22}$$

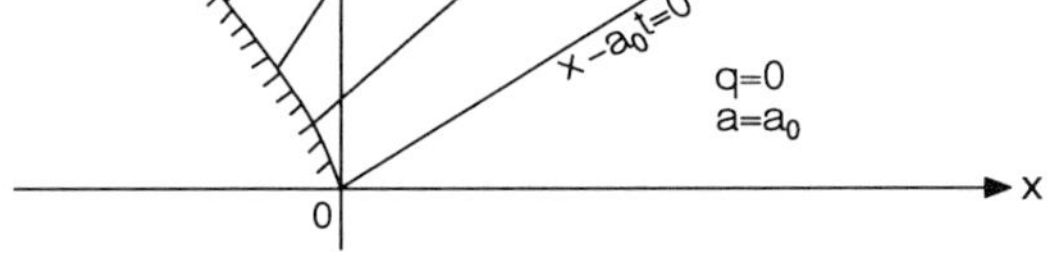

Fig. 3.1.2: When the piston path decelerates and attends a constant velocity after $t > \frac{1}{v}$, there are constant states on the two sides of the simple wave region.

The piston decelerates until $\frac{1}{v}$ when it reaches a velocity $-q_l$, then it continues to move with this constant velocity (Fig.3.1.2).

*The characteristic variable w here is different from that for a polytropic gas used later, see (3.1.68).

The solution consists of the constant state $q = 0, a = a_0$ in $x > a_0 t$; another constant state $q = -q_l, a = \left(a_0 - \frac{\gamma-1}{2} q_l\right) = a_l$ say, between the piston and simple wave i.e., $-q_l t + \frac{q_l}{2v} < x < (a_l - q_l)t - \frac{q_l}{2v}$ for $t > \frac{1}{v}$, and a simple wave in the domain bounded by the characteristics $x = a_0 t$, $x = -\frac{q_l}{2v} + \left(a_0 - \frac{\gamma+1}{2} q_2\right) t$ and the piston path $x = -\frac{1}{2} q_l v t^2$.

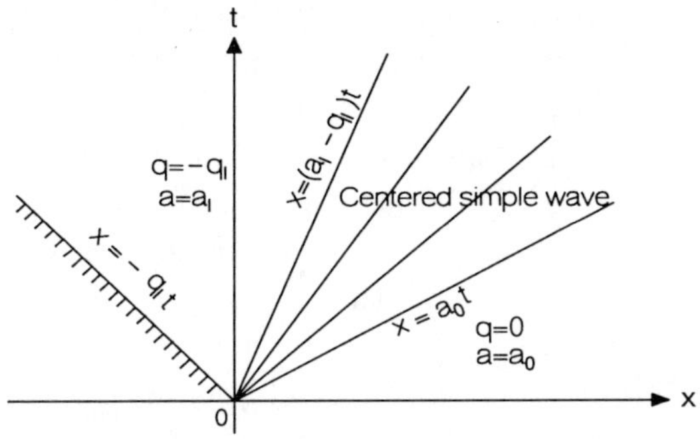

Fig. 3.1.3: when $v \to \infty$, the simple wave in Fig.3.1.2 approaches a centered simple wave.

An interesting limiting solution of the above solution is obtained by taking the limit as $v \to \infty$ as shown in Fig. 3.1.3. The starting points of all the diverging C_I family of characteristics collapse to the origin 0. The solution now consists of a *centered simple wave* bounded by two constant states on the two sides of it. It is easier to get this centered simple wave solution by substituting

$$q = Q(w) \quad , \quad w = \frac{x}{t} \tag{3.1.23}$$

in the equation (3.1.21) and noting that $Q' \neq 0$ in S. This gives $Q = \frac{2}{\gamma+1}(\frac{x}{t} - a_0)$. Therefore, the solution of the piston withdrawing with a constant velocity $-q_l$ is given by (note: $a_l = a_0 - \frac{\gamma-1}{2} q_l$)

$$q = \begin{cases} -q_l \;, & -q_l t < x \leq (a_l - q_l)\, t \\ \frac{2}{\gamma+1}(\frac{x}{t} - a_0), & (a_l - q_l)\, t < x \leq a_0 t \\ 0 \;, & a_0 t < x < \infty \end{cases} \tag{3.1.24}$$

$$a = \begin{cases} a_0 - \frac{\gamma-1}{2} q_l & , \quad -q_l t < x \le (a_l - q_l)\, t \\ \frac{2}{\gamma+1} a_0 + \frac{\gamma-1}{\gamma+1} \frac{x}{t} & , \quad (a_l - q_l)\, t < x \le a_0 t \\ a_0 & , \quad a_0 t < x < \infty \end{cases} \tag{3.1.25}$$

This is one of the rare examples of an exact solution in a very simple explicit form representing an important physical phenomenon. It is important also form another aspect. When the piston speed $-q_l$ is such that $-\frac{2}{\gamma-1} a_0 < -q_l < 0$ i.e., $a_l > 0$, there is always a constant state separating the piston and the tail of the simple wave on $x = (a_0 - \frac{\gamma+1}{2} q_l)t$. When the piston velocity $-q_l$ is equal to $-\frac{2}{\gamma-1} a_0$, there is a *complete* simple wave in the sense that at the tail of the simple wave i.e., at the piston, the sound velocity a (and hence density and pressure) vanish. When the piston withdraws even faster i.e., $-q_l < -\frac{2}{\gamma-1} a_0$, there is no more change in the simple wave but a vacuum is created between the piston and the tail of the simple wave.

3.1.2 Simple wave in one space dimension

Consider a reducible hyperbolic system of n equations

$$A(\mathbf{u}) \frac{\partial \mathbf{u}}{\partial t} + B(\mathbf{u}) \frac{\partial \mathbf{u}}{\partial x} = 0 \tag{3.1.26}$$

Definition 3.1.1 A simple wave solution of (3.1.26) is a genuine solution in a domain S such that all components u_i of $\mathbf{u}$ can be expressed in terms of a single function $w(x, t) \in C^1(S)$:

$$\mathbf{u}(x, t) = \mathbf{U}(w(x, t)) \tag{3.1.27}$$

Substituting (3.1.27) in (3.1.26) and assuming that $w_x \neq 0$, we get

$$[B(\mathbf{U}) - (w_t/w_x) A(\mathbf{U})] \frac{d\mathbf{U}}{dw} = 0 \tag{3.1.28}$$

For a nontrivial solution of $\frac{d\mathbf{U}}{dw}$ we require that $-w_t/w_x = c$ i.e.,

$$w_t + c(\mathbf{U}) w_x = 0 \ \text{ in } S \tag{3.1.29}$$

where c is an eigenvalue of (3.1.26). Since $\mathbf{U}$ is a function of w alone and does not explicitly contain x and t, $c(\mathbf{U})$ is a function of w alone. Therefore, the equation (3.1.29) implies that in S, w=constant i.e., $\mathbf{U}$ = constant along each line, $x - c(\mathbf{U})t$ = constant. The straight lines $x - c(\mathbf{U})t$ = constant, are characteristic curves of the system (3.1.26). The derivative $\frac{d\mathbf{U}}{dw}$ is parallel to the corresponding right eigenvector:

$$\frac{d\mathbf{U}}{dw} = \mathbf{r}(\mathbf{U}) \tag{3.1.30}$$

This is a system of n ordinary differential equations and it can be integrated with $n-1$ arbitrary constants ($n-1$, because it is an autonomous system): $\tilde{\pi} = (\pi_1, \pi_2, \ldots, \pi_{n-1})$. Thus,

$$\mathbf{U} = \mathbf{U}(w, \tilde{\pi}) \tag{3.1.31}$$

where $\tilde{\pi}$ remains constant in S. Assuming that the mapping

$$\mathbf{u} = \mathbf{U}(w, \tilde{\pi}), \qquad (w, \pi_1, \ldots, \pi_{n-1}) \rightarrow (u_1, \ldots, u_n) \tag{3.1.32}$$

is one to one, we can express $w, \pi_1, \ldots, \pi_{n-1}$ as functions of $u_1, \ldots,$ u_n

$$w = w(\mathbf{u}), \qquad \pi_\delta = \pi_\delta(\mathbf{u}), \quad \delta = 1, 2, \ldots, n-1 \tag{3.1.33}$$

Since w and $\pi_1, \ldots, \pi_{n-1}$ are independent variables, $\frac{\partial \pi_\delta}{\partial w} = 0$. But,

$$\frac{\partial \pi_\delta}{\partial w} = \sum_{i=1}^{n} \frac{\partial \pi_\delta}{\partial u_i} \frac{\partial u_i}{\partial w}$$

Using (3.1.30) we get

$$\mathbf{r}.\nabla_u \pi_\delta = 0, \quad \delta = 1, 2, \ldots, n-1 \tag{3.1.34}$$

This relation can be treated as a first order nonlinear partial differential equation

$$r_1(\mathbf{u})\pi_{u_1} + r_2(\mathbf{u})\pi_{u_2} + \ldots + r_n(\mathbf{u})\pi_{u_n} = 0 \tag{3.1.35}$$

From the theory of a single first order quasilinear partial differential equations, it follows that (3.1.35) has a set of $n-1$ independent solutions $\tilde{\pi} = (\pi_1, \ldots, \pi_{n-1})$ and any other solution π_n can be expressed in terms of these solutions i.e., $\pi_n = \pi_n(\pi_1, \ldots, \pi_{n-1})$. This leads to the definition of a new set of variables (Lax (1957)).

Definition 3.1.2 A solution of (3.1.35) is called a *Riemann invariant* of the characteristic field associated with the eigenvalue $c(\mathbf{u})$.

It follows that there are only $n-1$ independent Riemann invariants associated with a given eigenvalue $c(\mathbf{u})$ of the hyperbolic system (3.1.26). Any other Riemann invariant associated with $c(\mathbf{u})$ can be expressed in terms of these $n-1$ functions. An example of Riemann invariants in a system of more than two equations is given at the end of this subsection.

We are now in a position to extend the Definition 2.1.1 of a characteristic variable for linear hyperbolic system to a quasilinear system.

Definition 3.1.3 A function $w(\mathbf{u})$ such that

$$\frac{\partial(w, \pi_1, \pi_2, \ldots, \pi_{n-1})}{\partial(u_1, u_2, \ldots, u_n)} \neq 0 \tag{3.1.36}$$

in a domain of $\mathbf{u}$-space, is called *characteristic variable of the characteristic field* associated with the eigenvalue $c(\mathbf{u})$.

In a simple wave associated with the eigenvalue c, all Riemann invariants $\pi_1, \pi_2, \ldots, \pi_{n-1}$ have constant values everywhere but the characteristic variable w has different constant values on different members of the characteristic curves $x - c(\mathbf{u})t$ = constant, which are straight lines. This is true for each one of the n eigenvalues $c_1, c_2, \ldots, c_n$ which, for simplicity, we assume to be distinct. Thus, associated with the kth eigenvalue $c_k(k = 1, 2, \ldots, n)$ we can have a *simple wave of the kth family* in which the kth characteristic variable $w^{(k)}$ is constant along the straight line characteristics of the kth field i.e. $x - c_k(w^k, \pi_1^{(k)}, \ldots, \pi_{n-1}^{(k)})t$ = constant and, in which, the kth Riemann invariants $\tilde{\pi}^k = (\pi_1^{(k)}, \ldots, \pi_{n-1}^k)$ are constant throughout the simple wave region in the $(x,t)-$plane.† In what follows, we shall discuss a simple wave the kth family, however, we shall suppress the superscript k from w and $\tilde{\pi}$. The value of $\tilde{\pi}$ in the simple wave can be determined from the boundary of the simple wave, except for this

†Unlike the case of linear equations, in general, it is not possible to reduce a hyperbolic system of more than two quasilinear equations to a canonical form in which in each equation derivatives of only one characteristic variable appears Lax(1963), see also the comment at the end of this subsection.

$\tilde{\pi}$ plays no role. Therefore, we shall not show the dependence of c_k on $\tilde{\pi}$ and we shall write $c_k(w, \tilde{\pi}) \equiv c_k(w)$.

The equation governing the evolution of the characteristic variable w in the simple wave is a single first order quasilinear partial differential equation (3.1.29), where $\mathbf{U} = \mathbf{U}(w)$. Writing $c(\mathbf{U}(w))$ simply as $c(w)$, we write (3.1.29) as

$$w_t + c(w)w_x = 0 \tag{3.1.37}$$

where c is a known function of w as we found in the example (3.1.21).

Consider a constant value w_0 of w along a particular characteristic curve of the kth family in a kth simple wave. In the neighbourhood of this characteristic curve, we have

$$c(\mathbf{U}(w)) = c_0 + \left\{\langle \nabla_u \, c, \frac{d\mathbf{U}}{dw}\rangle\right\}_0 (w - w_0) + 0\{(w - w_0)^2\}$$

Using the result that $\frac{d\mathbf{U}}{dw}$ is parallel to $\mathbf{r}_0$, we find that, to the first order in $(w - w_0)$, the characteristic velocity c in (3.1.37) varies linearly with w if $\{\langle \nabla_u c, \mathbf{r}\rangle\} \neq 0$.

Definition 3.1.4 If

$$\langle \nabla_u c, \mathbf{r}\rangle \neq 0, \quad \forall \mathbf{u} \in D_u \tag{3.1.38}$$

where D_u is a domain in $(u_1, u_2, \ldots, u_n)$-space, we say that the characteristic field under consideration is *genuinely nonlinear*. Otherwise, the characteristic field is called *linearly degenerate*.

If a characteristic field is genuinely nonlinear, the shape of the pulse in a simple wave will deform with time (possibly leading to the formation of a shock). By retaining only the terms up to the first order in $w - w_0$ in the Taylor's expansion of c, we find that for a small amplitude simple wave the equation (3.1.37) can be approximated by the Burgers' equation. Note that in the case of equations (2.1.12 - 14) governing the one-dimensional motion of a polytropic gas, the characteristic fields corresponding to the eigenvalues $c_1 = q - a, c_3 = q + a$ are genuinely nonlinear, whereas the corresponding to $c_2 = q$ is linearly degenerate.

The following two results, which are simple to prove (Courant and Friedrichs, (1948) for a system of two equations; and Lax (1963) for a general system), show that a simple wave occurs in a most natural way. We state these results without proof.

Theorem 3.1.1 If a section of a kth characteristic carries constant value of $\mathbf{u}$, then in a region adjacent to this section, the solution is either a constant state or a simple wave of kth family.

Theorem 3.1.2 The solution in a region adjacent to a region of constant state is a simple wave solution.

It is quite easy to produce a simple wave of kth family from an initial value problem or a boundary value problem (see section 3.1.4).

Now we proceed to discuss a special class of simple wave solutions, namely the *centered simple wave*, which is of great importance in the solution of a fundamental problem called Riemann's problem. A *centered simple wave*, more appropriately called a *centered* rarefaction wave, is a genuine solution of (3.1.26) for $t > t_0$, in which $\mathbf{u}$ depends only on the ratio $(x-x_0)/(t-t_0)$; (x_0, t_0) being the center of the wave. Choosing $(x-x_0)/(t-t_0)$ to be the variable w in (3.1.27), it follows that such a solution is a simple wave with one of the n families of characteristics, say kth, being straight lines represented by $(x - x_0)/(t - t_0) =$ constant, all of which pass through the point (x_0, t_0). Now we state a theorem without proof.

Theorem 3.1.3 Let a constant state $\mathbf{u}_l$ on left[‡] be connected to another constant state $\mathbf{u}_r$ on right by a centered simple wave of the kth family, then

$$c_k(\mathbf{u}_l) < c_k(\mathbf{u}_r) \tag{3.1.39}$$

Noting that the characteristic curves through the origin in Figure 3.1.3 are given by $x = c_k(\mathbf{u})t$, we see that the inequality (3.1.39) is a necessary condition for the existence of a centered simple wave for $t > 0$.

The inequality (3.1.39) is of great importance to us. Let us explain it with a particular choice of the parameter δ:

$$\delta = c_k(\mathbf{u}_r) - c_k(\mathbf{u}_l) \tag{3.1.40}$$

In general, for an arbitrary simple wave (not necessarily centered) the parameter δ could have both positive and negative values.

‡With respect to the reader.

However, the inequality (3.1.40) permits only positive values of δ, i.e., "rarefaction simple waves" in a centered wave.

Example from gas dynamics

We first mention a few general results in fluid flows. The system of 3 equations (2.1.12 - 14) together represent conservation of mass, momentum and energy in a continuous flow of the gas. They are equivalent to the three conservation laws (2.1.8) with (2.1.20), which are valid also for discontinuous solutions. The equations (2.1.12 - 14) imply $(p\rho^{-\gamma})_t + u(p\rho^{-\gamma})_x = 0$ which states that $p\rho^{-\gamma}$ is constant as we move with a fluid element. $p\rho^{-\gamma}$ is a function of the entropy and hence the entropy of a fluid element also remains constant in a continuous flow as we move with the fluid. The system (3.1.1) of two equations is derived from (2.1.12 - 14) by making an additional assumption that the entropy has the same constant value throughout the fluid flow.

For the equations (2.1.12-14), we shall find the expressions for a pair of Riemann invariants corresponding to the third eigenvalue $c_3 = q + a$. The Riemann invariants satisfy (3.1.35) which, in this case, reduces to

$$\frac{\rho}{a}\pi_\rho + \pi_q + \rho a \pi_p = 0 \tag{3.1.41}$$

Using the theory of characteristics for the first order equation (3.1.41) and the expression $a^2 = \gamma p/\rho$, we find a pair of independent solutions

$$\pi_1 = \frac{1}{2}q - \frac{a}{\gamma - 1}, \quad \pi_2 = p\rho^{-\gamma} \tag{3.1.42}$$

These form a pair of Riemann invariants we are looking for. We note that π_1 is the Riemann invariant π used in the previous subsection. But, now we get a new Riemann invariant π_2, the function of entropy discussed above. It also follows that $\tilde{\pi} \equiv (\pi_1, \pi_2) =$ constant leads to the same simple wave which we discussed in the section 3.1.1.

There is an important difference between the system of three equations (2.1.12-14) and its particular case (3.1.1) of two equations. In the case of the former, the compatibility condition (2.1.19) can not be expressed in terms of a characteristic variable w, however, this can be done for the latter, where the compatibility condition can be expressed in the form (3.1.11 or 20) i.e.,

$$\left(\frac{\partial}{\partial t} + (q+a)\frac{\partial}{\partial x}\right) w = 0, \quad w = \frac{q}{2} + \frac{a}{\gamma - 1} \tag{3.1.43}$$

In the case of a simple wave of the system (3.1.1) we get a relation π=constant and that of the system (2.1.12 - 14) of three equations we get two relations $\tilde{\pi} = (\pi_1, \pi_2)$ = constant so that both equations (3.1.43) and (2.1.19) can be reduced to the same single first order partial differential equation (3.1.21) in one dependent variable. In this case (3.1.21) is of the form (3.1.37) with q as the characteristic variable. Similarly, we can get an equation for the characteristic variable $w = q + \frac{a_0}{\gamma - 1}$.

3.1.3 Simple wave in multi-dimensions

As explained in the beginning of this section, a simple wave is a plane or one-dimensional wave of a reducible system of equations. In this section, we shall look for such waves as solutions of equations in more than two independent variables

$$A(\mathbf{u})\mathbf{u}_t + B^{(\alpha)}(\mathbf{u})\mathbf{u}_{x_\alpha} = 0 \tag{3.1.44}$$

where matrices A and $B^{(\alpha)}$ are functions only of a dependent variable $\mathbf{u} \in I\!R^n$.

We look for a solution of the form

$$\mathbf{u}(t, \mathbf{x}) = \mathbf{u}(t, \xi), \quad \xi = n_1 x_1 + n_2 x_2 + \ldots + n_m x_m, \quad |\mathbf{n}| = 1 \tag{3.1.45}$$

where $n_1, n_2, \ldots, n_m$ are constants. The equation (3.1.44) becomes

$$A(\mathbf{u})\mathbf{u}_t + B(\mathbf{u})\mathbf{u}_\xi = 0 \tag{3.1.46}$$

where the matrix

$$B(\mathbf{u}) = n_\alpha B^{(\alpha)} \tag{3.1.47}$$

depends also on the m parameters $n_1, n_2, \ldots, n_m$ satisfying $|\mathbf{n}| = 1$.

The system (3.1.46) is a reducible system in two independent variables ξ and t. Following the discussion in the previous section, a simple wave solution of (3.1.46) is given by

$$u(\xi, t) = \mathbf{U}(w, \mathbf{n}, \tilde{\pi}) \tag{3.1.48}$$

where the scalar w satisfies the equation

$$w_t + c(\mathbf{U})w_x = 0 \tag{3.1.49}$$

and the components of the $n-1$ vector $\tilde{\pi}$

$$\tilde{\pi} \equiv \tilde{\pi}(\mathbf{n}) = (\pi_1, \pi_2, \ldots, \pi_{n-1}) \tag{3.1.50}$$

are $n-1$ Riemann invariants depending on $\mathbf{n}$. The simple wave velocity $c(\mathbf{U}) = c(\mathbf{u}, \mathbf{n})$ is an eigenvalue satisfying

$$\det[B - cA] \equiv \det[n_\alpha B^{(\alpha)} - c(\mathbf{u}, \mathbf{n})A] = 0 \tag{3.1.51}$$

which is exactly the same as (2.3.11). Thus c is an eigenvalue of (3.1.45) with parameters $n_1, n_2, \ldots, n_m$.

All results of the previous section on the simple wave are valid for this simple wave in which the role of the x-coordinate is played by the variable ξ, which is the distance measured in the direction of the constant vector $\mathbf{n}$.

Example from gas dynamics

Consider the Euler equations of motion (2.3.16 - 18). Substituting (3.1.45) in these equations we get

$$\rho_t + q_N \rho_\xi + \rho q_{N\xi} = 0 \tag{3.1.52}$$

$$\mathbf{q}_t + q_N \mathbf{q}_\xi + \frac{1}{\rho}\nabla p = 0 \tag{3.1.53}$$

$$p_t + q_N p_\xi + \rho a^2 q_{N\xi} = 0 \tag{3.1.54}$$

where

$$q_N = \langle \mathbf{n}, \mathbf{q} \rangle \quad , \quad \mathbf{n} = \text{ constant vector} \tag{3.1.55}$$

is the fluid velocity normal to the planes given by $\xi \equiv \langle \mathbf{n}, \mathbf{x} \rangle =$ constant. Taking the inner product of (3.1.53) with $\mathbf{n}$ gives

$$q_{Nt} + q_N q_{N\xi} + \frac{1}{\rho} p_\xi = 0 \tag{3.1.56}$$

The set of 3 equations (3.1.52, 54 and 56) are exactly the same as the system of equations (2.1.12 - 14), when we identify q_N with q. Corresponding to the eigenvalue

$$c = q_N + a = \langle \mathbf{n}, \mathbf{q} \rangle + a \tag{3.1.57}$$

of this set, the two Riemann invariants are

$$\pi_1 = \frac{1}{2}\mathbf{q}_N - \frac{a}{\gamma - 1}, \quad \pi_2 = p\rho^{-\gamma} \tag{3.1.58}$$

To find a complete set of Reimann invariants of (2.3.16 - 18), we take a right eigenvector $\mathbf{r}$ corresponding to the eigenvalue (3.1.57)

$$\mathbf{r} = (\rho/a, n_1, n_2, n_3, \rho a)^T \tag{3.1.59}$$

and solve the equation (3.1.35) i.e.,

$$\frac{\rho}{a}\pi_\rho + n_1 dq_1 + n_2 dq_2 + n_3 dq_3 + \rho a dp = 0 \tag{3.1.60}$$

The characteristic equations of (3.1.60) are

$$a\frac{d\rho}{\rho} = \frac{dq_1}{n_1} = \frac{dq_2}{n_2} = \frac{dq_3}{n_3} = \frac{dp}{\rho a} \tag{3.1.61}$$

We note that each segment of this ratio is also equal to dq_N, since $|\mathbf{n}| = 1$. Therefore, two first integrals of this system are

$$\pi_1 = \frac{1}{2}q_N - \frac{a}{\gamma - 1} \quad , \quad \pi_2 = p\rho^{-\gamma} \tag{3.1.62}$$

Other first integrals of (3.1.61) can be obtained by solving

$$\frac{dq_\alpha}{n_\alpha} = dq_N \quad , \quad \alpha = 1, 2, 3 \tag{3.1.63}$$

which gives $\pi'_{1\alpha} = q_\alpha - n_\alpha q_N \quad , \ \alpha = 1, 2, 3$. From the expressions of π_1 and $\pi'_{1\alpha}$ we get some more first integrals $\pi_{1\alpha}$, where

$$\pi_{1\alpha} = (n_\alpha \pi_1 + \frac{1}{2}\pi_{1\alpha}) = \frac{1}{2}q_\alpha - \frac{n_\alpha a}{\gamma - 1} \tag{3.1.64}$$

We have a wide choice of a set of Riemann invariants – any set of four independent functions from the set $\pi'_{1\alpha}, \pi_{1\alpha}, \pi_2$ $(\alpha = 1, 2, 3)$ can be chosen to be a set of Riemann invariants. However, we can not include all three π'_{11}, π'_{12} and π'_{13} in a set since $\sum_{\alpha=1}^{3} n_\alpha \pi'_{1\alpha} = 0$. Keeping symmetry in mind, we choose

$$\pi_{1\alpha} = \frac{1}{2}q_\alpha - \frac{n_\alpha a}{\gamma - 1}, \ \alpha = 1, 2, 3; \quad \pi_2 = p\rho^{-\gamma} \tag{3.1.65}$$

as a set of four independent Riemann invariants corresponding to the eigenvalue (3.1.57). There is also a wide choice for the characteristic variable w, it can be chosen to be any function of ρ, q_α, p which can not be expressed as a function of $\pi_{11}, \pi_{12}, \pi_{13}, \pi_2$. We choose

$$w = \frac{a_0}{\rho_0}(\rho - \rho_0) \tag{3.1.66}$$

where a_0 and ρ_0 are two constants. With this choice of w, it is simple to express $\rho, \mathbf{q}$ and p in terms of w and $\tilde{\pi} = (\pi_{11}, \pi_{12}, \pi_{13}, \pi_2)$:

$$\rho = \rho_0(1 + w/a_0), \quad p = \pi_2 \rho_0^\gamma (1 + w/a_0)^\gamma, \ q_\alpha = 2\pi_{1\alpha} + \frac{2n_\alpha}{\gamma - 1} A(w) \tag{3.1.67}$$

where

$$a = A(w) \equiv \sqrt{\gamma \pi_2 \rho_0^{\gamma-1}} (1 + w/a_0)^{(\gamma-1)/2} \tag{3.1.68}$$

Multiplying the system (2.3.16-18) by a left eigenvector $l = [0, \rho a \mathbf{n}, 1]$ with $\mathbf{n}$ constant, we get

$$\rho a \{q_{Nt} + (q_N + a) q_{N\xi}\} + \{p_t + (q_N + a) p_\xi\} = 0 \tag{3.1.69}$$

Substituting (3.1.67 - 68) and

$$q_N = 2\left(\sum_{\alpha=1}^{3} n_\alpha \pi_{1\alpha}\right) + \frac{2}{\gamma - 1} A(w) \tag{3.1.70}$$

in (3.1.69), we derive the partial differential equation governing the 3-D simple waves in a polytropic gas as

$$w_t + \left\{2\left(\sum_{\alpha=1}^{3} n_\alpha \pi_{1\alpha}\right) + \frac{\gamma + 1}{\gamma - 1} A(w)\right\} \langle \mathbf{n}, \nabla \rangle w = 0 \tag{3.1.71}$$

or

$$\frac{dw}{dt} = 0 \tag{3.1.72}$$

where

$$\frac{d}{dt} = \frac{\partial}{\partial t} + \left\{2\left(\sum_{\alpha=1}^{3} n_\alpha \pi_{1\alpha}\right) + \frac{\gamma + 1}{\gamma - 1} A(w)\right\} \langle \mathbf{n}, \nabla \rangle \tag{3.1.73}$$

The equation (3.1.71) or equations (3.1.72 - 73) imply that the amplitude w of the wave remains constant along the rays of the plane waves with constant normal $\mathbf{n}$:

$$\frac{d\mathbf{x}}{dt} = \mathbf{n}\left\{2\left(\sum_{\alpha=1}^{3} n_\alpha \pi_{1\alpha}\right) + \frac{\gamma+1}{\gamma-1}A(w)\right\} \tag{3.1.74}$$

Since w is assumed to be constant on the wavefront at any time, the second set of ray equations (2.4.13) are automatically satisfied in the form

$$\frac{d\mathbf{n}}{dt} = 0 \tag{3.1.75}$$

The equation (3.1.71), with constant values of $\mathbf{n}$ and $\tilde{\pi}$, gives a 3-D simple wave. Such a wave is one-dimensional, propagating in the direction of $\mathbf{n}$. A high-frequency wave, which we shall discuss in this monograph, is a simple modification of such a wave in which $\mathbf{n}$ and $\tilde{\pi}$ vary slowly.

3.1.4 An initial value problem leading to a kth simple wave

Consider the reducible hyperbolic system (3.1.26) in two independent variables and assume that the eigenvalues satisfy

$$c_1 < c_2 < \ldots < c_n \tag{3.1.76}$$

for all $\mathbf{u}$ under consideration. We now consider an initial value problem for the system with the initial condition

$$\mathbf{u}(x,0) = \begin{cases} \mathbf{u}_0 & , \quad x > 0 \\ \mathbf{u}_1(x) & , \quad x \le 0 \end{cases} \tag{3.1.77}$$

where $\mathbf{u}_0$ is constant and $\mathbf{u}_1$ is a smooth function with an additional restriction that all the $n-1$ Riemann invariants $\pi_1, \pi_2, \ldots, \pi_{n-1}$ of the kth family of characteristics satisfy

$$\tilde{\pi}(\mathbf{u}_1(x)) = \tilde{\pi}(\mathbf{u}_0) \quad , \quad x < 0 \tag{3.1.78}$$

Thus, this initial data $\tilde{\pi} = (\pi_1, \pi_2, \ldots, \pi_{n-1})$ is constant to the left of the point $x = 0$ even though $\mathbf{u}_1$ need not be constant. Now three cases arise.

Case (a) Let $\mathbf{u}_1$ be also a constant but the data has a discontinuity at $x = 0$ i.e., $[u(x,0)]_{x=0} = \mathbf{u}_0 - \mathbf{u}_1 \neq 0$. When this discontinuity does not satisfy the Lax entropy condition for the kth shock,

$$c_{k-1}(\mathbf{u}_1(0)) < S < c_k(\mathbf{u}_1(0)), \;\; c_k(\mathbf{u}_0) < S < c_{k+1}(\mathbf{u}_0) \qquad (3.1.79)$$

suitably modified for $k = 1$ and n, the initial discontinuity at $x = 0$ will immediately get resolved into a kth centered simple wave. In this way we have produced a kth centered simple wave running into a constant state $\mathbf{u}_0$:

$$\mathbf{u}(x,t) = \mathbf{a} + \mathbf{b}\frac{x}{t} \;\; , \;\; c_k(\mathbf{u}_1(0)) < \frac{x}{t} < c_k(\mathbf{u}_0) \qquad (3.1.80)$$

where

$$\mathbf{a} = \frac{c_k(\mathbf{u}_0)\mathbf{u}_1(0) - c_k(\mathbf{u}_1(0))\mathbf{u}_0}{c_k(\mathbf{u}_0) - c_k(\mathbf{u}_1(0))}, \;\; \mathbf{b} = \frac{(\mathbf{u}_0 - \mathbf{u}_1(0))}{c_k(\mathbf{u}_0) - c_k(\mathbf{u}_1(0))} \qquad (3.1.81)$$

The first derivatives of $\mathbf{u}$ will, in general, be discontinuous across the leading and trailing fronts of the simple waves i.e., across $x = c_k(\mathbf{u}_0)t$ and $x = c_k(\mathbf{u}_1(0))t$, respectively.

Case (b) If the smooth function $\mathbf{u}_1(x)$ is not a constant function but $[\mathbf{u}(x,0)]_{x=0} = 0$, the solution will have a simple wave of the kth family bounded on the right by the kth characteristic $x = c_k(\mathbf{u}_0)t$. The solution will be valid for all time if all characteristics of the kth family starting from the points $x < 0$ on $t = 0$ diverge. Otherwise, the solution will be valid only until a finite critical time.

Case (c) Let us assume now that $\mathbf{u}_1(x)$ is not a constant function $[\mathbf{u}(x,0)]_{x=0} \neq 0$, and the Lax entropy condition (3.1.79) is not satisfied. The initial data now leads to a simple wave in the domain $x < t\ c_k(\mathbf{u}_1(0))$, a centered simple wave given by (3.1.80 - 81) and a constant state in the domain $x > t\ c_k(u_0)$. This solution may also be only locally valid depending on the convergence of the kth family of characteristics starting from the negative values of x at $t = 0$.

We have shown above how to produce, at least locally in t, a simple wave of the kth family running into a known constant state. We could have produced the most general form of this simple wave covering the whole of strip $0 < t < t_c$ of the (x,t)-plane by choosing a piecewise smooth initial data $\mathbf{u}(x,0)$ for $x \in I\!R$ such that the kth Riemann invariants $\tilde{\pi}(\mathbf{u}(x,0)) =$ constant and the discontinuities in $\mathbf{u}(x,0)$ do not satisfy Lax's entropy condition. The solution will have a number of centered simple waves each bounded by more general simple waves on the either side.

3.2 High-frequency approximation, wavefront, Huygens' method and Fermat's principle

In this section we shall first give a formal definition of a wavefront which requires high-frequency approximation. We shall show that a wavefront in a continuous solution of a hyperbolic system satisfies the characteristic partial differential equation. Using this result, we shall discuss Huygens' method of wavefront construction for a hyperbolic system of linear equations. This method, initially stated for light waves, is one of the most powerful methods in the theory of wave propagation. We find that the original statement of Huygens, proposed more than 300 years earlier, remains valid and unchanged for a wave governed by a general hyperbolic system provided we extend the meaning of *spherical waves* appropriately. We shall show this to be true also for a nonlinear wavefront and a shock front in subsequent chapters. In the last three sections of this Chapter we shall discuss another powerful mathematical tool for wave propagation, Fermat's principle, and its application to the derivation of weakly nonlinear ray theory in the simplest case.

3.2.1 Definition of a wavefront

Waves form an all pervading phenomena. They involve transfer of energy from one part of a medium to another part without transfer of material particles. When we use such a general idea for the definition of waves, we may not be in a position to identify some special propagating surfaces which we would like to call wavefronts. Intuitively, a wavefront can be recognized as a propagating surface across which abrupt or rapid changes in the state of a medium takes place. We first extend this observation into a formal definition of a wavefront.

Definition 3.2.1 When the state $\mathbf{u}(\mathbf{x}, t), \mathbf{x} \in I\!R^m, t \in I\!R$ of a continuum medium is expressed in the form

$$\mathbf{u} = \mathbf{u}\left(\frac{\phi}{\epsilon}, \frac{\psi_1}{\epsilon_1}, \frac{\psi_2}{\epsilon_2}, \ldots, \frac{\psi_p}{\epsilon_p}\right) \tag{3.2.1}$$

where ϕ and ψ_α are functions of $\mathbf{x}, t$; $\epsilon > 0$, $\epsilon_\alpha > 0$ and

$$\frac{\epsilon}{\epsilon_\alpha} << 1 \ , \ \alpha = 1, 2, \ldots, p \tag{3.2.2}$$

we call the surface

$$\Omega_t : \phi(\mathbf{x}, t) = \text{ constant} \ , \ t = \text{ fixed} \tag{3.2.3}$$

in $(\mathbf{x})$-space to be a wavefront. When the constant in (3.2.3) takes all real values, we get a one parameter family of wavefronts.

The above definition for a wavefront involves an approximation: there is a more *rapid* change in the solution of the system of governing equations as we cross a wavefront transversely as compared to more gradual changes as we move along the wavefront. Thus, we can see a *short wavelength* variation in the value of the solution at a given time when we examine the state (represented by $\mathbf{u}$) of the continuum system in the direction transverse to a wavefront. We also notice it as a *high-frequency* variation in $\mathbf{u}$ with respect to time when we remain at a fixed point during the period the wavefront transverses the point. One of the reasons for the slow variation with respect to variables $\psi_1, \psi_2, \ldots, \psi_p$ could be that the state of the ambient medium in which the wave is propagating may vary with length and time scales associated with the larger parameters $\epsilon_1, \epsilon_2, \ldots, \epsilon_p$.

We can visualize a wavefront without any approximation provided the condition (3.2.2) becomes $\epsilon/\epsilon_\alpha = 0$, $\alpha = 1, 2, \ldots, p$. This can happen when there is a surface of discontinuity $\phi = 0$ in the state $\mathbf{u}$ (such as a shock front) or in the partial derivatives of $\mathbf{u}$ (in which case $\phi = 0$ becomes a characteristics surface, see section 2.5). The condition $\epsilon/\epsilon_\alpha = 0$, $\alpha = 1, 2, \ldots, p$, is also satisfied when $\phi =$ constant surfaces are plane and the state of the medium, in which the wave is running, is uniform i.e., $\epsilon_\alpha = \infty$ for $\alpha = 1, 2, \ldots, p$. The last condition is satisfied by the multi-dimensional simple wave discussed in subsection 3.1.3 provided there are constant states on the two sides, say for $\phi < \phi_l$ and $\phi > \phi_r$, of the simple wave. The high-frequency assumption (3.2.2) is also satisfied if we perturb the simple wave solution such that the unit normal $\mathbf{n}$ and the Riemann invariants $\tilde{\pi}$ vary slowly compared to the variation of $\mathbf{u}$ with respect to the *phase function* $\phi = \langle \mathbf{n}, \mathbf{x} \rangle$. In any case, the high frequency approximation implies that all state variables can be expressed approximately in terms of a single variable. We have seen it to be true for a simple wave solution and we shall show it to be true approximately in all other cases.

3.2.2 Huygens' method of wavefront construction

Much before the nature of the light waves was understood, Christiaan Huygens found in 1676-78 a method of construction of successive positions of a wavefront starting from its initial position. This, known as Huygens' principle [§] in physics, appeared in 1690 in *"Traite' de la Lumiere"* and can be stated in a very simple language: *all points of a wavefront can be considered to be point sources of spherical secondary wavelets and after a time t the new position of the wavefront is an envelope of these secondary wavelets.* Huygens' method does not tell about which of the two envelopes is to be chosen as the wavefront. Both are possible but once the direction of propagation is initially decided depending on the source which created the wavefront, we have to follow that direction. Today, this method is the source of derivation of many properties like laws of reflection and refraction of waves in a first course on light. Huygens' method, stated more than three centuries ago only for light waves, is valid also for the waves governed by a general system of hyperbolic equations (Courant and Hilbert, 1962). We present in this section its proof for a linear system and later, we shall show it to be valid also for a nonlinear wavefront and a shock front.

We shall show in the next chapter that successive positions of a wavefront $\Omega_t : \phi(\mathbf{x}, t) = 0$ in a system governed by the linear hyperbolic system (2.5.1), is given by the characteristic partial differential equation

$$Q(\mathbf{x}, t; \nabla\phi, \phi_t) \equiv \det(A\phi_t + B^{(\alpha)}\phi_{x_\alpha}) = 0 \tag{3.2.4}$$

[§] In mathematics, by Huygens's principle, we mean today a different property of wave propagation (Courant and Hilbert (1962)).

If we solve $\phi(\mathbf{x}, t) = 0$ for t in the form $t = \psi(\mathbf{x})$ so that $\phi(\mathbf{x}, \psi(\mathbf{x})) = 0$, the wavefront Ω_t at time t is represented by

$$\Omega_t : \ \psi(\mathbf{x}) \ - t = 0 \tag{3.2.5}$$

Since $\phi_{x_\alpha} = -\phi_t \psi_{x_\alpha}$ and Q is a homogeneous function of ϕ_t and $\nabla\phi$, ψ satisfies the equation

$$\bar{Q}(\mathbf{x}, \psi; \nabla\psi) \equiv Q(\mathbf{x}, \psi(\mathbf{x}); \nabla\psi, -1) = \det(B^{(\alpha)}\psi_{x_\alpha} - A) = 0 \tag{3.2.6}$$

Let $P_0(\mathbf{x}^0 = (x_1^0, x_2^0, \dots, x_m^0))$ be a point on the wavefront Ω_{t_0} at the time $t = t_0$. The coordinates of the point P_0 can be expressed in terms of the surface coordinates $(\eta_1, \eta_2, \dots, \eta_{m-1})$

$$\mathbf{x}^0 = \mathbf{x}^0(\tilde{\eta}) \, , \quad \tilde{\eta} = (\eta_1, \eta_2, \dots, \eta_{m-1}) \tag{3.2.7}$$

The unit normal $\mathbf{n}^0(\tilde{\eta})$ to Ω_{t_0} can be determined from (3.2.7). Our aim is to construct Ω_t knowing the position of Ω_{t_0}. We assume that Ω_t corresponds to the kth characteristic family so that the appropriate mode of propagation is fixed once and for all and no ambiguity as to which of the two or many envelopes is to be considered.

Consider the solution of the ray equations (2.4.6 - 7) with $c = c_k(\mathbf{x}, t; \mathbf{n}_0)$, the kth eigenvalue, with initial values $\mathbf{x} = \mathbf{x}^0$ at $t = t_0$ as given in (3.2.7) and with $\mathbf{n} = \mathbf{n}_0$, where $\mathbf{n}_0$ is an arbitrary unit vector ($\mathbf{n}^0$ being only a particular value of $\mathbf{n}_0$ obtained from (3.2.7) at $\mathbf{x}^0$). This gives a set of rays of the kth family

$$\mathbf{x} = \mathbf{x}(t, \mathbf{x}^0, t_0, \mathbf{n}_0) \tag{3.2.8}$$

which, as $\mathbf{n}_0$ varies subject to $|\mathbf{n}_0| = 1$, generates a characteristic conoid at the point $(\mathbf{x}^0, t_0)$. We denote the equation of the convex hull of the characteristic conoid in the form

$$t - t_0 = w(\mathbf{x}; \mathbf{x}^0, t_0) \tag{3.2.9}$$

Since (2.5.1) is hyperbolic, for t = constant, this equation represents a *spherical wave front* (a closed surface) centered at the point $\mathbf{x}^0$ on the wavefront at the time t_0. This spherical wave may even pass through $\mathbf{x}^0$ as is the case for the eigenvalue $c = \langle \mathbf{n}, \mathbf{q} \rangle$ of the Euler's equations (2.3.16-18).

For the wave equation, the rays are lines given by (2.2.14) with $\mathbf{n} = \mathbf{n}_0$. The spherical wavefronts are spheres of radius $a_0(t - t_0)$, which, according to the original statement of Huygens' wavefront construction, touches the wavefront Ω_t at the tip of the ray, which is in the direction $\mathbf{n}^0$ to Ω_{t_0} at $\mathbf{x}^0$.

We consider now two cases:

Case (a) When A and $B^{(\alpha)}$ are independent of t, the function w is independent of t_0 and satisfies the partial differential equation (3.2.6). Now w is a m parameter family of complete integrals of (3.2.6), the parameters being components of $\mathbf{x}^0$. When $\mathbf{x}^0$ lies on Ω_{t_0}, we get an $m-1$ parameter subfamily of complete integrals with parameters $\eta_1, \eta_2, \ldots, \eta_{m-1}$. From the theory of first order partial differential equations, it follows that the envelope of this $m-1$ parameter family will also be a solution of the (3.2.6). We denote this solution by $\psi(\mathbf{x})$. Then

$$t - t_0 = \psi(\mathbf{x}) \tag{3.2.10}$$

represents the wavefront Ω_t which coincides with Ω_{t_0} at time t_0. Thus, we have proved the Huygens' method of wavefront construction at the time t, the wavefront Ω_t is given by the envelope of the *spherical wavefronts* of *radius* $t - t_0$, defined in terms of the metric determined by the given hyperbolic system, whose centers are the points of the wavefront Ω_{t_0} at time t_0.

Case (b) When A and $B^{(\alpha)}$ depend on t, the coefficients in the partial differential equation for ψ now depend explicitly on t. Therefore the function w satisfies (3.2.6) not as a partial differential equation but on the surface (3.2.9) i.e., when t is expressed in terms of $\mathbf{x}$. Therefore, we take the equation of the characteristic conoid at the point $(\mathbf{x}^0, t_0)$ in the space-time (the locus of (3.2.8) as $\mathbf{n}_0$ varies) in the form

$$W(\mathbf{x}, t; \mathbf{x}^0, t_0) = 0 \tag{3.2.11}$$

where the function W satisfies (3.2.4) as a partial differential equation. Moreover, the spherical wavefront function w in (3.2.9) depends on t_0 and

$$W(\mathbf{x}, t_0 + w(\mathbf{x}; \mathbf{x}^0, t_0); \mathbf{x}^0, t_0) = 0 \tag{3.2.12}$$

The solution W depends on $m-1$ parameters $\tilde{\eta}$ in $\mathbf{x}^0$. We form the envelope of the conoids (3.2.11) with respect to the parameters $\eta_1, \eta_2, \ldots, \eta_{m-1}$, the envelope being represented by

$$\phi(\mathbf{x}, t; t_0) = 0 \tag{3.2.13}$$

From the theory of first order partial differential equations it follows that the envelope also satisfies the equation (3.2.4) and would give the wavefront Ω_t at time t. This wavefront Ω_t will be the envelope of the spherical wavefronts (3.2.9). Thus, we have deduced the Huygens' method of wavefront construction with the help of *spherical wavefronts* which depends not only on $\mathbf{x}^0$ on Ω_{t_0} but also on t_0.
The construction of the *spherical wavefront* (3.2.9) with the help of rays establishes also the role of rays: the wavefront Ω_t can be obtained as the locus of the tip of rays starting from the points $\mathrm{P}_0(\mathbf{x}^0)$ with the unit normal $\mathbf{n}^0$ of Ω_{t_0}. The spherical wavefront at $\mathbf{x}^0$ touches the wavefront Ω_t at this tip. The characteristic conoid at $(\mathbf{x}^0, t_0)$ will touch the characteristic surface $\Omega : \phi(x,t) = 0$ in space-time along the integral curves (3.2.8) (with $\mathbf{n}_0 = \mathbf{n}^0$) of the ray equations (2.4.6 - 7). Thus, the Huygens' method of construction of the wavefront is equivalent to a construction with the help of rays.
It is an irony that Huygens' method was rejected by all 18th century scientists and even far into the 19th century. Acceptance of Huygens' method required abandoning the idea that rays have much intrinsic physical significance and very few people were willing to do so (Lang, 1992). As we have shown above, construction of the wavefront with the help of spherical wavefronts is equivalent to that with the help of the rays. Our proof is valid only when the spherical wavefronts have centres at interior points of the wavefront Ω_{t_0} and when the spherical wavefronts and rays are not intercepted by an obstacle. It is known that the ray theory fails to explain diffraction, which is bending of light (or waves) around an obstacle such as the edge of a slit. However, Huygens' method using the spherical wavefront is valid even when a wavefront passes round corners. Fresnel's successful application of Huygens' method in 1819 to explain diffraction phenomenon was responsible for the acceptance of the wave theory of light and Huygens' important method. In this monograph, we discuss wavefront propagation only in free space, where the construction of the front Ω_t at a later time t can be achieved with the help of an envelope of spherical wavefronts as well as locus of the tip of a ray as the starting point of the ray moves on the initial wavefront Ω_{t_0}.

3.2.3 Huygens' method and ray theory

Because of the development of the theory of partial differential equations, we understand the relation between a wavefront and the rays associated with it. However, Huygens' method does not postulate the existence of rays – even in the construction of the spherical wavefront. Let us assume that the matrices A and $B^{(\alpha)}$ do not explicitly depend on t and we further assume the existence of the spherical wavefronts satisfying the equation (3.2.6) without any reference to rays. With only this much data we shall now deduce from Huygens' method the simpler method of wavefront construction using rays obtained from the characteristic partial differential equation (3.2.4). We first establish that *the rays obtained* from *the characteristic curves of the equations (3.2.4) and (3.2.6) are the same.* This result is expected since (3.2.6) is also a partial differential equation for ψ which gives the wavefront Ω_t in the form $\psi = t$. However, it is not obvious.

Instead of the characteristic partial differential equation (2.4.1), we take the alternative equation (2.4.10) satisfied by the phase function $\phi(\mathbf{x}, t)$:

$$Q(\mathbf{x}, t; \nabla\phi, \phi_t) \equiv (\mathbf{l}A\mathbf{r})\phi_t + (\mathbf{l}B^{(\alpha)}\mathbf{r})\phi_{x_\alpha} = 0 \tag{3.2.14}$$

The characteristic curves of this equation are given by the bicharacteristic (or ray) equations (2.4.6-7).

The equation satisfied by $\psi(\mathbf{x})$ is

$$\bar{Q}(\mathbf{x}, \psi; \nabla\psi) \equiv \mathbf{l}B^{(\alpha)}\mathbf{r}\psi_{x_\alpha} - \mathbf{l}A\mathbf{r} = 0 \tag{3.2.15}$$

The characteristic curves of this equation are given by

$$\frac{dx_\alpha}{d\sigma'} = \frac{1}{2}\bar{Q}_{\psi_{x_\alpha}} = \frac{1}{2}\mathbf{l}B^{(\alpha)}\mathbf{r} \tag{3.2.16}$$

and

$$\frac{d\psi_{x_\alpha}}{d\sigma'} = -\frac{1}{2}\bar{Q}_{x_\alpha} - \frac{1}{2}\psi_{x_\alpha}\bar{Q}_\psi \tag{3.2.17}$$

$$\frac{d\psi}{d\sigma'} = \frac{1}{2}\psi_{x_\alpha}\bar{Q}_{\psi_{x_\alpha}} \tag{3.2.18}$$

The eigenvectors $\mathbf{l}$ and $\mathbf{r}$ depend on $\mathbf{x}, t = \psi$ and $\mathbf{n}$ (i.e. $\nabla\psi$). Since $\mathbf{l}(B^\alpha\psi_{x_\alpha} - A) = 0$ and $(B^{(\alpha)}\psi_{x_\alpha} - A)\mathbf{r} = 0$, it follows that

$$\bar{Q}_{x_\alpha} = \mathbf{l}(B^{(\beta)}_{x_\alpha}\psi_{x_\beta} - A_{x_\alpha})\mathbf{r} \tag{3.2.19}$$

The unit normal $\mathbf{n}$ and the eigenvalue c are given in terms of the first derivatives of ψ by

$$\mathbf{n} = \frac{\nabla\psi}{|\nabla\psi|} \quad , \quad c = \frac{1}{|\nabla\psi|} \tag{3.2.20}$$

Using the result (3.2.17) and following the procedure of the derivation of (2.4.7) with $\mathbf{p}$ replaced by $\nabla\psi$, we get

$$\frac{dn_\alpha}{d\sigma'} = -\frac{1}{2}\mathbf{l}\left\{n_\beta\left(-c\frac{\partial A}{\partial\eta_\beta^\alpha} + n_\gamma\frac{\partial B^{(\gamma)}}{\partial\eta_\beta^\alpha}\right)\right\}\mathbf{r} \tag{3.2.21}$$

Using the expression for $\bar{Q}$ in (3.2.18) and noting that $t = \psi$, we get

$$\frac{dt}{d\sigma'} = \frac{d\psi}{d\sigma'} = \frac{1}{2}(\mathbf{l}B^{(\alpha)}\mathbf{r})\psi_{x_\alpha} = \frac{1}{2}\mathbf{l}A\mathbf{r} \tag{3.2.22}$$

where we have used the equation $\bar{Q} = 0$. This result, when combined with (3.2.16) and (3.2.21), lead to the bicharacteristic equations (2.4.6) and (2.4.7)

Thus, we have shown the expected result that *the rays of the hyperbolic system are exactly the same as the characteristic curves of the partial differential equation (3.2.6).* These rays in $(\mathbf{x})$-space are given in parametric form $\mathbf{x} = \mathbf{x}(t)$ with t as the parameter. But if we view the rays in space-time they are the bicharacteristic curves of the hyperbolic system or the Monge curves of the characteristic equation (3.2.6) since $t = \psi(\mathbf{x})$ is an integral surface of (3.2.6).

We now assume that the wavefront $\Omega_t : t - t_0 = \psi(\mathbf{x})$ is obtained as the envelope of $m-1$ parameter family of spherical wavefronts $t - t_0 = w(\mathbf{x}, \mathbf{x}^0)$ with $\mathbf{x}^0 = \mathbf{x}^0(\eta_1, \eta_2, \ldots, \eta_{m-1})$ lying on the wavefront Ω_{t_0} at time t_0. When we look at these surfaces in $(\mathbf{x}, \psi)$-space or $(\mathbf{x}, t)$-space, $t - t_0 = \psi(\mathbf{x})$ represents the integral surface Ω of (3.2.6) and $t - t_0 = w(\mathbf{x}; \mathbf{x}_0)$ represents an $m-1$ parameter family of integral surfaces (characteristic conoids) which envelops Ω in this $(m+1)$ dimensional space. We can show (Courant and Hilbert (1962), page 104) that the curves of contact of a member of the complete integral surface (in $(\mathbf{x}, \psi)$ or $(\mathbf{x}, t)$-space) of (3.2.6) and the envelope Ω in $(\mathbf{x}, \psi)$ or $(\mathbf{x}, t)$-space is a Monge curve (called a characteristic curve in Courant and Hilbert - we reserve the name *characteristic curve* for the curves in the space of independent variables) of the partial differential equation (3.2.6). Now, as shown above, these Monge curves are nothing but the bicharacteristic curves of the original hyperbolic system. Thus, *the wavefront* Ω_t *obtained by Huygens' method can also be obtained as the locus of the tips (at time* t*) of the rays starting from the various points of the wavefront* Ω_{t_0}.

3.2.4 Fermat's principle

Twenty-seven years before Huygens stated his method of wavefront construction, another powerful principle was proposed by Fermat in 1650. It concerned itself with finding the path which light must take in travelling from one point of space to another point. The fact that nature *economizes its resources* during an evolution of a system is of profound importance in developing either techniques or theories in applications of mathematics. The *economy in resources* leads to minimization (or finding a stationary value) of certain quantities associated with the system. In the case of the propagation of light, this quantity is the transit time between two fixed points in space. Fermat's principle states that *a light ray going from one point* P_0 *to another point* P_1 *in space chooses a path such that the time of transit is minimum (or more precisely stationary) with respect to small variations in the path.*

The Fermat's principle defines the ray joining the two points P_0 and P_1. In this approach the wavefront is a derived concept. For Huygens' method, the wavefront plays the main role and the concept of rays can be derived from it: as the loci (when t varies) of the points of contact of the spherical wavefronts (3.2.9) with the wavefront (or the envelope) Ω. We showed that these rays are merely the rays obtained from the bicharacteristic curves of the hyperbolic system. An important mathematical problem is to show that the ray obtained by Fermat's principle is the same as the ray derived from Huygens' method. Like Huygens' method, Fermat's principle is also a very important tool for ray tracing in many practical problems such as those found in geophysics. In the next two sections we wish to discuss equivalence of the rays obtained from Fermat's principle and Huygens' method for an isotropic wave governed by the wave equation.

Before we take the particular case of the wave equation, we give a general formulation of Fermat's principle for a wave propagation in $(x_1, \ldots, x_m)$-space. We assume that at any point $P(\mathbf{x})$ on a wavefront

Ω_t at time t with unit normal $\mathbf{n}$, there exists a ray velocity $\frac{d\mathbf{x}}{dt} = \boldsymbol{\chi} \equiv (\chi_1, \ldots, \chi_m)$ such that $\boldsymbol{\chi} = \boldsymbol{\chi}(\mathbf{x}, t, \mathbf{n})$. The refractive index of the medium is inversely proportional to the ray speed $\Lambda = |\boldsymbol{\chi}|$, which, in general, depends on $\mathbf{x}, t$ and $\mathbf{n}$. Assume that it is possible to solve the relation $\boldsymbol{\chi}(\mathbf{x}, t, \mathbf{n}) = \boldsymbol{\chi}$ for $\mathbf{n}$ and write $\mathbf{n} = \mathbf{n}(\mathbf{x}, t, \chi)$. In an isotropic medium the two vectors $\mathbf{n}$ and χ are in the same direction. Let us write the equation of a path between the two fixed points $P_0(\mathbf{x}^0)$ and $P_1(\mathbf{x}^1)$ in the form

$$\mathbf{x} = \mathbf{x}(\sigma); \quad \mathbf{x}(0) = \mathbf{x}^0, \ \mathbf{x}(1) = \mathbf{x}^1 \tag{3.2.23}$$

where σ is a parameter. For example, if x_1 varies monotonically on this path, we may choose $\sigma = (x_1 - x_1^0)/(x_1^1 - x_1^0)$. Once the path is chosen we may move along this path with the speed Λ (which depends on $\mathbf{x}, t$ and the direction $\boldsymbol{\chi}$ of the path), then the arrival time t at any position P on the path will depend on σ (and also on t_0)

$$t = t(\sigma, t_0); \quad t(0, t_0) = t_0, \ t(1, t_0) = t_1 \tag{3.2.24}$$

The time of transit $I = t_1 - t_0$ from P_0 to P_1 is given by

$$I = \int_{t_0}^{t_1} \frac{|\mathbf{x}'|}{\Lambda} dt = \int_0^1 \frac{|\dot{\mathbf{x}}|}{\Lambda} d\sigma \tag{3.2.25}$$

where we used the symbols

$$\mathbf{x}' \equiv \frac{d\mathbf{x}}{dt} \tag{3.2.26}$$

and

$$\dot{\mathbf{x}} = \frac{d\mathbf{x}}{d\sigma} \tag{3.2.27}$$

and Λ is a function of $\mathbf{x}, \mathbf{x}'$ and t or $\mathbf{x}, \dot{\mathbf{x}}$ and t.

According to Fermat's principle, for a path from P_0 to P_1 to be a ray, it should be chosen so that the first variation of $I = t_1 - t_0$ should be zero with respect to small variations in the path. In all applications of Fermat's principle, the medium was assumed to be stationary i.e., Λ was independent of $t : \Lambda = \Lambda(\mathbf{x}, \mathbf{x}')$. A formulation of Fermat's principle for a nonstationary medium is given in Courant and Hilbert (1962) on page 117 (see the expression of F). In this monograph, we shall use the second expression for I in (3.2.25) i.e.,

$$I = \int_0^1 F(\mathbf{x}, \dot{\mathbf{x}}, t) d\sigma \tag{3.2.28}$$

where

$$F(\mathbf{x}, \dot{\mathbf{x}}, t) = \frac{|\dot{\mathbf{x}}|}{\Lambda(\mathbf{x}, \dot{\mathbf{x}}, t)} \tag{3.2.29}$$

In this variational problem, the boundary values $\mathbf{x}(0) = \mathbf{x}^0$ and $\mathbf{x}(1) = \mathbf{x}^1$ are fixed. Since we need to make $I = t_1 - t_0$ to be stationary, the boundary values $t_0 = t(0)$ and $t_1 = t(1)$ are free.

The question of the equivalence of rays obtained from Fermat's principle and those from the bicharacteristic curves of the governing equations is not yet settled in the most general situation. The equivalence has been shown for elasticity equations by Epstein and Sniatycki (1992) and for Euler's equations of a polytropic gas by Prasad and Russo (1993). We shall discuss this question in the next two sections for the simplest case i.e., in the case of the wave propagation in an isotropic medium, where the governing partial differential equation is the wave equation

$$u_{tt} - a^2 \nabla u = 0 \tag{3.2.30}$$

In this case the rays are orthogonal to the wavefront i.e., $\mathbf{n} = \mathbf{x}' \, |\mathbf{x}'|$, $|\mathbf{x}'| \equiv \Lambda = a(\mathbf{x}, t)$. These results imply

$$\mathbf{x}' = \mathbf{n} a \tag{3.2.31}$$

We discuss the case of a stationary medium and medium with a time dependent refractive index separately.

3.2.5 Fermat's principle in a stationary medium

The variational problem

$$I = \int_0^1 F(\mathbf{x}, \dot{\mathbf{x}}) d\sigma, \quad F(\mathbf{x}, \dot{\mathbf{x}}) = \frac{|\dot{\mathbf{x}}|}{a(\mathbf{x})} \tag{3.2.32}$$

has been discussed in all standard text books. The boundary values $\mathbf{x}^0$ and $\mathbf{x}^1$ at $\sigma = 0$ and $\sigma = 1$ are fixed. The condition that I be stationary is that $\mathbf{x}(\sigma)$ must satisfy the Euler's equations

$$\frac{d}{d\sigma}\left(\frac{\partial F}{\partial \dot{x}_\alpha}\right) = \frac{\partial F}{\partial x_\alpha} \tag{3.2.33}$$

These equations lead to

$$\frac{dn_\alpha}{d\sigma} = \frac{n_\alpha}{a}\frac{da}{d\sigma} - \frac{|\dot{\mathbf{x}}|}{a}a_{x_\alpha}, \quad \alpha = 2, \ldots, m \tag{3.2.34}$$

where we have used $\dot{x}_\alpha/|\dot{\mathbf{x}}| = n_\alpha$. We have written the Euler equations only for $\alpha = 2, \ldots, m$ since we have indicated in the previous section a choice of σ as $\sigma = (x_1 - x_1^0)/(x_1^1 - x_1^0)$. Since $\dot{\mathbf{x}} = d\mathbf{x}/d\sigma$, we replace the differentiation with respect to σ by t along the ray (see (3.2.26-27)) and get

$$\frac{dn_\alpha}{dt} = \frac{n_\alpha}{a}\frac{da}{dt} - \frac{|\mathbf{x}'|}{a}a_{x_\alpha}\ , \ \alpha = 2, \ldots, m \tag{3.2.35}$$

Since $|\mathbf{x}'| = a$ along a ray and *the medium is assumed to be stationary so that* $\frac{d}{dt} = a\sum_{\alpha=1}^m n_\alpha \frac{\partial}{\partial x_\alpha}$, the last equation becomes

$$\frac{dn_\alpha}{dt} = -\left(\frac{\partial}{\partial x_\alpha} - n_\alpha\langle \mathbf{n}, \nabla\rangle\right)a, \ \ \alpha = 2, \ldots, m \tag{3.2.36}$$

Differentiating $n_1 = \sqrt{1 - \sum_{\alpha=2}^m n_\alpha^2}$ and using the above equations we get

$$\frac{dn_1}{dt} = \frac{1}{n_1}\sum_{\alpha=2}^m n_\alpha \frac{\partial a}{\partial x_\alpha} - \frac{n_2^2 + \ldots + n_m^2}{n_1}\langle \mathbf{n}, \nabla\rangle a \tag{3.2.37}$$

Using $n_2^2 + \ldots + n_m^2 = 1 - n_1^2$, in the last term, we can show that the equation (3.2.36) is valid also for $n = 1$. The equations for n_α can be written together in the vector form

$$\frac{d\mathbf{n}}{dt} = -(\nabla - \mathbf{n}\langle \mathbf{n}, \nabla\rangle)a \tag{3.2.38}$$

Thus, the ray equations obtained from Fermat's principle in an isotropic stationary inhomogeneous medium with a given sound speed a, form a system of equations (3.2.31) and (3.2.38). These are exactly the same as the ray equations derived from the bicharacteristic equations of the wave equation (3.2.30).

3.2.6 Fermat's principle in a nonstationary medium

The variational problem is

$$I = \int_0^1 F(\mathbf{x}, t, \dot{\mathbf{x}}, \dot{t}) d\sigma, \quad F = \frac{|\dot{\mathbf{x}}|}{a(\mathbf{x}, t)} \tag{3.2.39}$$

where we note that $\dot{t}$ does not appear in the expression form F. As mentioned earlier, the boundary values of $\mathbf{x}$ at $\sigma = 0$ and $\sigma = 1$ are fixed but $t(0) = t_0$ and $t(1) = t_1$ are free. The Euler equations are

$$\frac{d}{d\sigma}\left(\frac{\partial F}{\partial \dot{x}_\alpha}\right) = \frac{\partial F}{\partial x_\alpha}, \quad \alpha = 2, 3, \ldots, m \tag{3.2.40}$$

and

$$\frac{d}{d\sigma}\left(\frac{\partial F}{\partial \dot{t}}\right) = \frac{\partial F}{\partial t} \Rightarrow 0 = -\frac{|\dot{\mathbf{x}}|}{a^2} a_t \tag{3.2.41}$$

with the natural boundary conditions (Courant and Hilbert, 1953)

$$F_{\dot{t}} = 0 \text{ at } \sigma = 0 \text{ and } \sigma = 1 \tag{3.2.42}$$

The boundary conditions (3.2.42) are automatically satisfied since $\dot{t}$ does not appear in F. However, the equation (3.2.41) implies that $a_t = 0$ i.e., the medium is stationary. Thus, the Fermat's principle in terms of the variational problem (3.2.39) is not properly formulated for a nonstationary medium.

The inconsistency observed in the above formulation of Fermat's principle is due to the following reason. Fermat's principle was first formulated for a stationary medium in which any two points could be connected by a ray and hence also the end points P_0 and P_1 could be connected by a ray. The formulation (3.2.28 - 29) given in Courant and Hilbert (1962) for two arbitrary points $\tilde{P}_0(\mathbf{x}^0, t_0)$ and $\tilde{P}(\mathbf{x}^1, t_1)$ in space-time requires examination whether $\tilde{P}_0$ and $\tilde{P}_1$ can also be connected by a ray. This is not true for an arbitrary pair $(\tilde{P}_0, \tilde{P}_1)$ in space-time. This was observed by Kovner (1990) followed by a demonstration by Nityananda and Samuel (1992) in general relativity who restricted the points $\tilde{P}_0$ and $\tilde{P}_1$ to points which can be joined by null curves. The point noted by Kovner is relevant not only in general relativity but in a system governed by any hyperbolic system. We should consider only those points $\tilde{P}_0$ and $\tilde{P}_1$ in space-time which can be connected by a bicharacteristic curve i.e., the $\tilde{P}_1$

should be a point on the forward characteristic conoid of the point $\tilde{P}_0$. When $\tilde{P}_0$ and $\tilde{P}_1$ are such points, it is now clear that the value of I should be stationary with respect to paths which lie on the characteristic conoid at $\tilde{P}_0$; such paths need not be bicharacteristics but the path which makes I stationary should turn out to be the bicharacteristic curve joining $\tilde{P}_0$ and $\tilde{P}_1$.

Now we give a general formulation of the Fermat's principle, which we call the *extended Fermat's principle* in a nonstationary medium governed by a hyperbolic system. Let the equation of the characteristic conoid of the point $\tilde{P}_0(\mathbf{x}^0, t_0)$ be given by

$$t = \phi(\mathbf{x}) \tag{3.2.43}$$

and let $\tilde{P}_1(\mathbf{x}^1, t_1)$ be such that $t_1 = \phi(\mathbf{x}^1)$. We now define a function $\bar{F}$ on the characteristic conoid by

$$\bar{F}(\mathbf{x}, \dot{\mathbf{x}}) = \frac{|\dot{\mathbf{x}}|}{\bar{a}(\mathbf{x})}, \quad \bar{a}(\mathbf{x}) = a(\mathbf{x}, \phi(\mathbf{x})) \tag{3.2.44}$$

Extended Fermat's principle says that a ray is defined to be a path which makes the integral

$$I = \int_0^1 \bar{F}(\mathbf{x}, \dot{\mathbf{x}}) d\sigma, \ \bar{F} = \frac{|\dot{\mathbf{x}}|}{\bar{a}(\mathbf{x})} \tag{3.2.45}$$

stationary with respect to variations in the paths, which now obviously lie on the characteristic conoids at the point $\tilde{P}_0$.

The Euler's equations corresponding to the variational problem (3.2.45) are

$$\frac{d}{d\sigma}\left(\frac{\partial \bar{F}}{\partial \dot{x}_\alpha}\right) = \frac{\partial \bar{F}}{\partial x_\alpha}, \quad \alpha = 2, 3, \ldots, m \tag{3.2.46}$$

As in the case of the stationary medium in section 3.2.5, these equations lead to

$$\frac{d\mathbf{n}}{dt} = -(\nabla - \mathbf{n}\langle \mathbf{n}, \nabla \rangle)\bar{a} \equiv -\mathbf{L}\bar{a} \tag{3.2.47}$$

where $\mathbf{n} = \nabla\phi/|\nabla\phi|$ and $\mathbf{L}$ is defined by (2.4.21). Since the operator $\mathbf{L} = \nabla - \mathbf{n}\langle \mathbf{n}, \nabla \rangle$ represents a tangential derivative with respect to the characteristic conoid $t = \phi$ so that $\mathbf{L}\phi = 0$, we get

$$\mathbf{L}\bar{a} = \mathbf{L}a(\mathbf{x}, \phi(\mathbf{x})) = \mathbf{L}a(\mathbf{x}, t) + a_t \mathbf{L}\phi = \mathbf{L}a \tag{3.2.48}$$

Therefore, the rays, which make the integral I stationary, are given by

$$\frac{d\mathbf{x}}{dt} = \mathbf{n}a\,, \quad \frac{d\mathbf{n}}{dt} = -\mathbf{L}a \tag{3.2.49}$$

which are again the ray equations obtained from the bicharacteristic equations of the wave equation (3.2.30). Thus we find that the extended Fermat's principle in terms of the variational formulation (3.2.45) is not only well-posed for a nonstationary medium but gives correct rays. It is now a simple matter to use the rays to construct successive positions of a wavefront, which we present below.

Consider now a surface Ω_0 in $(\mathbf{x})$-space and let $\mathbf{n}^0$ denote the unit normal of Ω_0 at a point $P_0(\mathbf{x}^0)$ on Ω_0. Solving the system of equations (3.2.49) with the initial condition $\mathbf{x} = \mathbf{x}^0, \mathbf{n} = \mathbf{n}^0$ at $t = 0$ we get

$$\mathbf{x} = \mathbf{x}(t, \mathbf{x}^0)\,, \quad \mathbf{n} = \mathbf{n}(t, \mathbf{x}^0) \tag{3.2.50}$$

The locus of the points $\mathbf{x}$ at a fixed time t, when $\mathbf{x}^0$ varies on Ω_0, is a surface Ω_t. This surface is the wavefront at time t having initial position Ω_0. The ray (3.2.50) from the point P_0 at time $t = 0$ is the ray associated with the common bicharacteristic curve on the characteristic conoid at $\tilde{P}_0(\mathbf{x}^0, 0)$ and the characteristic surface Ω in space-time whose t-level surfaces are Ω_t. The section of the characteristic conoid by $t=$ constant is the spherical wavefront which touches Ω_t at the point $P(\mathbf{x}, t)$ given by $\mathbf{x} = \mathbf{x}(t, \mathbf{x}^0)$ in (3.2.50). The unit normal to the wavefront at the point P is given by the second expression in (3.2.50).

We can use the same procedure to formulate the extended Fermat's principle in the general case i.e., the case (3.2.28 - 29), where the ray speed Λ, obtained from the ray velocity χ of a hyperbolic system, depends explicitly on t.

3.2.7 Weakly nonlinear ray theory (WNLRT) in an isotropic medium using Fermat's principle

Consider the propagation of small amplitude waves in the form of a continuous pulse in an isotropic medium in high-frequency approximation so that the disturbed region due to the waves can be characterized by a one parameter family of wavefronts. We assume that the state ahead of the leading wavefront to be a constant state with sound velocity a_0. The velocity a of propagation of any one of these wavefronts would differ from the constant value a_0 by a small quantity $\epsilon a_0 k\tilde{w}$ so that

$$a = a_0(1 + \epsilon k\tilde{w}) \tag{3.2.51}$$

where k is a nonlinearity constant of the medium. The perturbed amplitude $\tilde{w}$ in the wave would depend on $\mathbf{x}$ as well as t so that in this case we come across a situation in which the refractive index or the front velocity is time-dependent. Now we consider the propagation of one of these one parameter family f nonlinear wavefronts. We denote this wavefront by Ω_t.

In terms of the unit normal $\mathbf{n}$ of Ω_t, the ray velocity in the isotropic medium is given by

$$\frac{d\mathbf{x}}{dt} = \boldsymbol{\chi} \equiv \mathbf{n}a_0(1 + \epsilon k\tilde{w}) \tag{3.2.52}$$

We now use the extended Fermat's principle with $F = |\dot{\mathbf{x}}|/\{a_0(1 + \epsilon k\tilde{w}(\mathbf{x}, t))\}$ and derive from the second equation in (3.2.49)

$$\frac{d\mathbf{n}}{dt} = -\epsilon a_0 k\mathbf{L}\tilde{w} \tag{3.2.53}$$

(3.2.52 - 53) form a system of ray equations for a weakly nonlinear ray theory. Since $\langle \mathbf{n}, \mathbf{L}\rangle = 0$, equation (3.2.53) implies that $\mathbf{n}$ satisfies the consistency condition $|\mathbf{n}| = 1$ for all time if it is initially so. Thus, there are 5 independent ray equations for 6 quantities: 3 components of $\mathbf{x}$, only 2 of the 3 components of $\mathbf{n}$ and w i.e., the system of equations (3.2.52 - 53) is under determined. To close the system we need a transport equation for w or the equation for the propagation of energy along the rays. In many physical systems, such as gas dynamics, the energy crossing per unit ray tube area A is proportional to the square of the amplitude w. In the case of a continuous wave profile along a ray (as is the case of a nonlinear wave which is not a shock wave), the energy propagating in a ray tube is conserved. Hence, from the consideration of conservation of energy in a ray tube, we get

$$\frac{d}{dt}(Aw^2) = 0 \tag{3.2.54}$$

Using relations (2.2.22 - 23), we get the following transport equation for w along a ray

$$\frac{dw}{dt} = \Omega a_0 \tilde{w} \tag{3.2.55}$$

where Ω, the mean curvature of the nonlinear wavefront is given in terms of its unit normal $\mathbf{n}$ by the expression (2.2.22).

Equations (3.2.52 - 53 and 55) form the complete set of equations of the WNLRT. A formal derivation of the energy conservation law (3.2.54) along a nonlinear ray tube and its extension to a general hyperbolic system forms a difficult mathematical problem which will be discussed in the next chapter. This problem is quite simple for linear waves where one derives the transport equation along linear rays, Sommerfeld and Runge (1911). However, for WNLRT we need to have a perturbation scheme in which the amplitude appears in the coefficients of the leading order approximate equation so that the w appears in the eikonal equation itself (Prasad, 1975, 1994, 2000). The use of the extended Fermat's principle presented here makes the derivation of the ray equations (3.2.52 - 53) not only very simple but justifies the perturbation scheme used in the derivation of the WNLRT in sections 4.3 and 4.4, where the transport equation for a general hyperbolic system has been derived.

3.3 Kinematics of a propagating curve

In this section we continue the discussion of results which are geometric in nature and which do not depend on dynamic equations of the medium. We discuss first a few general properties of linear wavefront propagation and then present a new set of conservation laws in two space dimensions (Morton, Prasad and Ravindran (1992), and Prasad (1995)) based on the conservation of distance in two independent directions. These conservation laws lead in a natural way to a new phenomenon of *kinks*, which are images in the (x, y)-plane of geometric shocks in the ray coordinate system (these coordinates will be introduced in section 3.3.2). The idea of kinks was first introduced by Whitham (1957), who called them *shock-shocks* as they appeared in his theory of shock propagation. Our discussion in this section shows that a kink is a geometric result which is common to more general propagating surfaces.

3.3.1 Caustic, wavefront folding and some other general properties

Singularities on wavefronts are very common physical phenomena.

A kink appears on the surface of tea or coffee in a cup when the cup is placed on a suitable place relative to a source of light. It also appears when the rays of light pass through a gravitational lens causing a cosmological phenomenon. In both these examples the converging rays envelope a surface, called caustic, on which the successive positions of the wavefront have a cusp type of singularity. Usually, a caustic starts with a cusp type of singularity, called arête, and its two branches (in two-space dimensions) bound a region in which the wavefront folds and crosses itself (Fig.3.3.1). An interesting example of a caustic appears during the propagation a two-dimensional wavefront, which is initially given by

$$\left.\begin{array}{lll} y^2 = 4x & , & 0 \le x \le 1 \\ y = \pm(x+1) & , & x > 1 \end{array}\right\} \tag{3.3.1}$$

The central part of the initial wavefront is a parabola extended by its tangents beyond $x > 1$. It has a continuously turning tangent even at the points $(1, \pm 2)$.

We take the wavefront to be the one which satisfies the characteristic equation (2.2.2) with $a_0 = 1$. The ray, starting from a point represented by $(\eta^2, 2\eta)$ for $0 \le \eta \le 1$ and $(2\eta - 1, 2\eta)$ for $\eta > 1$, on the upper part of the initial wavefront, is given by

$$\left.\begin{array}{l} x = \eta^2 + \frac{t}{\sqrt{1+\eta^2}} \ , \ y = 2\eta - \frac{\eta t}{\sqrt{1+\eta^2}} \ \text{for } 0 \le \eta \le 1 \\ \text{and} \\ x = 2\eta - 1 + \frac{1}{\sqrt{2}} t \ , \ y = 2\eta - \frac{1}{\sqrt{2}} t \ \text{for } \eta > 1 \end{array}\right\} \tag{3.3.2}$$

respectively. The caustic i.e., the envelope of the rays, is represented parametrically by

$$x = 2 + 3\eta^2 \ , \ y = -2\eta^3 \ , \ 0 \le \eta \le 1 \tag{3.3.3}$$

The arête of this caustic is at (2,0) (Fig.3.3.1). The two branches of the caustic end at a finite distance from the arête, the lower branch, enveloped by the rays from the upper part of the initial wavefront extends from (2,0) to (5,-2).

The wavefront reaches the arête at $t = 2$. A parametric representation of equations of one half of the wavefront at any time t is given by (3.3.2) with t = constant and η as the parameter. For $2 < t < 4\sqrt{2}$, this half has a cusp type of singularity where $\frac{dx}{d\eta} = 0, \frac{dy}{d\eta} = 0$ i.e. at (x_c, y_c) where

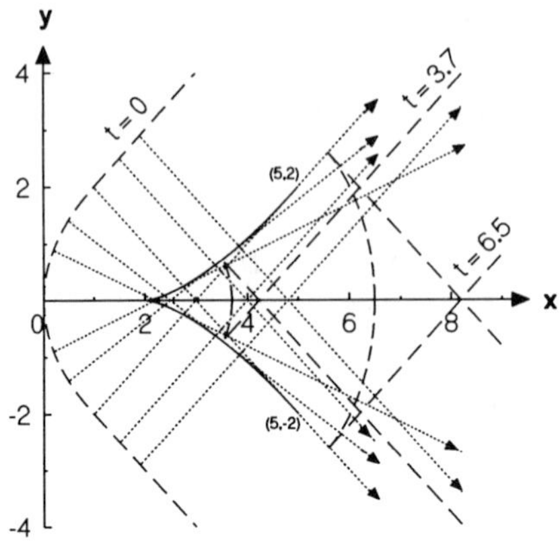

Fig. 3.3.1: Linear wavefront propagation in an isotropic homogeneous medium with speed of propagation unity.
- - - - - - : wavefront, ——— : caustic, ... : rays

$$x_c = \left\{\left(\tfrac{t}{2}\right)^{2/3} - 1\right\} + \left(\tfrac{1}{2}t\right)^{2/3}, y_c = \left\{\left(\tfrac{t}{2}\right)^{2/3} - 1\right\}^{\frac{1}{2}} \left\{2 - \left(\tfrac{1}{2}t^2\right)^{\frac{1}{3}}\right\},$$
$$\text{for } 2 < t < 4\sqrt{2} \tag{3.3.4}$$

For $t > 4\sqrt{2}$, this cusp type of singularity moves to (x_c', y_c') beyond a point where the two parts of the wavefront, given by the two different expressions in (3.3.1), meet

$$x_c' = 1 + \frac{t}{\sqrt{2}}, y_c' = 2 - \frac{t}{\sqrt{2}}, \text{ for } t > 4\sqrt{2} \tag{3.3.5}$$

An interesting part of this example is the existence of a cusp type of singularity for the wavefront for $t > 4\sqrt{2}$ even though there is no caustic. This singularity on the wavefront results from the discontinuity in the curvature of the initial wavefront at the point B.

In a very special case when all rays converge to a point, the caustic degenerates onto a point which is called *focus*.

We now present another interesting result associated with a linear wavefront propagation. This concerns the appearance of singularities on a wavefront as the wavefront enters into a caustic region and the resolution of these singularities as the wavefront emerges out of the caustic region.

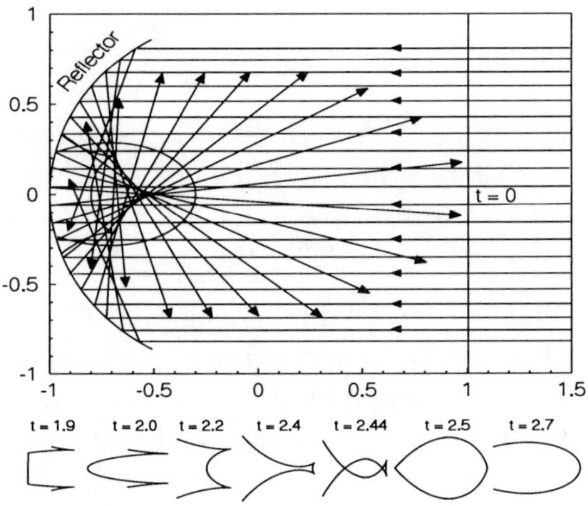

Fig. 3.3.2: Caustic on the surface of tea in a cup, neglecting the multiple reflections of the rays; and seven typical shapes of the wavefront after a plane wavefront is reflected.

Fig. 3.3.2 represents the successive positions of a plane wavefront after it is reflected from the interior surface of a circular cylinder (neglecting multiple reflections). The brightly illuminated side of the caustic is the domain bounded by the reflecting surface and the caustic. Soon after the reflection, at a time $t < t_c$, the wavefront develops a pair of cusps both of which approach the arête at a critical time t_c after which the singularities disappear and the wavefront becomes smooth at a time $t > t_c$. This represents a phenomenon in which the transition at the arête takes place in reverse order of that depicted in Fig 3.3.1 when the smooth wavefront meets the arête. This is an important observation and verifies that the propagation of a linear wavefront is a reversible process.

We note here three fundamental properties of a linear wavefront propagation.

1. *Self propagation.* This means that a linear wavefront is determined by the information only on the wavefront at any previous time and is not influenced by the wavefronts which follow or precede it. For a linear wavefront, the information required is simply the position (and hence the geometry) of the wavefront and not the amplitude of the wavefront.

2. *Local determinacy.* This means that the motion of an arbitrary small arc of a linear wavefront is independent of the neighbouring arcs.

3. *Reversibility in time.* According to linear propagation, where the amplitude of the wave has no influence, if a wavefront Ω_{t_1}, at time t_1 leads to a wavefront Ω_{t_2} at time $t_2(> t_1)$, then by reversing the direction of propagation velocity we can get Ω_{t_1} from Ω_{t_2}.

We shall examine later in this monograph which of the above three properties are valid for a nonlinear wavefront and a shock front.

In 1957, Whitham developed a theory of shock front propagation using intuitive arguments and discovered a new type of singularity on the front. Since such a singularity was an image in the (x, y)-plane of a shock of the equations of his *shock dynamics in a ray coordinate system*, he called it *shock-shock* and interpreted it as the trace of a triple shock interaction (the third shock is missing in this theory) on the shock front. The shock strength and the direction of the normal to the shock are discontinuous across this singularity and the singularity physically appears as a *kink*. The first experimental result, showing formation and propagation of a kink on a shock front in a gaseus medium, was obtained by Sturtevant and Kulkarni (1976) although the Mach reflection and the triple shock interaction are phenomena that were observed long ago (see Courant and Friedrichs (1948)). Even today, it is a great mathematical challange to verify the various experimentally observed properties of the flow field containing these kinks (Tabak and Rosales (1994); Hunter (1997)). Kink is a singularity which appears not only on a shock front but also on a nonlinear wavefront. It is a more general geometrical property associated with a moving curve in two-dimensions or a moving surface in three-dimensions. It manifests itself on a moving curve or surface when a very special type of dynamical property of a medium is available. This special dynamical property of the medium is the genuine nonlinearity of the mode of propagation under consideration.

The basic kinematical equations of a propagating surface, derived on simple geometrical considerations, have been available in the form of differential equations in a ray coordinate system for a very long time (Thomas (1961)). Since kinks are shocks in this coordinate system, the partial differential equations are inadequate to describe the kinks, for which we need physically realistic conservation forms. Appropriate conservation laws for the propagation of a two-dimensional wavefronts in an isotropic medium were first derived by Morton, Prasad and Ravindran (1992) and for a three-dimensional wavefront by Giles, Prasad and Ravindran (1996). In the next section we shall present an extension of the work of Morton, Prasad and Ravindran based on the work of Prasad (1995).

Singularities on wavefronts were first discovered by Huygens in 1654. In a theory of linear partial differential equations, singularities have been studied in detail (Gårding (1980)) but these are related to the characteristic of multiplicity higher than one. We are interested in problems in which the multiplicity of the characteristic is uniformly equal to one. General classification of a singularity on the wavefront is provided by the catastrophe theory, Arnol'd (1993). However, we are interested not in classification but in the formation and propagation of kinks starting from a given wavefront at an initial time. Only the kink type of singularities seems to appear on a nonlinear wavefront and a shock front of moderate intensity.

3.3.2 Ray coordinate system and kinematical conservation laws

Let Ω_t be a curve, representing a wavefront which occupies different positions at different times. For discussion in this section, a wavefront refers also to a shock front. The distinction between a wavefront across which state variables are continuous and a shock front needs to be taken into account when we consider dynamical equations. Associated with a wavefront Ω_t, there exists at every point on it a ray velocity $\boldsymbol{\chi} = (\chi_1, \chi_2)$. This gives a one parameter family of curves, called rays, each one of which is traced by a point on Ω_t moving with the velocity $\boldsymbol{\chi}$. Expression for $\boldsymbol{\chi}$ can be obtained only from the properties of the medium in which Ω_t propagates and depends also on the unit normal $\mathbf{n} = (n_1, n_2)$ of Ω_t. We write the equation of Ω_t in the form

$$\Omega_t : x = x(\xi, t), \;\; y = y(\xi, t) \tag{3.3.6}$$

where constant values of t give the positions of the propagating curve Ω_t at different times and ξ = constant represents a ray. In the case of isotropic wave propagation, rays are orthogonal to the family of wavefronts Ω_t. When the equation of Ω_t is represented in terms of a *ray coordinate system* (ξ, t) as in (3.3.6), the ray velocity $\boldsymbol{\chi} = (\chi_1, \chi_2)$ is given by

$$(x_t, y_t) = (\chi_1, \chi_2) \tag{3.3.7}$$

C, the speed of propagation of Ω_t is

$$C = < n, \boldsymbol{\chi} > \equiv n_1\chi_1 + n_2\chi_2 \tag{3.3.8}$$

and the component T of the ray velocity in the direction of the tangent to Ω_t is

$$T = -n_2\chi_1 + n_1\chi_2 \tag{3.3.9}$$

Consider the curves Ω_t and Ω_{t+dt}. Let P$'$ and Q$'$ be the positions at time $t + dt$ of P and Q, respectively on two rays at a distance $gd\xi$ on Ω_t (Fig.3.3.3). $PN = Cdt$ is the normal displacement of Ω_t in time dt. Thus, g is the metric associated with ξ and C is the metric associated with t in the ray coordinate system. If the coordinates of $Q^{'}$ are $(x + dx, y + dy)$ then (dx, dy) is a displacement in the (x, y) plane corresponding to a displacement $(d\xi, dt)$ in the ray coordinate plane, so that

$$\begin{aligned} dx &= (Cdt)n_1 - (gd\xi + Tdt)n_2 \\ dy &= (Cdt)n_2 + (gd\xi + Tdt)n_1 \end{aligned} \tag{3.3.10}$$

Let θ be the angle which the normal to Ω_t makes with the x-axis $(n_1 = \cos\theta, n_2 = \sin\theta)$. The above relation gives the Jacobian matrix of the transformation from (ξ, t)-plane to (x, y)-plane

$$\begin{pmatrix} x_\xi & x_t \\ y_\xi & y_t \end{pmatrix} = \begin{pmatrix} -g\sin\theta & C\cos\theta - T\sin\theta \\ g\cos\theta & C\sin\theta + T\cos\theta \end{pmatrix} \tag{3.3.11}$$

The Jacobian, i.e., the determinant of the Jacobian matrix, is $-gC$, which shows that the transformation between (ξ, t) and (x, y) is non-singular as long as g and C are non-zero and finite.

A little care is required while interpreting the transformation between (x, y)-plane and the ray coordinate plane of (ξ, t). In this case

$$\frac{\partial}{\partial t} = (Cn_1 - Tn_2)\frac{\partial}{\partial x} + (Cn_2 + Tn_1)\frac{\partial}{\partial y}, \quad \frac{\partial}{g\partial\xi} = -n_2\frac{\partial}{\partial x} + n_1\frac{\partial}{\partial y} \tag{3.3.12}$$

where we note that $\frac{\partial}{\partial t}$ the time-rate of change along a ray, is same as $\frac{d}{dt} = \frac{\partial}{\partial t} + \chi_1 \frac{\partial}{\partial x_1} + \chi_2 \frac{\partial}{\partial x_2}$ in (x, y, t)-space.

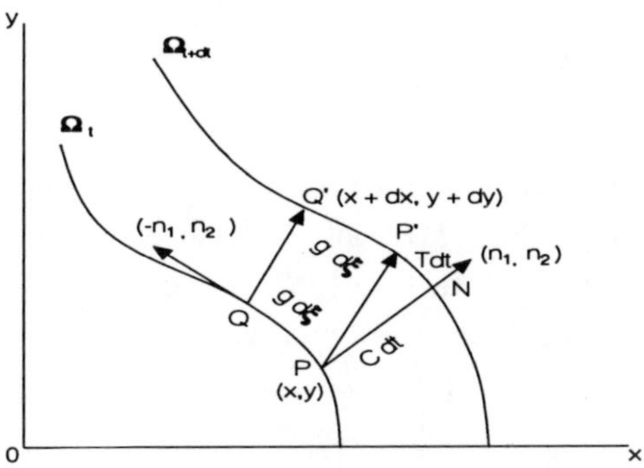

Fig. 3.3.3: P' and Q' are the positions on Ω_{t+dt} of P and Q, respectively, on two rays at a distance $gd\xi$ on Ω_t.

Following Morton, Prasad and Ravindran (1992), we derive a pair of relations in the conservation form by equating $x_{\xi t}$ to $x_{t\xi}$ and $y_{\xi t}$ to $y_{t\xi}$:

$$(g \sin\theta)_t + (C\cos\theta - T\sin\theta)_\xi = 0 \tag{3.3.13}$$

$$(g \cos\theta)_t - (C\sin\theta + T\cos\theta)_\xi = 0 \tag{3.3.14}$$

We call (3.3.13 - 14) *kinematical conservation laws.* From these, we deduce the following partial differential equations

$$g_t = C\theta_\xi + T_\xi \tag{3.3.15}$$

$$\theta_t = -\frac{1}{g}C_\xi + \frac{1}{g}T\theta_\xi \tag{3.3.16}$$

as kinematical relations for any propagating curve Ω_t. Equations (3.3.15 - 16) or their conservation forms (3.3.13 - 14) represent a system of two equations involving four quantities g, θ, C and T. For a wavefront, which can be defined only in the high frequency limit, the quantities C and T can be expressed in terms of an amplitude w of the wavefront Ω_t and its unit normal $\mathbf{n}$. Therefore, to get a determined system of equations, we must add to these equations another evolution equation for w. Such an equation turns out to be a transport equation along a ray, which we shall discuss later in this chapter.

We shall now show the equivalence of the equations (3.3.15 - 16) to the ray equations (or the bicharacteristic equations) of a hyperbolic system

$$A\mathbf{u}_t + B^{(1)}\mathbf{u}_x + B^{(2)}\mathbf{u}_y + F = 0 \tag{3.3.17}$$

when Ω_t is taken to be the projection on the (x, y)-plane of the section of the characteristic surface $\Omega : \phi(x, y, t) = 0$ by $t =$ constant plane. Here $\mathbf{u}$ and $F(x, y, t, \mathbf{u})$ are n dimensional column vectors and $A(x, y, t, \mathbf{u},)$, $B^{(1)}(x, y, t, \mathbf{u},)$ and $B^{(2)}(x, y, t, \mathbf{u},)$ are n x n matrices. In this case C is an eigenvalue c, $C = c = -\phi_t/|\nabla\phi|$ and $\mathbf{n} = \nabla\phi/|\nabla\phi|$. The ray equations (2.4.6-7) become

$$\frac{dx}{dt} = \frac{\mathbf{l}B^{(1)}r}{\mathbf{l}A\mathbf{r}} \equiv \chi_1, \quad \frac{dy}{dt} = \frac{\mathbf{l}B^{(2)}\mathbf{r}}{\mathbf{l}A\mathbf{r}} \equiv \chi_2 \tag{3.3.18}$$

and

$$\frac{d\theta}{dt} = \frac{1}{g(\mathbf{l}A\mathbf{r})}\,\mathbf{l}\left(c\frac{\partial A}{\partial \xi} - n_1\frac{\partial B^{(1)}}{\partial \xi} - n_2\frac{\partial B^{(2)}}{\partial \xi}\right)\mathbf{r} \tag{3.3.19}$$

where $\mathbf{l}$ and $\mathbf{r}$ are left and right null vectors of $A\phi_t + B^{(1)}\phi_x + B^{(2)}\phi_y$. Since

$$c(\mathbf{l}A\mathbf{r}) = \mathbf{l}(n_1B^{(1)} + n_2B^{(2)})\mathbf{r}, \quad n_1 = \cos\theta, n_2 = \sin\theta$$

$$C_\xi = -\frac{1}{\mathbf{l}A\mathbf{r}}\mathbf{l}\,(cA_\xi - n_1B^{(1)}_\xi - n_2B^{(2)}_\xi)\mathbf{r} + \frac{1}{\mathbf{l}A\mathbf{r}}\mathbf{l}\,(-n_2B^{(1)} + n_1B^{(2)})\mathbf{r}\theta_\xi \tag{3.3.20}$$

where we have used

$$cA\mathbf{r} = (n_1B^{(1)} + n_2B^{(2)})\mathbf{r}, \quad \mathbf{l}cA = n_1B^{(1)} + n_2B^{(2)}$$

Using (3.3.9), (3.3.18), and noting that $\frac{d}{dt}$ becomes the partial derivative $\frac{\partial}{\partial t}$ in (ξ, t)-plane, we find that (3.3.19) reduces to (3.3.16) with $c = C$. To deduce (3.3.15) from (3.3.18), we differentiate $g^2 = x_\xi^2 + y_\xi^2$ with respect to t to get

$$gg_t = x_\xi x_{\xi t} + y_\xi y_{\xi t} = x_\xi\, x_{t\xi} + y_\xi\, y_{t\xi}$$

Substituting in the above the expressions for x_t and y_t from (3.3.18) and noting that $x_\xi/g = -n_2$, $y_\xi/g = n_1$ we get

$$\begin{aligned} g_t &= -n_2(\chi_1)_\xi + n_1(\chi_2)_\xi \\ &= (-n_2\chi_1 + n_1\chi_2)_\xi + (n_1\chi_1 + n_2\chi_2)\theta_\xi \end{aligned}$$

which is the relation (3.3.15).

We can deduce a pair of kinematic equations in conservation form also in (x, y)-plane. The Jacobian of the transformation from (x, y)-plane to (ξ, t)-plane is given by

$$\begin{pmatrix} \xi_x & \xi_y \\ t_x & t_y \end{pmatrix} = \begin{pmatrix} -\frac{1}{gC}(C\sin\theta + T\cos\theta) & \frac{1}{gC}(C\cos - T\sin\theta) \\ \frac{1}{C}\cos\theta & \frac{1}{C}\sin\theta \end{pmatrix} \tag{3.3.21}$$

Equating the partial derivatives ξ_{xy} to ξ_{yx} and t_{xy} to t_{yx}, we get the conservation forms

$$\left(\frac{\cos\theta - (T/C)\sin\theta}{g}\right)_x + \left(\frac{\sin\theta + (T/C)\cos\theta}{g}\right)_y = 0 \tag{3.3.22}$$

$$\left(\frac{\sin\theta}{C}\right)_x - \left(\frac{\cos\theta}{C}\right)_y = 0 \tag{3.3.23}$$

These conservation forms with T=0 were obtained by Whitham (1957). However, they are not suitable to study the propagation of the singularities. We shall show later that the two sets (3.3.13 - 14) and (3.3.22 - 23) are equivalent in the sense that both lead to the same jump relations across a kink.

3.3.3 Two types of singularities and jump conditions across a kink

We start with an observation which we can take as a basic assumption: rays are neither lost nor created. Thus, the variable ξ introduced on a wavefront varies continuously even if a singularity appears on a wavefront, which may get folded or suddenly bent. Now two cases arise.

Case a: In the first case, θ, g, C and T remain smooth functions satisfying the partial differential equations (3.3.15 - 16) but the Jacobian $\partial(x, y)/(\xi, t)$ given by the determinant of the matrix (3.3.11) vanishes in (ξ, t)-plane. This can happen at an isolated point in the

(ξ, t)-plane or along a curve m in the (ξ, t)-plane. We consider only the situation when the Jacobian $= -gC$ vanishes along a curve m. The situation is again complicated. $-gC$ can vanish when C vanishes, an example of which is a sonic line in a steady flow of a gas (section 5.1) where not only C but T also vanishes for wavefronts which are orthogonal to the stream lines. We restrict our consideration to the case in which the Jacobian vanishes due to the vanishing of the metric g and that $C \neq 0$. The image of m in the (x, y)-plane is an edge. Consider now a point P on m. It can be proved (Courant and Friedrichs, 1948) that at P there exists an exceptional direction in the (ξ, t)-plane such that the image of any curve passing through P in the exceptional direction has a cusp on the edge in the (x, y)-plane. This direction given by $(\dot{\xi}, \dot{t})$ is the right null vector of the Jacobian matrix at P. Since $g = 0$ at P, we get $\dot{\xi} \neq 0, \dot{t} = 0$ showing that the wavefront (t = constant) itself is in the exceptional direction at the points P of m. Thus, the wavefront at any t remains in a domain only on one side of the edge and has a cusp on it. A ray is not tangential to the wavefront in (ξ, t)-plane. Hence, the ray through P is not in the exceptional direction and is mapped into a curve which is tangential to the edge in the (x, y)-plane. This shows that when $g = 0$ and $C \neq 0$, the edge is an envelope of the rays i.e., a caustic and a wavefront Ω_t has a cusp type of singularity on t. Such singularities appear frequently in linear wave propagation. This discussion does not rule out that a cusp will not appear on a nonlinear wavefront.

Case b: We shall show in section 6.1.2 that the equations (3.3.15 to 16) along with the transport equation for the amplitude in high frequency approximation, imply that the quantities θ, g, C and T, though continuous functions of ξ at $t = 0$, may become discontinuous at a point L $(\xi_p(t), t)$. For a fixed t, limits of these quantities (denoted by subscripts $-$ and $+$) as we approach L from lower and higher values of ξ to $\xi_p(t)$ are finite so that the jumps $[\theta] = \theta_+ - \theta_-$, $[g] = g_+ - g_-$, $[C] = C_+ - C_-$ and $[T] = T_+ - T_-$ are also finite. Since the jump $[\theta]$ is found to be non-zero, the curve Ω_t suffers a sudden change in the tangent direction at a point P which is the image in the (x, y)-plane of L in the (ξ, t)-plane. Because $[\theta] \neq \pi$ (which can be seen from the results in a particular case – curves S_1 and S_2 in Fig. 6.2.4), the point P is not a cusp of Ω_t but is a new type of singularity, which we call a *kink*. When we move on Ω_t the Jacobian $-gC$ of the transformation also suffers a finite jump as we cross a kink and neither of the values $(-gC)_-$ and $(-gC)_+$ is zero. Fig.3.3.4 represents a kink phenomenon in the (x, y)-plane. The kink at P on Ω_t occupies a position Q' on Ω_{t+dt}. The kink path PQ' separates two states represented by subscripts + and −. PN' and $Q'N$ are the normals from P and Q' to the −ve side of Ω_{t+dt} and +ve side of Ω_t, respectively. $N'P' = T_-dt$ is the displacement of a ray along Ω_{t+dt} due to the tangential velocity T_- and $g_-d\xi$ is the distance along Ω_{t+dt} between P' and Q$'$. Corresponding quantities $g_+d\xi$ and T_+dt on the positive side are PQ and QN, respectively.

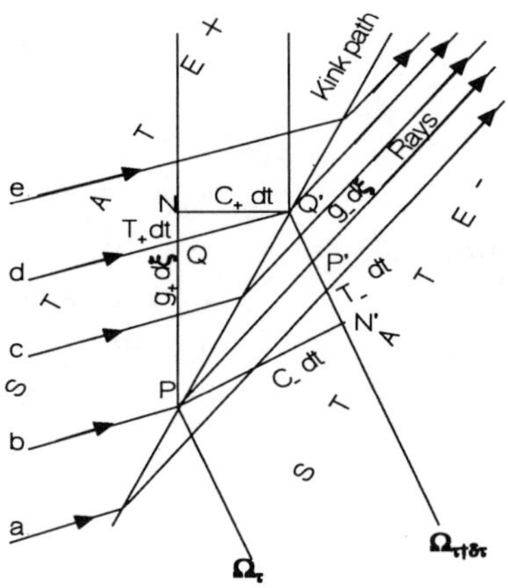

Fig. 3.3.4: Geometry of wavefront and rays on the two sides of a kink.

Using the Pythagoras theorem we get

$$(g_+\, d\xi + T_+dt)^2 + (C_+dt)^2 = (g_-d\xi + T_-dt)^2 + (C_-dt)^2 \quad (3.3.24)$$

which shows that the kink velocity $K = d\xi_p/dt$ satisfies the quadratic equation

$$(g_-^2 - g_+^2)K^2 + 2(g_-T_- - g_+T_+)K + (C_-^2 - C_+^2) + (T_-^2 - T_+^2) = 0 \quad (3.3.25)$$

This equation has real roots if

$$(g_-T_+ - g_+T_-)^2 + (C_+^2 - C_-^2)(g_-^2 - g_+^2) > 0 \quad (3.3.26)$$

The kink velocity K in the (ξ, t)-plane can also be deduced from the conservation forms (3.3.13 - 14). The jump relations obtained from these are

$$-K[g\sin\theta]+[C\cos\theta-T\sin\theta]=0,\ K[g\cos\theta]+[C\sin\theta+T\cos\theta]=0 \tag{3.3.27}$$

Eliminating θ from these two we get the equation (3.3.25). This shows that the two conservation laws (3.3.13 - 14) represent conservation of distance in the (x, y)-plane as explained below and hence are physically realistic conservation laws.

We shall now make a precise statement on the conservation of distance and give a more explicit proof for it. The jump relations (3.3.27) imply *conservation of distance* in x and y directions (and hence in any arbitrary direction) in the sense that *the vector displacement* $(d\mathbf{r})_k$ *of a kink in an infinitesimal time interval* dt *when computed in terms of variables* $(g_-, C_-, T_-, \theta_-)$ *and* $(g_+, C_+, T_+, \theta_+)$ *on the two sides of the kink path are the same.* This explicit proof was given by Giles, Prasad and Ravindran (1996) in a more general context of the propagation of a three-dimensional nonlinear wavefront but in an isotropic medium. In the two-dimensional case (3.3.10) gives expressions for the displacement $(d\mathbf{r})_k$ in terms of quantities on both sides of the kink

$$\begin{aligned}(d\mathbf{r})_k &= \{(\cos\theta_-, \sin\theta_-)C_- + (-\sin\theta_-, \cos\theta_-)T_-\}dt \\ &+ \{(-\sin\theta_-, \cos\theta_-)g_-\}d\xi \\ &= \{\cos\theta_+, \sin\theta_+)C_+ + (-\sin\theta_+, \cos\theta_+)T_+\}dt \\ &+ \{(-\sin\theta_+, \cos\theta_+)g_+\}d\xi\end{aligned} \tag{3.3.28}$$

Taking the first component of the above relation, dividing by dt and using $\frac{d\xi}{dt} = K$, we get the first jump relation in (3.3.27). Similarly, the second jump relation in (3.3.27) also follows.

Eliminating K from (3.3.27), we get the following *Hugoniot curve*

$$\cos(\theta_- - \theta_+) = \frac{C_-g_- + C_+g_+}{C_-g_+ + C_+g_-} + \frac{g_+T_- - g_-T_+}{g_+C_- + g_-C_+}\sin(\theta_- - \theta_+) \tag{3.3.29}$$

Let S be the slope of the path of the kink in (x, y)-plane i.e., $S = (dy/dx)_{kink}$. The jump relations derived from conservation laws (3.3.22 - 23) give the following expressions for S:

$$S = \frac{C_-\cos\theta_+ - C_+\cos\theta_-}{C_+\sin\theta_- - C_-\sin\theta_+} \tag{3.3.30}$$

$$S = \frac{g_+ \sin\theta_- - g_- \sin\theta_+ + (T_- g_+/C_-)\cos\theta_- - (T_+ g_-/C_+)\cos\theta_+}{g_+ \cos\theta_- - g_- \cos\theta_+ + (T_+ g_-/C_+)\sin\theta_+ - (T_- g_+/C_-)\sin\theta_-} \tag{3.3.31}$$

respectively. Equating the two expressions for S, we get the relation (3.3.29). Without any loss of generality, we assume $\theta_- = 0$, which can be achieved by choosing the direction of the x-axis to be instantaneously coincident with the normal of front in the $-$ state. This only simplifies steps required to derive some results below. Then the kink velocity K from the first relation in (3.3.27) satisfies

$$K = (C_+ \cos\theta_+ - T_+ \sin\theta_+ - C_-)/(g_+ \sin\theta_+) \tag{3.3.32}$$

In (3.3.10), we take the differentials dx and dy along the kink line in the + state, then the kink velocity $S = (dy/dx)_{\text{kink}}$ is given by

$$S = \frac{(C_+ \sin\theta_+ + T_+ \cos\theta_+) + g_+ K \cos\theta_+}{(C_+ \cos\theta_+ - T_+ \sin\theta_+) - g_+ K \sin\theta_+}, \quad K = \frac{d\xi_p}{dt}$$

Substituting the expression for K from (3.3.32) we get

$$S = (C_+ - C_- \cos\theta_+)/(C_- \sin\theta_+) \tag{3.3.33}$$

which is exactly the same as the expression (3.3.30) with $\theta_- = 0$. These results show that the two sets of conservation laws (3.3.13 - 14) and (3.3.22 - 23) are equivalent in the sense that both lead to the same Hugoniot relation across a kink and both give the same kink line in the (x, y)-plane.

The conservation laws (3.3.13 - 14) are consistent with all geometrical features which may result from the propagation of the two wings of the front meeting at a kink. As above, assume that there are two straight wings, W_-^0 and W_+^0 with $(C_-, \theta_- = 0)$ and (C_+, θ_+) respectively. We further assume that initially at the origin in (x, y)-plane. At a later time t, when the two wings propagate independently, they are given by

$$W_- \ : \ x = C_- t \tag{3.3.34}$$

and

$$W_+ \ : \ x\cos\theta_+ + y\sin\theta_+ = C_+ t \tag{3.3.35}$$

Note that T_+ does not appear in this equation. The point of intersection of W_- and W_+ is $(C_- t,\ (C_+ - C_- \cos\theta_+)t/\sin\theta_+)$, which should be the position of the kink at time t. This can be verified by noting that the shape of the line joining the origin and this point is $(C_+ - C_- \cos\theta_+)t/(C_- \sin\theta_+)$ which is the same as the expression (3.3.33). Thus we get a very simple method of determining of the position of a kink *"calculate the positions of the two portions of the front on the two sides of the kink as if they propagate independently and determine the point of interaction."* This method, although it can not be continued accurately for a long time in a numerical computation, was indeed used by Kevlahan (1996).

It is possible to get one more interesting result using only the kinematical conditions. Consider a wavefront with x-axis as the line of symmetry (Fig.3.3.5) with two kinks P_1 and P_2 joining a straight disk (with θ_-, C_-, T_-) of the wavefront and two wings (with $\mp\theta_+$, C_+, T_+). The question arises: "Do the kinks move away from or tend to approach one another?" To answer this, we note that the slope S of the upper kink path in (x, y)-plane is negative if $S < 0$. We consider the whole configuration of the wavefront to be moving in the +ve x-direction, then $C_- > 0$ and since $\theta_- = 0$ and $\theta_+ < 0$. (3.3.33) with $S < 0$ gives

$$C_- \cos\theta_+ < C_+ \tag{3.3.36}$$

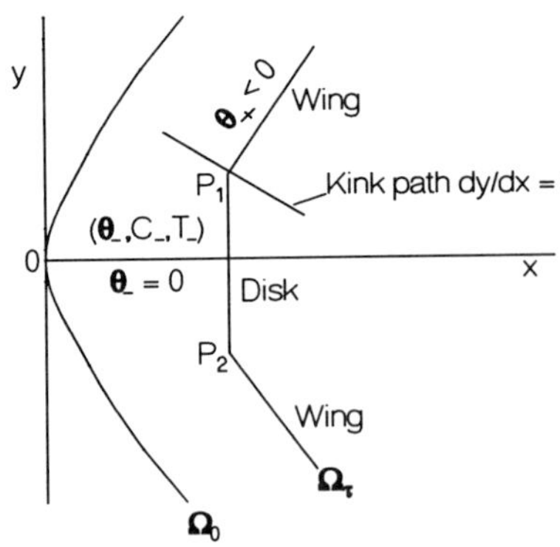

Fig. 3.3.5: The two kinks approach one another if $S < 0$.

Setting $\theta_- = 0$ in (3.3.29) and then eliminating θ_+ from (3.3.29) and (3.3.36), we get required necessary and sufficient conditions for the two kinks to approach each other. This condition takes a particularly simple form in the case of the wave propagation in an isotropic medium i.e., if $T = 0$. Eliminating θ_+ between (3.3.29) with $\theta_- = 0$ and $T = 0$ and (3.3.36) we get

$$C_- < C_+ \tag{3.3.37}$$

This condition was first derived by Kevlahan (1996).

For a discontinuous solution, different conservation laws with ξ, t as independent variables are, in general, not equivalent. Hence *(3.3.13 - 14) are the only physically realistic conservation laws, which conserve distance as explained above.*

3.3.4 Kinematical compatibility conditions on a surface of discontinuity in multi-dimensions

The results of the section 3.3.2 on the kinematical conservation laws have been extended for a propagating surface in three-dimensional space by Giles, Prasad and Ravindran (1996). They have derived the conservation laws and explicitly shown that these conservation laws represent conservation of distance in two independent directions on the propagating surface with a kink curve on it. We shall not discuss these recent developments but present some other results (Thomas, 1961; Truesdell and Toupin, 1960), which are required for further development of a theory to study propagation of discontinuities along rays associated with a propagating surface.

Let $\Omega : \phi(\mathbf{x}, t) = 0$ be the locus in space-time of a propagating m-dimensional surface $\Omega_t : \phi(\mathbf{x}, t) = 0,\ t =$ constant. We assume that the function ϕ is smooth. The unit normal $\mathbf{n}$ of the surface Ω_t and its velocity of propagation (see (2.2.3)) are given by

$$\mathbf{n} = \nabla\phi/|\nabla\phi| \text{ and } c = -\phi_t/|\nabla\phi| \tag{3.3.38}$$

Let Ω be a surface of discontinuity of a function $\mathbf{u}(\mathbf{x}, t)$ and its derivatives, which we assume to tend to finite limits as they approach the surface Ω from either side. We call such discontinuities as *discontinuities of the first kind.* A tangential derivative $\frac{\partial}{\partial T}$ on Ω_t of any function is completely determined by the distribution of the function on Ω_t. If we apply this to the limiting values of $\mathbf{u}$ on both sides of Ω_t and take the difference, we get *Hadamard's lemma*

$$\frac{\partial}{\partial T}[\mathbf{u}] = \left[\frac{\partial \mathbf{u}}{\partial T}\right] \tag{3.3.39}$$

where, as in section 1.3, the symbol [ ] implies jump of a quantity as we cross Ω_t from one side of Ω to the other side. Let $\frac{\partial}{\partial n}$ represent the normal derivative on Ω_t i.e., $\frac{\partial}{\partial n} = \langle \mathbf{n}, \nabla \rangle$ then

$$\left[\frac{\partial \mathbf{u}}{\partial n}\right] = \sum_{\alpha=1}^{m} n_\alpha \left[\frac{\partial \mathbf{u}}{\partial x_\alpha}\right] \tag{3.3.40}$$

These are general results. In particular, if the function $\mathbf{u}$ is continuous across Ω, $[\mathbf{u}] = 0$ which implies that $[\partial \mathbf{u}/\partial T] = 0$. Now we note that for each α, β from $1, 2, \ldots, m$, the differential operators $\phi_{x_\alpha} \frac{\partial}{\partial t} - \phi_t \frac{\partial}{\partial x_\alpha}$ and $\phi_{x_\alpha} \frac{\partial}{\partial x_\beta} - \phi_{x_\beta} \frac{\partial}{\partial x_\alpha}$ are tangential derivatives on Ω. Therefore, we get

$$\phi_{x_\alpha}[\mathbf{u}_t] - \phi_t[\mathbf{u}_{x_\alpha}] = 0,\ \phi_{x_\alpha}[\mathbf{u}_{x_\beta}] - \phi_{x_\beta}[\mathbf{u}_{x_\alpha}] = 0 \text{ when } [\mathbf{u}] = 0 \tag{3.3.41}$$

Using (3.3.38), we express these results in the form

$$[\mathbf{u}_t] = -c\left[\frac{\partial \mathbf{u}}{\partial n}\right],\quad \left[\mathbf{u}_{x_\beta}\right] = n_\beta \left[\frac{\partial \mathbf{u}}{\partial n}\right],\quad \text{when } [\mathbf{u}] = 0 \tag{3.3.42}$$

3.4 Breakdown of the continuity of a solution of a quasilinear system

In the very first section of this monograph we started with the discussion of the most important property of a solution of equations with genuine nonlinearity. Even if a solution itself remains finite, its derivatives may tend to infinity in a finite time called, *critical time* i.e., the continuity of an initially continuous solution may break down. The problem of determination of the critical time for the blow up of the derivatives of smooth solutions of systems of more than one equation, especially in three or more independent variables, is not easy. For some results on a special pair of equations (namely p-system) in two independent variables, one may consult Smoller (1983), chapter 20, section A.

A related problem, discussion of growth and decay of discontinuities in the derivatives of a solution, is a much simpler problem – this is more so if the state ahead of the characteristic surface across which the discontinuity exists is known. The problem in this form for two independent variables was dealt with in a very simple way by Whitham (1959a) by writing an expansion of the solution behind the curve of discontinuity (i.e., the characteristic curve). Pack (1960) discussed this problem more rigorously in regards to Euler equations of a compressible gas for a radially symmetric flow and mentioned an interesting conclusion of Burton (1893) for a flow in a space of arbitrary dimension m. This result is physically meaningful only when $m = 1, 2$ and 3. However, it beautifully brings out the result of competition between the shock formation tendency of genuine nonlinearity and its opposing effect of the increase in the surface area of a wavefront with diverging rays. In the next section, we shall use Whitham's method to derive these results.

The general problem of evolution of discontinuities in the first derivatives of a solution of a quasilinear hyperbolic system in any number of independent variables was solved by Varley and Cumberbatch (1965). We shall present a generalization of the results of their work in section 3.4.2 for a curved wavefront (across which the discontinuities exist) running into a known state.

3.4.1 Combined effect of genuine nonlinearity and geometrical divergence

In section 2.2.4 we showed that geometrical convergence and divergence of rays have significant effect on the growth and decay of the amplitude of a singularity (in that case, a discontinuity in a second derivative) of a solution of the wave equation. Genuine nonlinearity also causes growth and decay in the amplitude of a singularity. In this section we shall study their combined effect when both processes are simultaneously present in the equations. The effect is best seen by considering radially symmetric isentropic motion of a polytropic gas (see also section 3.1.1)

$$A\mathbf{u}_t + B\mathbf{u}_r + C = 0 \qquad (3.4.1)$$

where

$$\mathbf{u} = \begin{bmatrix} \rho \\ q \end{bmatrix}, \quad A = I, \quad B = \begin{bmatrix} q & \rho \\ \frac{a^2}{\rho} & q \end{bmatrix}, \quad C = \begin{bmatrix} \alpha \rho q \\ 0 \end{bmatrix}, \quad \alpha = 0, 1, 2, \ldots \tag{3.4.2}$$

and the sound velocity a satisfies $a^2 = \frac{dp}{d\rho}$, $p = A\rho^\gamma$ where A and γ are constant. The eigenvalues of the system are $q - a$ and $q + a$. Here $\alpha = 0, 1, 2, \ldots, m, \ldots$ correspond to radially symmetric flow in one, two, three, $\ldots, m+1, \ldots$ dimensional space.

Let us consider a diverging (i.e., moving away from the origin) wave running into a uniform state given by $\mathbf{u}_0 = (\rho_0, 0)^T$. Assume that the state of the gas varies continuously from the uniform state ahead to the disturbed one behind. The leading wavefront is given by $\Omega : r - a_0 t = \text{constant} = r_0$, say. We further assume that the first order derivatives of $\mathbf{u}$ suffer discontinuity of the first kind across Ω. To derive the transport equation for the amplitude of the discontinuity, we use Whitham's method, in which the solution behind the leading wavefront can be expressed in the form

$$\mathbf{u} = \mathbf{u}_0 + (t - r/a_0)\mathbf{u}_1(r) + \frac{1}{2}(t - r/a_0)^2 \mathbf{u}_2(r) + \ldots \tag{3.4.3}$$

Here $\mathbf{u}_1$ represents the value of $\mathbf{u}_t$ just behind the leading wavefront Ω. Since $\mathbf{u}_t = 0$ in the uniform state ahead of $\Omega, \mathbf{u}_1$ represents the jump in $\mathbf{u}_t$ across Ω.

Substituting (3.4.3) in (3.4.1) and equating coefficients of powers of $t - \frac{r}{a_0}$ on the left equal to zero, we get

$$\rho_1 = (\rho_0/a_0)q_1 \tag{3.4.4}$$

$$\rho_2 - (\rho_0/a_0)q_2 - 2\rho_1 q_1/a_0 + \rho_0 q_1' + \alpha\rho_0 q_1/r = 0 \tag{3.4.5}$$

and

$$-(a_0/\rho_0)(\rho_2 - (\rho_0/a_0)q_2) - q_1^2/a_0 - (\gamma - 2)(a_0/\rho_0^2)\rho_1^2 + (a_0^2/\rho_0)\rho_1' = 0 \tag{3.4.6}$$

Eliminating ρ_1, ρ_2 and q_2 from the above relations we get

$$\frac{dq_1}{dr} + \frac{\alpha}{2r}q_1 - \frac{\gamma+1}{2a_0^2}q_1^2 = 0 \tag{3.4.7}$$

As explained above, q_1 represents the jump in the time rate of change of the fluid velocity q across the leading wavefront. (3.4.7) is the final transport equation for q_1.

The solution of the equation (3.4.7) satisfying $q_1 = q_{10}$ at $r = r_0$ is

$$\frac{1}{q_1} = \left(\frac{r}{r_0}\right)^{\alpha/2} \left\{\frac{1}{q_{10}} + \frac{\gamma+1}{(2-\alpha)a_0^2} r_0\right\} - \frac{\gamma+1}{(2-\alpha)a_0^2} r, \text{ for } \alpha \neq 2 \quad (3.4.8)$$

and

$$\frac{1}{q_1} = r\left\{\frac{1}{q_{10} r_0} - \frac{\gamma+1}{2a_0^2} \log\frac{r}{r_0}\right\}, \quad \text{for } \alpha = 2 \quad (3.4.9)$$

The solution given above is valid also for a converging (i.e., moving toward the origin) wave and can be obtained proceeding exactly in the same way except that we need to write the expansion (3.4.3) in powers of $t + r/a_0$ instead of $t - r/a_0$. An expanding compression wave implies jump $q_1 > 0$ and a converging compression wave $q_1 < 0$. As long as q_1 is finite (either in a compression or expansion wave), the velocity q of the gas (and so density ρ and pressure p) is continuous across the leading wavefront $r \pm a_0 t = r_0$. A shock wave appears at the leading wavefront when q_1 tend to infinity i.e., at a point $r = R$ given by

$$R^{1-\alpha/2} = r_0^{1-\alpha/2}\left\{1 + \frac{(2-\alpha)a_0}{(\gamma+1)q_{10}r_0}\right\} \quad \text{for } \alpha \neq 2 \quad (3.4.10)$$

and

$$R = r_0 \exp\left\{\frac{2a_0^2}{(\gamma+1)q_{10}r_0}\right\}, \quad \text{for } \alpha = 2 \quad (3.4.11)$$

where $r_0 > 0$ (without loss of generality, we can make this assumption even when $\alpha = 0$).

Consider now the two cases:

Case a: Diverging wave.

When the diverging wave is an expansion wave, $q_{10} < 0$. Expressions (3.4.10 - 11) show that in all cases i.e., in one, two, three, four ... dimensions $R < r_0$ which is not possible because the diverging wave moves in positive direction from r_0. Thus, in this case, q_1 can not become infinite and no shock is ever formed at the leading wavefront of an expansion wave.

When the diverging wave is a compression wave $q_{10} > 0$, expressions (3.4.10 - 11) for $\alpha = 0, 1, 2$ give values of $R > r_0$. Thus every diverging compression wave ends into a shock wave in one, two and three-dimensions. Whenever $2 < \alpha < 2+(\gamma+1)q_{10}r_0/a_0$ is satisfied, a shock is again formed in every compression wave in this mathematically (only mathematically and not physically) meaningful higher dimensional space.

To understand the significance of the purely mathematical result stated at the end of the last paragraph, we note that there are two physical processes present in the equation (3.4.7). The second term containing α represents the effect of geometric increase or decrease in the area of $\alpha + 1$ dimensional sphere and the third term containing q_1^2 represents the effect of the genuine nonlinearity present in the equations (3.4.1 - 2). For $\alpha = 0, 1, 2$ the genuine nonlinearity dominates over the geometric decay for a diverging compression wave and a shock is eventually formed. For $\alpha = 3, 4, \dots,$ whenever $\alpha < 2 + (\gamma + 1)q_{10}r_0/a_0$ the genuine nonlinearity continues to dominate but when $\alpha > 2 + (\gamma + 1)q_{10}r_0/a_0$ the geometric damping on the diverging wave is too strong. It can be easily checked that in this case q_1 remains finite even when the leading wavefront reaches infinity.

Case b: Converging wave

As expected, for every converging wave (with an exception of the case $\alpha = 0$ i.e., plane wave) q_1 tends to infinity as r tends to zero. Hence we consider below only the case when q_1 blows up at a non-zero value of r.

When we have converging compression wave $q_{10} < 0$, expressions (3.4.10 - 11) show that for all values of α, $R < r_0$. Thus, a shock is always formed.

For a space of sufficiently high dimension, we may think that a shock may be formed at a finite distance less than r_0 due to rapid convergence at the head of an expansion wave. But this is not possible since $q_1 \to \infty$ would violate the entropy condition. For example, (3.4.10 - 11) give $R > r_0$ for $\alpha = 0, 1, 2$ and

$$2 < \alpha < 2 + \frac{(\gamma + 1)q_{10}r_0}{a_0} \tag{3.4.12}$$

When $\alpha > 2 + (\gamma + 1)q_{10}r_0/a_0$, (3.4.10) does not give any positive value of R.

3.4.2 Transport equation for discontinuities in derivatives for a system in multi-dimensions

Consider the hyperbolic system of equations in the form

$$A(\mathbf{x},t,\mathbf{u})\mathbf{u}_t + B^{(\alpha)}(\mathbf{x},t,\mathbf{u})\mathbf{u}_{x_\alpha} + C(\mathbf{x},t,\mathbf{u}) = 0 \qquad (3.4.13)$$

where $A, B^{(\alpha)}$ and C are smooth functions of their arguments. We assume that $\mathbf{u}$ is continuous across a surface $\Omega : \phi(\mathbf{x},t) = 0$ but its first derivatives have discontinuities of the first kind across Ω.

We also assume that the solution on one side of Ω, say the direction in which the normal $\mathbf{n}$ points, is known and is denoted by $\mathbf{u}_0(\mathbf{x},t)$.

In this section we shall first show that Ω is a characteristic surface and then we shall derive a transport equation to study growth or decay of the strength of the discontinuities in the first derivatives of $\mathbf{u}$ along the bicharacteristics on Ω.

We assume ϕ itself to be smooth and introduce a new coordinate system

$$(\mathbf{x},t) \to (\mathbf{x}',\phi); \quad \mathbf{x}' = \mathbf{x}, \phi = \phi(\mathbf{x},t) \qquad (3.4.14)$$

Then

$$\frac{\partial}{\partial t} = \phi_t \frac{\partial}{\partial \phi}, \quad \frac{\partial}{\partial x_\alpha} = \frac{\partial}{\partial x'_\alpha} + \phi_{x_\alpha}\frac{\partial}{\partial \phi} \qquad (3.4.15)$$

and

$$\frac{\partial}{\partial \phi} = \frac{1}{\phi_t}\frac{\partial}{\partial t}, \quad \frac{\partial}{\partial x'_\alpha} = \frac{\partial}{\partial x_\alpha} - \frac{\phi_{x_\alpha}}{\phi_t}\frac{\partial}{\partial t} = \frac{\partial}{\partial x_\alpha} + \frac{n_\alpha}{c}\frac{\partial}{\partial t} \qquad (3.4.16)$$

where $\mathbf{n} = \nabla\phi/|\nabla\phi|$ and $c = -\phi_t/|\nabla\phi|$. As noted in section 2.2.3, $\frac{\partial}{\partial x'}$ represents a derivative in the direction of a tangent to the surface Ω in space time and $\frac{\partial}{\partial \phi}$ represents it in a transversal direction. The equation (3.4.13) transforms to

$$(A\phi_t + \phi_{x_\alpha}B^{(\alpha)})u_\phi + B^{(\alpha)}u_{x'_\alpha} + C = 0 \qquad (3.4.17)$$

We make a slight departure from the usual convention by writing (3.4.17) in a mixed coordinate system $(\mathbf{x}',t)$

$$\left(A - \frac{1}{c}n_\alpha B^{(\alpha)}\right)\mathbf{u}_t + B^{(\alpha)}\mathbf{u}_{x'_\alpha} + C = 0 \qquad (3.4.18)$$

Since $\mathbf{u}$ is continuous on Ω, the jump $[\mathbf{u}_{x'_\alpha}] = 0$. But according to our assumption, the transversal derivative $\mathbf{u}_\phi$ or $\mathbf{u}_t$ are discontinuous across Ω i.e., $[\mathbf{u}_t] = \phi_t[\mathbf{u}_\phi] \neq 0$. Further jumps of $A, B^{(\alpha)}, C, c$ and $\mathbf{n}$ across Ω are zero. Taking jump of (3.4.18) across Ω we get

$$\left(n_\alpha B^{(\alpha)} - cA\right)_0 [\mathbf{u}_t] = 0 \tag{3.4.19}$$

where the subscript 0 represents the value when $\mathbf{u} = \mathbf{u}_0$ and we define the jump

$$[\mathbf{u}_t] = \mathbf{u}_{0t} - (\mathbf{u}_t)_l \tag{3.4.20}$$

with the subscript l denoting the value on the negative side of the direction of $\mathbf{n}$ i.e., the state behind the wavefront.

Since $[\mathbf{u}_t] \neq 0$, this equation implies that on Ω

$$\det\left(n_\alpha B^\alpha - cA\right) = |\nabla\phi| \det\left(\phi_{x_\alpha} B^{(\alpha)} + \phi_t A\right) = 0 \tag{3.4.21}$$

i.e., we get an extension of the result of the section 2.5, *the surface of discontinuity Ω of the first derivatives is a characteristic surface when* $[\mathbf{u}] = 0$. In the section 2.5 for a linear system $[\mathbf{u}] = 0$ is not required. We assume that Ω is a characteristic of the kth family.¶ We denote, as usual, the left and right eigenvectors on Ω_t corresponding to the kth eigenvalue c (assumed to be simple) by $\mathbf{l}_0$ and $\mathbf{r}_0$ (note that we have suppressed the superscript k on $\mathbf{l}$ and $\mathbf{r}$ and the subscript k on c). The equation (3.4.19) implies

$$[\mathbf{u}_t] = \tilde{w}_1\, \mathbf{r}_0 \tag{3.4.22}$$

where $\tilde{w}_1$ is a scalar defined on Ω. Next we derive the transport equation for $\tilde{w}_1$.

As in section 2.5, we could have worked with the function w_1 related to the jump in u_ϕ instead of that in u_t. The transport equation for w_1 would have been simpler than (3.4.27) below for $\tilde{w}_1$. The jump $[u_t]$ has a better physical interpretation than $[u_\phi]$. $[u_t]$ is related to the jump $\left[\frac{\partial u}{\partial n}\right]$ in the normal derivative by the relation (3.3.42), where as $[u_\phi]$ contains a term ϕ_t, which has no simple physical interpretation.

¶We have already shown in section 3.1.4 how to produce kth family of simple waves for a reducible system of equations.

Differentiating (3.4.13) with respect to t we get

$$A\mathbf{u}_{tt} + B^{(\alpha)}\frac{\partial \mathbf{u}_t}{\partial x_\alpha} + (A_t + (\nabla_u A \cdot \mathbf{u}_t))\mathbf{u}_t$$
$$+(B_t^{(\alpha)} + (\nabla_u B^{(\alpha)} \cdot \mathbf{u}_t))\mathbf{u}_{x_\alpha} + (C_t + (\nabla_u C \cdot \mathbf{u}_t)) = 0 \quad (3.4.23)$$

where we have used $\mathbf{u}_{x_\alpha t} = \mathbf{u}_{tx_\alpha}$ in the second term. We note that the partial derivatives with respect to x'_α and t do not commute i.e., $u_{x'_\alpha t} \neq u_{tx'_\alpha}$. Using (3.4.16), we write this equation in terms of derivatives in a mixed coordinate system $(\mathbf{x}', t)$:

$$\left(A - \frac{n_\alpha}{c}B^{(\alpha)}\right)\mathbf{u}_{tt} + B^{(\alpha)}\frac{\partial \mathbf{u_t}}{\partial x'_\alpha} + A_t\mathbf{u}_t + B_t^{(\alpha)}\left(\mathbf{u}_{x'_\alpha} - \frac{n_\alpha}{c}\mathbf{u}_t\right)$$
$$+C_t + (\nabla_u C \cdot \mathbf{u}_t) + (\nabla_u A \cdot \mathbf{u}_t)\mathbf{u}_t - \frac{n_\alpha}{c}(\nabla_u B^{(\alpha)} \cdot \mathbf{u}_t)\mathbf{u}_t$$
$$+(\nabla_u B^{(\alpha)} \cdot \mathbf{u}_t)\mathbf{u}_{x'_\alpha} = 0 \quad (3.4.24)$$

We note that all functions appearing in (3.4.24) except $\mathbf{u}_t, \mathbf{u}_{tx'_\alpha}$ and $\mathbf{u}_{tt}$ are continuous on Ω. We also note that a jump of a product fg of two functions on Ω can be expressed in the form

$$[fg] = -[f][g] + f_0[g] + g_0[h] \quad (3.4.25)$$

Taking a jump of the equation (3.4.24) across Ω, we get the following relation

$$\left(A - \frac{n_\alpha}{c}B^{(\alpha)}\right)_0[\mathbf{u}_{tt}] + B_0^{(\alpha)}\frac{\partial}{\partial x'_\alpha}[\mathbf{u}_t] + A_t[\mathbf{u}_t]$$
$$-\frac{n_\alpha}{c_0}B_{t_0}^{(\alpha)}[\mathbf{u}_t] + ((\nabla_u C)_0 \cdot [\mathbf{u}_t]) + ((\nabla_u B^{(\alpha)})_0 \cdot [\mathbf{u}_t])\mathbf{u}_{0x'_\alpha}$$
$$+((\nabla_u A)_0 \cdot \mathbf{u}_{0t})[\mathbf{u}_t] + ((\nabla_u A)_0 \cdot [\mathbf{u}_t])\mathbf{u}_{0t}$$
$$-\frac{n_\alpha}{c_0}\left\{((\nabla_u B^{(\alpha)})_0 \cdot \mathbf{u}_{0t})[\mathbf{u}_t] + ((\nabla_u B^{(\alpha)})_0 \cdot [\mathbf{u}_t])\mathbf{u}_{0t}\right\}$$
$$-((\nabla_u A)_0 \cdot [\mathbf{u}_t])[\mathbf{u}_t] + \frac{n_\alpha}{c_0}((\nabla_u B^{(\alpha)})_0 \cdot [\mathbf{u}_t])[\mathbf{u}_t] = 0 \quad (3.4.26)$$

where n_α is the unit normal of Ω_t.

Premultiplying this equation by $\mathbf{l}_0$ (when the first term vanishes) and substituting (3.4.22) for $[\mathbf{u}_t]$, we get

$$\frac{d\tilde{w}_1}{d\sigma} + \left\{\left(\mathbf{l}_0 B_0^{(\alpha)}\frac{\partial \mathbf{r}_0}{\partial x'_\alpha}\right) + M\right\}\tilde{w}_1 + K\tilde{w}_1^2 = 0 \quad (3.4.27)$$

where $d/d\sigma$ is related to the time rate of change d/dt along the rays associated with the wavefront Ω_t by

$$\frac{d}{dt} = \frac{1}{\mathbf{l}_0 A_0 \mathbf{r}_0} \frac{d}{d\sigma} = \left(\mathbf{l}_0 B_0^{(\alpha)} \mathbf{r}_0\right) \frac{\partial}{\partial x'_\alpha} \tag{3.4.28}$$

$$\begin{aligned} M = {} & \mathbf{l}_0\{A_{t0}\mathbf{r}_0 + ((\nabla_u A)_0 \cdot \mathbf{u}_{0t})\mathbf{r}_0 + ((\nabla_u A)_0 \cdot \mathbf{r}_0)\mathbf{u}_{0t}\} \\ & + \mathbf{l}_0 \left((\nabla_u B^{(\alpha)})_0 \cdot \mathbf{r}_0)(\mathbf{u}_{0x'_\alpha} - \frac{n_\alpha}{c_0}\mathbf{u}_{0t}) \right. \\ & \left. - \frac{n_\alpha}{c_0}\{B_{t0}^{(\alpha)} + ((\nabla_u B^{(\alpha)})_0 \cdot \mathbf{u}_{0t})\}\mathbf{r}_0 \right) + \mathbf{l}_0((\nabla_u C)_0 \cdot \mathbf{r}_0) \end{aligned} \tag{3.4.29}$$

and

$$K = \mathbf{l}_0\{\frac{n_\alpha}{c_0}((\nabla_u B^{(\alpha)})_0 \cdot \mathbf{r}_0) - ((\nabla_u A)_0, \mathbf{r}_0)\}\mathbf{r}_0 \tag{3.4.30}$$

(3.4.27) is the final form of the transport equation for the amplitude $\tilde{w}_1$ of the jump $[\mathbf{u}_t]$ across a wavefront Ω_t running into a known state $\mathbf{u}_0$. The transport equation (3.4.27) can be solved along a ray only if the ray path associated with the wavefront Ω_t is known from the solution of the equations (2.4.6 - 7). Since $\mathbf{u}$ is continuous across Ω_t on which the tangential derivatives appear in (2.4.7), the right hand sides of the ray equations (2.4.6 - 7) are to be evaluated at $\mathbf{u}_0(\mathbf{x}, t)$ which is the value of $\mathbf{u}$ on Ω_t. Thus, the position $\mathbf{x}(t)$ of a point moving on a ray and the normal $\mathbf{n}(t)$ of the wavefront at that point can be determined by solving the $2n$ nonlinear ordinary differential equations (2.4.6-7) with $\mathbf{u}$ replaced by $\mathbf{u}_0$. Once $\mathbf{x}(t)$ and $\mathbf{n}(t)$ are determined along rays as functions of t and the surface coordinates on the initial wavefront Ω_0, all coefficients appearing in the transport equation (3.4.27) can be expressed in terms of t (or σ) and hence can be solved in principle. This gives the amplitude $\tilde{w}_1$ of the discontinuity in $\mathbf{u}_t$ across Ω_t along a ray if the value of $\tilde{w}_1$ is known at any location on the ray.

The left and right eigenvectors $\mathbf{l}_0$ and $\mathbf{r}_0$ are known only on Ω_t. Therefore, the coefficients $\mathbf{l}_0 B_0^{(\alpha)} \frac{\partial \mathbf{r}_0}{\partial x'_\alpha}$ in (3.4.27), which contain tangential derivatives on the wavefront Ω_t, can be evaluated. This term corresponds to the second term in the equation (3.4.7) for the Euler's equations of gas dynamics and in many applications it is found to be proportional to the mean curvature of the surface Ω_t. See Varley and Cumberbatch (1965) for details of two cases of wave propagation which are not radially symmetric.

Unlike the result in section 2.5, the equation (3.4.27) is the non-linear containing second power of $\tilde{w}$ on the right hand side. This implies that the discontinuity $[\mathbf{u}_t]$ may tend to infinity in finite time. This happens often in many interesting physical situations as we saw in the case of a compression wave in a polytropic gas.

The particular case of the transport equation of Varley and Cumberbatch is obtained by taking the matrix A to be the identity matrix and $B^{(\alpha)}$ and C to be independent of t. We further assume the state ahead of Ω_t to be steady: $\mathbf{u}_0(\mathbf{x})$. Many terms in (3.4.29-30) drop out. We first notice that

$$\frac{\partial \mathbf{r}_0}{\partial x'_\alpha} = \frac{\partial \mathbf{r}_0}{\partial x_\alpha} \ , \quad \mathbf{u}_{0x'_\alpha} = \mathbf{u}_{0x_\alpha} \tag{3.4.31}$$

The coefficients M and K in the transport equation now become

$$M = \mathbf{l}_0 \left(((\nabla_u B^{(\alpha)})_0 \cdot \mathbf{r}_0)\mathbf{u}_{0x_\alpha} + ((\nabla_u C)_0 \cdot \mathbf{r}_0) \right) \tag{3.4.32}$$

and

$$K = \mathbf{l}_0 \left(\frac{n_\alpha}{c_0} ((\nabla_u B^{(\alpha)})_0 \cdot \mathbf{r}_0) \right) \mathbf{r}_0 \tag{3.4.33}$$

3.5 Jump conditions on a curved shock

A shock wave is a beautiful example of a solution in high frequency approximation. As mentioned in section 3.2.1, the approximation is satisfied exactly across a shock front. In this section we shall present the jump conditions on a shock for a system of n first order equations in *conservation form* in $m+1$ independent variables. For simplicity we shall consider a system of n *conservation laws*

$$\frac{\partial H(\mathbf{u})}{\partial t} + \langle \nabla, \mathbf{F}(\mathbf{u}) \rangle = 0 \tag{3.5.1}$$

where n components of the *density* vector $\mathbf{H}$ and the flux vector $\mathbf{F}$ are independent of $\mathbf{x}$ and t and $\mathbf{u} : \mathbf{R}^4 \to \mathbf{R}^n$. We present here a derivation of the jump conditions following Maslov (1980).

For a *weak or generalized* solution of the system of conservation laws (3.5.1), the derivatives of $\mathbf{H}$ and $\mathbf{F}$ in the equation are treated as derivatives of a distribution or a generalized function (Gel'fand and Shilov (1964)). To be more specific, a weak solution of (3.5.1) is a bounded measurable function $\mathbf{u}$ which satisfies

$$\int_{\mathbf{R}^4} (\langle \varphi_t, \mathbf{H} \rangle + \langle \nabla\varphi, \mathbf{F} \rangle)\, d\mathbf{x}dt = 0 \tag{3.5.2}$$

for all test functions $\varphi : \mathbf{R}^4 \to \mathbf{R}^n$ such that $\varphi \in C_0^\infty(\mathbf{R}^4)$.

For a derivation of the jump conditions across a shock surface, we assume that a weak solution is represented by functions which are $C^\infty(\mathbf{R}^4)$ except for a smooth surface $\Omega : s(\mathbf{x}, t) = 0$. We also assume that the solution suffers only a discontinuity of the first kind on Ω, i.e., the limiting values of the solution and its partial derivatives, as we approach Ω from either side of it, exist and are bounded. Such a function g is represented in the form

$$g = g_0(\mathbf{x}, t) + H(s)g_1(\mathbf{x}, t) \tag{3.5.3}$$

where g_0 and $g_1 \in C^\infty(\mathbf{R}^4)$ and $H(s)$ is the Heaviside function. The jump in the function g on Ω as we cross it from the domain $s < 0$ into the domain $s > 0$ is g_1.

$$\text{i.e.,} \qquad [g] = g_1 \ \text{ on } \ \Omega \tag{3.5.4}$$

The representation (3.5.3) of a piecewise smooth function g is not unique since the function $g_0(= g \text{ in } s < 0)$ is uniquely determined only in the domain $s < 0$ and $g_0 + g_1(= g \text{ in } s > 0)$ is uniquely determined in the domain $s > 0$.

Let us denote the space of functions representable in the form (3.5.3) with $g_0, g_1 \in C^\infty(\mathbf{R}^4)$ by $\mathcal{R}_\Omega$. Then $\mathcal{R}_\Omega$ is closed under addition, multiplication and substitution as arguments in smooth functions. The last one implies that if $\psi : \mathbf{R} \to \mathbf{R}$ (or ψ: $\mathbf{R}^n \to \mathbf{R}^n$) be such that $\psi \in C^\infty(\mathbf{R})$ and $g \in \mathcal{R}_\Omega$, then $\psi(g) \in \mathcal{R}_\Omega$. Thus, for the solution of (3.5.1) with a shock manifold Ω, the *density* function $\mathbf{H}$ and the *flux* function $\mathbf{F}$ belong to $\mathcal{R}_\Omega$.

Let $\mathbf{e} = (e_1, e_2, e_3, e_4) \in \mathbf{R}^4$ be a constant vector and

$$\nabla = \left(\frac{\partial}{\partial x_1}, \frac{\partial}{\partial x_2}, \frac{\partial}{\partial x_3}\right) \quad \text{and} \quad \nabla_{xt} = \left(\nabla, \frac{\partial}{\partial t}\right) \tag{3.5.5}$$

By $\mathcal{D}_{e,\Omega}$, we denote a set of generalized functions obtained by differentiating a member of $\mathcal{R}_\Omega$ in the direction of $\mathbf{e}$, i.e., $h \in \mathcal{D}_{e,\Omega}$ if there exists a function $g \in \mathcal{R}_\Omega$ such that

$$h = \langle \mathbf{e}, \nabla_{x,t} \rangle g \tag{3.5.6}$$

The derivatives in (3.5.6) are to be interpreted as derivatives of a generalized function.

Now define a class $\mathcal{D}_\Omega$ of generalized functions representable as a finite sum of functions belonging to various sets $\mathcal{D}_{e,\Omega}$ for different vectors $\mathbf{e} \in \mathbf{R}^4$. Thus, $h \in \mathcal{D}_\Omega$ if there exists an integer $m > 0$, vectors $\mathbf{e}^1, \mathbf{e}^2, \ldots, \mathbf{e}^m$ and functions $g^1, g^2, \ldots, g^m$; $g^i \in \mathcal{R}_\Omega$ such that

$$h(\mathbf{x}, t) = \sum_{i=1}^{m} h^i(\mathbf{x}, t) \tag{3.5.7}$$

where $h^i =< \mathbf{e}^i, \nabla_{xt} > g^i \in \mathcal{D}_{e^i,\Omega}$.

It is now interesting to note that, for the weak solution under consideration, the left hand side of each equation of the system of conservation law (3.5.1) (written explicitly) is an element of $\mathcal{D}_{e,\Omega}$. For example, the conservation law for mass of a compressible fluid is of the form

$$\frac{\partial \rho}{\partial t} + \sum_{i=1}^{3} \frac{\partial (\rho q_i)}{\partial x_i} = 0$$

which can be written in the form

$$\langle \mathbf{e}^1, \nabla_{xt} \rangle g^1 + \langle \mathbf{e}^2, \nabla_{xt} \rangle g^2 + \langle \mathbf{e}^3, \nabla_{xt} \rangle g^3 + \langle \mathbf{e}^4, \nabla_{xt} \rangle g^4 = 0 \tag{3.5.8}$$

where

$$\mathbf{e}^1 = (1, 0, 0, 0), \ \ \mathbf{e}^2 = (0, 1, 0, 0), \ \ \mathbf{e}^3 = (0, 0, 1, 0), \ \ \mathbf{e}^4 = (0, 0, 0, 1) \tag{3.5.9}$$

and

$$g^1 = \rho q_1, \ \ g^2 = \rho q_2, \ \ g^3 = \rho q_3 \text{ and } g^4 = \rho \tag{3.5.10}$$

Thus the system of conservation laws (3.5.1) consists of a system of equations of the form

$$h(\mathbf{x}, t) \equiv \sum_{i=1}^{m} < \mathbf{e}^i, \ \nabla_{xt} > g^i = 0 \tag{3.5.11}$$

where $\mathbf{e}^i$ are constant vectors and $g^i \in \mathcal{R}_\Omega$, $i = 1, 2, \ldots, m$. Let H be the domain in space-time which corresponds to $s > 0$, i.e., the region behind the shock and let $\nu(\mathbf{x},t)\mid_\Omega$ be the unit normal to the surface Ω directed into the domain H; then

$$\nu(\mathbf{x},t)\Big|_\Omega = \frac{\nabla_{xt}s}{|\nabla_{xt}s|}\Big|_\Omega \tag{3.5.12}$$

Let us denote the shock front in the physical space (i.e., (x_1,x_2,x_3)-space) by Ω_t; then Ω_t is represented by $s(\mathbf{x},t)=0$ with t as a parameter. Let $C|_\Omega$ be the velocity of propagation of the shockfront and $\mathbf{N}|_\Omega$ be the unit normal to the shock front, then

$$C\Big|_\Omega = \mp\frac{s_t}{|\nabla s|}\Big|_\Omega, \qquad \mathbf{N}\Big|_\Omega = \pm\frac{\nabla s}{|\nabla s|}\Big|_\Omega \tag{3.5.13}$$

where the upper or lower sign is to be taken according to the choice of s.

Theorem 3.5.1 If the functions

$$g^i(\mathbf{x},t) = g_0^i(\mathbf{x},t)+H(s)g_1^i(\mathbf{x},t), \;\; i=1,2,\ldots,m$$

belonging to $\mathcal{R}_\Omega$ satisfy the conservation law (3.5.11), then the jumps $g_1^i|_\Omega$ in g^i across Ω satisfy

$$\sum_{i=1}^{m} g_1^i(\mathbf{x},t) < \mathbf{e}^i, \nu(x,t) > |_\Omega = 0 \tag{3.5.14}$$

and the generalized function $h(\mathbf{x},t)$ is representable in the form

$$h(\mathbf{x},t) = h_0(\mathbf{x},t)+H(s)h_1(\mathbf{x},t) \tag{3.5.15}$$

Proof Let $\varphi:\mathbf{R}^4\to\mathbf{R}$ belong to $C_0^\infty(\mathbf{R}^4)$. If the equation (3.5.11) is satisfied in the sense of distribution, then

$$\begin{aligned}
0 &= \int_{\mathbf{R}^4} h(\mathbf{x},t)\varphi(\mathbf{x},t)d\mathbf{x}dt \\
&= \sum_{i=1}^{m}\int_{\mathbf{R}^4} \varphi\langle\mathbf{e}^i,\nabla_{xt}\rangle g^i d\mathbf{x}dt \\
&= -\sum_{i=1}^{m}\int g^i\langle\mathbf{e}^i,\nabla_{xt}\rangle\varphi d\mathbf{x}dt, \quad \text{from definition of a weak solution}
\end{aligned}$$

$$= -\sum_{i=1}^{m}\Bigg[\int_{\mathbf{R}^4} g_0^i\langle \mathbf{e}^i, \nabla_{xt}\rangle \varphi d\mathbf{x}dt$$

$$+ \int_{\mathbf{R}^4} H(s(x,t)) g_1^i \langle \mathbf{e}^i, \nabla_{xt}\rangle \varphi d\mathbf{x}dt\Bigg]$$

Using Green's theorem to both terms (and noting that φ vanishes outside a closed bounded set and $H(s) = 0$ outside the domain $s > 0$), we get

$$\begin{aligned} 0 = & - \sum_{i=1}^{m}\Bigg[-\int_{\mathbf{R}^4} \varphi\langle \mathbf{e}^i, \nabla_{xt}\rangle g_0^i d\mathbf{x}dt \\ & - \int_{\mathbf{R}^4} \varphi H(s)\langle \mathbf{e}^i, \nabla_{xt}\rangle g_1^i d\mathbf{x}dt \\ & - \int_{\Omega} g_1^i \varphi \langle \mathbf{e}^i, \nu(\mathbf{x},t)\rangle d\mathbf{x}dt\Bigg] \end{aligned}$$

where ν is the unit normal to Ω directed into the domain $s > 0$. Then

$$\begin{aligned} 0 = & \int_{\Omega} \varphi \sum_{i=1}^{m}\Big\{ g_1^i \langle \mathbf{e}^i, \nu(\mathbf{x},t)\rangle \Big\} d\mathbf{x}dt \\ & + \int_{\mathbf{R}^4} \varphi \Big\{ \sum_{i=1}^{m} \langle \mathbf{e}^i, \nabla_{xt}\rangle g_0^i + H(s)\sum_{i=1}^{m}\langle \mathbf{e}^i, \nabla_{xt}\rangle g_1^i \Big\} d\mathbf{x}dt \end{aligned} \quad (3.5.16)$$

Since φ is an arbitrary function, it follows that

$$\sum_{i=1}^{m} g_1^i < \mathbf{e}^i, \nu(\mathbf{x},t) > \; = 0 \quad \text{on} \quad \Omega$$

which is the result (3.5.14). Now it follows that

$$\begin{aligned} & \int_{\mathbf{R}^4} h\ (\mathbf{x},t)\ \varphi(\mathbf{x},t) d\mathbf{x}dt \\ & \quad = \int_{\mathbf{R}^4} \varphi \Big\{ \sum_{i=1}^{m} < \mathbf{e}^i, \nabla_{xt} > g_0^i \\ & \quad + \; H(s)\sum_{i=1}^{m} < \mathbf{e}^i, \nabla_{xt} > g_1^i \Big\} d\mathbf{x}dt \end{aligned} \quad (3.5.17)$$

The equality is true for all test functions φ, hence we get

$$h(\mathbf{x},t) = h_0(\mathbf{x},t) + H(s)h_1(\mathbf{x},t) \tag{3.5.18}$$

where

$$h_0(\mathbf{x},t) = \sum_{i=1}^{m} \langle \mathbf{e}^i, \nabla_{xt}\rangle g_0^i\ , \quad h_1(\mathbf{x},t) = \sum_{i=1}^{m} \langle \mathbf{e}^i, \nabla_{xt}\rangle g_1^i \tag{3.5.19}$$

This completes the proof of the theorem.

Equation (3.5.14) is the well-known jump condition across a shock. A more familiar form of it is obtained by replacing the components of ν by $\mathbf{N}|_\Omega$ and $-C|_\Omega$ with the help of (3.5.13). This gives

$$-\left\{\sum_{i=1}^{m} g_1^i e_4^i C\right\}|_\Omega + \left\{\sum_{i=1}^{m} g_1^i (e_1^i N_1 + e_2^i N_2 + e_3^i N_3)\right\}|_\Omega = 0. \tag{3.5.20}$$

For the conservation law of mass in gas dynamics, i.e., for the equation (3.5.8), the relation (3.5.20) becomes

$$-\rho_1 C|_\Omega + \rho_1 q_{11} N_1 + \rho_1 q_{21} N_2 + \rho_1 q_{31} N_3 = 0 \tag{3.5.21}$$

Let us now write the jump relations (or RH conditions) for the vector conservation law (3.5.1) in a more familiar form

$$-C|_\Omega[\mathbf{H}(\mathbf{u})] + \sum_{i=1}^{m} N_i[\mathbf{F}(\mathbf{u})] = 0 \tag{3.5.22}$$

For a system of conservation laws in one spatial dimension

$$(\mathbf{H}(u))_t + (\mathbf{F}(u))_x = 0 \tag{3.5.23}$$

the RH conditions become

$$-S[\mathbf{H}] + [\mathbf{F}] = 0 \tag{3.5.24}$$

where $S = \dot{\mathbf{x}}(t)$ is the shock velocity in +ve x direction. This is an extension of the relation (1.3.6) for a single conservation law.

Chapter 4

Weakly nonlinear ray theory (WNLRT): derivation

4.1 A historical account

Extending Fermat's principle, we derived in section 3.2.7 the nonlinear ray equations and then using the conservation of energy along a ray tube, we derived the transport equation for the amplitude along the nonlinear rays for an isotropic wave propagation. This lead to a set of coupled equations of a weakly nonlinear ray theory for such a system. Extension of the theory to a more general system is quite difficult.

In this chapter we shall discuss derivation of the transport equation for the amplitude of a high frequency wave for a general hyperbolic system of quasilinear equations in multi-dimensions assuming amplitude to vary continuously (in fact as smoothly as required in the analysis) on a curved pulse which is characterized by a one parameter family of wavefronts. We shall take up three different derivations.

The first derivation (Choquet-Bruhat (1969)) in section 4.2 is formal and quite elegant but uses linear rays and hence is valid only over a *small* distance of propagation since these rays may deviate significantly from the exact rays of the nonlinear system. The second and the third derivations give the same transport equation as Choquet-Bruhat's theory but now along nonlinear rays so that the results obtained from it are valid over *very long* distances of propagation. The second derivation (Prasad (1975, 1994)) is simple and uses beautiful geometrical ideas following Gubkin (1958). The third derivation also credited to Prasad (2000) is formal and complex but helps in analysing clearly the order of magnitude of different terms in the approximate equations. It is also important because an elegant and computationally efficient shock ray theory can be derived from it. Derivation of shock ray equations and a new theory of shock dynamics are the subject matter in the last two chapters of this monograph.

A few words on the historical account of attempts on the calculation of the amplitude of a wave under high frequency approximation are in order here. Amplitude amplification due to convergence of rays (see also sections 2.2.4, 3.4.1 - 2) have been known for a long time. A formal derivation of this for the wave equation was given by Sommerfeld and Runge in (1911). Keller (1954) attempted to deal with the nonlinear effects and the geometrical effect together in order to discuss propagation of a curved weak shock. While following the propagation of weakly nonlinear waves along linear rays, Whitham (1956) proposed a nonlinearization technique following Lighthill (1949) which took into account the nonlinear deformation along linear rays of a curved pulse due to amplitude dependence of the speed of propagation in the direction of rays. This technique was used to calculate the sonic boom signature – an important application for which no other method was available. Almost simultaneously Whitham (1957, 1959) developed intuitive arguments for shock dynamics which gave finite amplitude of the shock in the region where caustic existed; the amplitude remained of the same order as that on the wavefront away from the caustic region. However, efforts to develop mathematical theories to understand the solution of linear equations in the caustic region continued (Buchal and Keller (1960) and Ludwig (1966)). In this attempt the amplitude in the next approximation did change from an infinite value to a finite but still remained too large to be realistic for small amplitude assumption to be valid for equations with genuine nonlinearity. A formal and systematic expansion procedure to deal with this type of nonlinearity in a general hyperbolic system was given by Choquet-Bruhat (1969) and independently by Parker (1969, 1971) who included many other physical processes in this discussion. Choquet-Bruhat's and Parker's

perturbation scheme leads to an eikonal equation (the eikonal equation turns, out to be the same as the characteristic equation for a hyperbolic system) which is independent of the wave amplitude w and therefore, gives a transport equation for the amplitude along the linear rays. The nonlinearity is taken into account by stretching the linear rays in the longitudinal direction due to dependence of the ray velocity on the wave amplitude. Thus, in essence, this theory is exactly the same as Whitham's nonlinearization technique (Whitham's shock dynamics is a very different theory) and we may refer to Whitham's, Choquet-Bruhat's and Parker's work by a short name – CPW theory.
The CPW theory does not satisfy Fermat's principle because when we apply the CPW theory to the nonlinear waves in a polytropic gas, the right hand side of (3.2.53) is absent. Alternative derivations of a generalization of (3.2.53), given in sections 4.2 and 4.3 in this chapter, require special perturbation schemes in which the amplitude of the wave appears in the eikonal equation itself (Prasad (1975, 1994, 2000)). This leads to a system of ray equations coupled with the transport equation. The equation (3.2.52) is exactly the same as that in CPW theory and implies the already mentioned effect of nonlinear longitudinal stretching of rays but the equation (3.2.53), derived from Fermat's principle, adds a new dimension to the propagation of a curved nonlinear wavefront. The equation (3.2.53) implies that the rays turn their direction or the wavefront rotates due to nonuniform distribution of the amplitude on the wavefront. This leads to a significant deviation of the nonlinear rays from linear rays over a length (and also time) scale on which the new perturbation scheme is valid. Extensive numerical results for converging nonlinear wavefronts show that the assumptions (high frequency and small amplitude) remain valid on distances much larger than the distance of the arête (i.e., the caustic region) from the initial wavefront. The two effects, elongation of rays contained in the equation (3.2.52) and the deviation of the rays from linear rays contained in the equation (3.2.53) are jointly responsible for the resolution of the caustic (Ravindran and Prasad (1985)) and the formation of kinks introduced in the section 3.3. This implies that this new weakly nonlinear ray theory (which we shall denote by WNLRT) gives a topologically different shape of a wavefront from that of the linear theory.

The difference between CPW theory and WNLRT disappears

when wave propagation in one space dimension is considered. In this case, rays are in the direction of one-dimensional propagation, say the x-axis and hence the question of lateral deviation of the rays from linear rays does not arise. The only affect of nonlinearity is the longitudinal stretching of the rays. We should have discussed the CPW type of perturbation scheme in one-space-dimension first but it is too simple, it will be taken up briefly in section 5.2, where it will be needed for application to study the stability of one-dimensional steady transonic flows. It was extensively used by Tanuity and his collaborators, who called it the *reductive perturbation method*, to derive KdV type of equations (Gȧrdner and Morikawa (1960), Tanuity and Wei (1968) and Tanuity (1974)).

With these brief remarks we now proceed to a description of Choquet-Bruhat's theory in the next section mainly to analyze the reasons which explain why the expansion involved in it fails to capture the correct nonlinear rays.

4.2 Derivation of CPW theory

We consider a system of first order quasilinear equations

$$A(\mathbf{u}, \mathbf{x}, t)\mathbf{u}_t + B^{(\alpha)}(\mathbf{u}, \mathbf{x}, t)\mathbf{u}_{x_\alpha} + \mathbf{C}(\mathbf{u}, \mathbf{x}, t) = 0 \tag{4.2.1}$$

where $\mathbf{u} \in I\!R^n$, $A \in I\!R^{n\times n}$, $B^{(\alpha)} \in I\!R^{n\times n}$ and $\mathbf{C} \in I\!R^n$. In this section, we do not assume that the system is hyperbolic. We assume that the system has one real simple eigenvalue $c(\mathbf{u}, \mathbf{x}, t)$, for all values of its arguments, which satisfies the usual characteristic equation

$$\det\left(n_\alpha B^{(\alpha)} - cA\right) = 0 \tag{4.2.2}$$

for $\mathbf{n} \in I\!R^m$.

We present here the simplest form of a derivation of the transport equation for an amplitude of a weakly nonlinear wave in high frequency approximation. This simple derivation which looks most natural, is important because it shows that we need to develop a special perturbation scheme to incorporate the first order amplitude correction in the eikonal equation or characteristic partial differential equation for the phase function.

We consider a small amplitude wave on a basic state $\mathbf{u}_0(\mathbf{x}, t)$ so that $\mathbf{u}_0$ satisfies the equation (4.2.1). Then we redefine a new function $\mathbf{v} = \mathbf{u} - \mathbf{u}_0$ so that $\mathbf{v} = 0$ is solution of a new first order quasilinear system $\mathbf{v} = 0$. Thus, without a loss of generality, we may assume $\mathbf{u} = 0$ to be a solution of (4.2.1) satisfying

$$\mathbf{C}(0, \mathbf{x}, t) = 0 \tag{4.2.3}$$

Let $0 < \epsilon << 1$ be a quantity of the order of the amplitude of the waves running into a state described by $\mathbf{u} = 0$. We assume

$$\mathbf{u} = \epsilon\, \mathbf{u}_1(\mathbf{x}, t, \theta^*) + \epsilon^2 \mathbf{u}_2(\mathbf{x}, t, \theta^*) + \ldots \tag{4.2.4}$$

where

$$\theta^* = \phi^*(\mathbf{x}, t)/\epsilon \tag{4.2.5}$$

Here ϕ^* is the phase function of the wave with $\phi^* = 0$ being one of the one parameter family of wavefronts under consideration. The significance of a superscript $*$ on ϕ and θ will become clear in section 4.4.3. Under the high frequency approximation in an ϵ-neighbourhood of the wavefront $\phi^* = 0$, ϕ^* is small and of the order ϵ. For these waves θ^* is of the order of unity. We further assume that the components of $\nabla\phi^*$ and ϕ^*_t are of order of one.

We substitute (4.2.4) in (4.2.1), expand all terms in powers of ϵ and equate various powers of ϵ on the left hand side equal to zero to get

$$O(1) \text{ terms}: \quad (A_*\phi^*_t + B^{(\alpha)}_* \phi^*_{x_\alpha})\mathbf{u}_{1\theta} = 0 \tag{4.2.6}$$

$$\begin{aligned} O(\epsilon) \text{ terms}: \quad & (A_*\phi^*_t + B^{(\alpha)}_* \phi^*_{x_\alpha})\mathbf{u}_{2\theta} \\ & + (A_*\mathbf{u}_{1t} + B^{(\alpha)}_* \mathbf{u}_{1x_\alpha} + (\nabla_u \mathbf{C})_* \mathbf{u}_1) \\ & + \langle \{\phi^*_t (\nabla_u A)_* \cdot \mathbf{u}_1 + \phi^*_{x_\alpha} (\nabla_u B^{(\alpha)})_* \cdot \mathbf{u}_1\},\ \mathbf{u}_{1\theta}\rangle = 0 \end{aligned} \tag{4.2.7}$$

where a subscript $*$ on a quantity represents its value at $\mathbf{u} = 0$ e.g.,

$$A_* = A(0, \mathbf{x}, t) \tag{4.2.8}$$

In order that (4.2.6) gives a non-zero solution for $\mathbf{u}_{1\theta}$, the phase function satisfies the *linearized* eikonal or characteristic partial differential equation

$$\det(A_*\phi^*_t + B^{(\alpha)}_* \phi^*_{x_\alpha}) = 0 \tag{4.2.9}$$

Since we have assumed that the system (4.2.1) has a real and simple eigenvalue $c(\mathbf{u}, \mathbf{x}, t)$, such a phase function ϕ^* exists and

$$c_* \equiv c(0, x, t) = -\phi_t^*/|\nabla\phi^*| \tag{4.2.10}$$

c^* satisfies

$$\det\left(n_{\alpha *}B_*^{(\alpha)} - c_* A_*\right) = 0 \tag{4.2.11}$$

with unit normal $\mathbf{n}_*$ of the *linearized* wavefronts given by the *linearized* phase function ϕ^* (see equation (2.4.5)) obtained in the basic state given by $\mathbf{u} = \mathbf{0}$. In terms of the left and right eigenvectors $\mathbf{l}_*$ and $\mathbf{r}_*$, the *linear* rays of the characteristic field c_* are given by (see equations (2.4.6 - 7)).

$$\frac{d\mathbf{x}_{\alpha *}}{dt} = \frac{\mathbf{l}_* B_*^{(\alpha)}\mathbf{r}_*}{\mathbf{l}_* A_* \mathbf{r}_*} \tag{4.2.12}$$

and

$$\frac{dn_{\alpha *}}{dt} = -\frac{1}{\mathbf{l}_* A_* \mathbf{r}_*}\mathbf{l}_*\left\{n_{\beta *}\left(-c_*\frac{\partial A_*}{\partial \eta_\beta^\alpha} + n_{\gamma *}\frac{\partial B_*^{(\gamma)}}{\partial \eta_\beta^\alpha}\right)\right\}\mathbf{r}_* = \Psi_{\alpha *} \tag{4.2.13}$$

A solution of (4.2.6) vanishing outside a small neighbourhood of $\phi_* = 0$ is given by

$$\mathbf{u}_1 = \tilde{w}(\mathbf{x}, t, \theta)\mathbf{r}_* \ , \quad \lim_{\theta\to\pm\infty} \tilde{w} = 0 \tag{4.2.14}$$

Premultiplying (4.2.7) by $\mathbf{l}_*$ so that the first term vanishes and substituting (4.2.14) we get

$$\frac{d^*\tilde{w}}{dt^*} + \mathcal{G}_*\tilde{w}\tilde{w}_{\theta^*} + \Omega_*\tilde{w} = 0 \tag{4.2.15}$$

where

$$\frac{d^*}{dt^*} = \frac{\partial}{\partial t} + \chi_{\alpha *}\frac{\partial}{\partial x_\alpha} \ , \quad \chi_{\alpha *} = \frac{\mathbf{l}_* B_*^{(\alpha)}\mathbf{r}_*}{\mathbf{l}_* A_* \mathbf{r}_*} \tag{4.2.16}$$

$$\mathcal{G}_* = \frac{1}{\mathbf{l}_* A_* \mathbf{r}_*}\mathbf{l}_*\left\{\phi_t^*(\nabla_u A)_* \cdot \mathbf{r}_* + \phi_{x_\alpha}^*(\nabla_u B^{(\alpha)})_* \cdot \mathbf{r}_*\right\}\mathbf{r}_* \tag{4.2.17}$$

$$\Omega_* = \frac{1}{\mathbf{l}_* A_* \mathbf{r}_*}\mathbf{l}_*\left\{A_*\frac{\partial \mathbf{r}_*}{\partial t} + B_*^{(\alpha)}\frac{\partial \mathbf{r}_*}{\partial x_\alpha}\right\} + \{\mathbf{l}_*(\nabla_u \mathbf{C})_* \cdot \mathbf{r}_*\} / \{(\mathbf{l}_* A_* \mathbf{r}_*)\} \tag{4.2.18}$$

The operator $\frac{d^*}{dt^*}$ represents the time rate of change along the linear rays.

(4.2.15) is the final form of the transport equation for the amplitude $\tilde{w}$ of the wave in CPW theory along linear rays given by (4.2.12 - 13). The linear rays are determined from the base state given by $\mathbf{u} = \mathbf{0}$ and do not depend on the solution $\tilde{w}$ representing the nonlinear wave. Once these rays have been drawn, $\mathcal{G}_*$ and Ω_* can be evaluated along a ray as functions of t. Given initial distribution of $\tilde{w}$ as a function θ^*, the equation (4.2.15) can be integrated along different rays starting from the various points of a pulse designated by different values of θ^*. In this way, the signature of a pulse can be determined as was done in the sonic boom problem by Rao (1956). The pulse deforms in the direction of a ray due to genuine nonlinearity present in the second term in (4.2.15) derived from the genuine nonlinearity of the characteristic field under consideration. Note that we did not make the assumption that the system (4.2.1) is hyperbolic. The linear ray geometry may have singularities like caustic where Ω_* tends to infinity. The amplitude of the wave will also tend to infinity at such points due to convergence of the rays. Thus, much before the geometric singularities of the linear rays appear, the CPW theory breaks down and we need a new theory to discuss the problem. This forms the aim of the discussion in the rest of this chapter.

4.3 A geometric derivation of (WNLRT)

CPW theory is also a weakly nonlinear ray theory, in which a quasilinear transport equation for the wave amplitude was derived along linear rays. By WNLRT we mean a nonlinear ray theory in which the transport equation is derived along the nonlinear rays. In this section we shall give a geometric derivation of WNLRT, which is due to Prasad (1975) and was inspired by the work of Gubkin (1958). As far as we know, this is the simplest and most beautiful derivation of the WNLRT.

In this section we assume for simplicity that the matrices $A, B^{(\alpha)}$ and the vector $\mathbf{C}$ are functions of $\mathbf{u} \in \mathbb{R}^n$ and $\mathbf{x} \in \mathbb{R}^m$ only. Unlike the case in the last section, the geometric derivation depends on the assumption that the system is hyperbolic. Hence, we consider a hyperbolic system of n quasilinear equations in the form

$$A(\mathbf{u}, \mathbf{x})\mathbf{u}_t + B^{(\alpha)}(\mathbf{u}, \mathbf{x})\mathbf{u}_{x_\alpha} + \mathbf{C}(\mathbf{u}, \mathbf{x}) = 0 \tag{4.3.1}$$

We use the usual summation convention for a repeated suffix except when the subscript or superscript is L and M. As pointed out in the last section, we could take the undisturbed basic solution $\mathbf{u}_0 = 0$, however we proceed in this section with $\mathbf{u}_0(\mathbf{x}) \neq 0$ satisfying

$$B^{(\alpha)}(\mathbf{u}_0, \mathbf{x})\mathbf{u}_{0x_\alpha} + \mathbf{C}(\mathbf{u}_0, \mathbf{x}) = 0 \tag{4.3.2}$$

4.3.1 WNLRT for a hyperbolic system

Since the system 4.3.1 is hyperbolic with t as a time-like variable, i.e., for an arbitrary set of real numbers $\{n_\alpha\}$, there are n real characteristic roots c_i (not necessarily distinct) of the equation

$$\det[n_\alpha B^{(\alpha)} - \lambda A] = 0 \tag{4.3.3}$$

and there exist n linearly independent left eigenvectors

$$\mathbf{l}^{(k)} \equiv (l_1^{(k)}, l_2^{(k)}, \ldots, l_n^{(k)})$$

which imply existence of n linearly independent right eigenvectors

$$\mathbf{r}^{(k)} = (r_1^{(k)}, r_2^{(k)}, \ldots, r_n^{(k)})^T$$

satisfying

$$\mathbf{l}^{(L)} n_\alpha B^{(\alpha)} = c_L \mathbf{l}^{(L)} A, n_\alpha B^{(\alpha)}\mathbf{r}^{(L)} = c_L A\mathbf{r}^{(L)}, L = 1, 2, \ldots, n \tag{4.3.4}$$

If the equation of the characteristic surface Ω corresponding to the characteristic velocity c_k be denoted by $\phi^{(k)}(\mathbf{x}, t) =$ constant, the derivatives $\nabla\phi^{(k)}$ and $\phi_t^{(k)}$ are proportional to the unit normal $\mathbf{n}$ of the wavefront Ω_t and $-c_k$.

Let us introduce a new set of $m+1$ independent variables $(x'_\alpha, \phi^{(L)})$, where

$$\phi^{(L)} = \phi^{(L)}(x_\alpha, t), \quad x'_\alpha = x_\alpha, \tag{4.3.5}$$

L being a fixed number taken from the set $(1, 2, \ldots, n)$. The system reduces to

$$\left(A\phi_t^{(L)} + B^{(\alpha)}\phi_{x_\alpha}^{(L)}\right)\frac{\partial \mathbf{u}}{\partial \phi^{(L)}} + B^{(\alpha)}\frac{\partial \mathbf{u}}{\partial x'_\alpha} + \mathbf{C}_i = 0 \tag{4.3.6}$$

Note that $\mathbf{l}^{(L)}(A\phi_t^{(L)} + B^{(\alpha)}\phi_{x_\alpha}^{(L)}) = 0$, so premultiplying this result by $\mathbf{l}^{(L)}$ we get

$$\mathbf{l}^{(L)}B^{(\alpha)}\frac{\partial \mathbf{u}}{\partial x'_\alpha} + \mathbf{l}^{(L)}\mathbf{C} = 0 \tag{4.3.7}$$

which is the compatibility condition along the characteristic surface. $\partial/\partial x'_\alpha$ is a tangential derivative in the characteristic surface $\phi^{(L)} =$ constant and is given by

$$\frac{\partial}{\partial x'_\alpha} = \frac{\partial}{\partial x_\alpha} + \frac{n_\alpha}{c_L}\frac{\partial}{\partial t} \tag{4.3.8}$$

If the equation of a bicharacteristic curve lying in the characteristic surface $\phi^{(L)}(\mathbf{x}, t) =$ constant be written as

$$\mathbf{x} = \mathbf{x}(\sigma_L), \quad t = t(\sigma_L) \tag{4.3.9}$$

from the lemma on bicharacteristic (section 2.4) we can suitably choose σ_L such that

$$\frac{d\mathbf{x}}{d\sigma_L} = \mathbf{l}^{(L)}B^{(\alpha)}\mathbf{r}^{(L)}, \quad \frac{dt}{d\sigma_L} = \mathbf{l}^{(L)}A\mathbf{r}^{(L)} \tag{4.3.10}$$

The directional derivative in the direction of the bichatacteristic curve is given by

$$\frac{d}{d\sigma_L} \equiv \frac{dt}{d\sigma_L}\frac{\partial}{\partial t} + \frac{dx_\alpha}{d\sigma_L}\frac{\partial}{\partial x_\alpha} = (\mathbf{l}^{(L)}B^{(\alpha)}\mathbf{r}^{(L)})\frac{\partial}{\partial x'_\alpha} \tag{4.3.11}$$

We consider a perturbation

$$\mathbf{v} = \mathbf{u} - \mathbf{u}_0 \tag{4.3.12}$$

on the given steady state such that the amplitude $|\mathbf{v}|$ is of the order of a small quantity δ. From (4.3.7) we get

$$\mathbf{l}^{(L)}B^{(\alpha)}\frac{\partial \mathbf{v}}{\partial x'_\alpha} + \bar{F}^{(L)} = 0 \tag{4.3.13}$$

where

$$\bar{F}^{(L)} = \mathbf{l}^{(L)}\mathbf{C} + \mathbf{l}^{(L)}B^{(\alpha)}\frac{\partial \mathbf{u}_0}{\partial x_\alpha} \tag{4.3.14}$$

Expanding the functions $\mathbf{C}_i$ and $B^{(\alpha)}$ about the steady solution $\mathbf{u}_0$ we get

$$\mathbf{C} + B^{(\alpha)}\frac{\partial \mathbf{u}_0}{\partial x_\alpha} = F_0\mathbf{v} + O(\delta^2) \tag{4.3.15}$$

where the matrix F_0 is defined by

$$F_0\mathbf{v} = (\nabla_u \mathbf{C})_0 \cdot \mathbf{v} + ((\nabla_u B^{(\alpha)})_0 \cdot \mathbf{v})\frac{\partial \mathbf{u}_0}{\partial x_\alpha} \tag{4.3.16}$$

and the subscript 0 on a quantity represents its value evaluated in solution $\mathbf{u}_0$ for example

$$B_0^{(\alpha)} = B^{(\alpha)}(\mathbf{x}, \mathbf{u}_0(\mathbf{x})), \quad \mathbf{l}_0 = \mathbf{l}(\mathbf{x}, \mathbf{u}_0, \mathbf{n}) \text{ etc.} \tag{4.3.17}$$

Since $\mathbf{n}$ is the unit normal of the exact nonlinear wavefront, $\mathbf{l}_0$ depends also on the perturbation. From (4.3.14) we get

$$\bar{F}^{(L)} = \mathbf{l}_0^{(L)}(F_0\mathbf{v}) + O(\delta^2) \tag{4.3.18}$$

Since the set $\{\mathbf{r}^{(1)}, \mathbf{r}^{(2)}, \ldots, \mathbf{r}^{(n)}\}$ of right eigenvectors is linearly independent we can replace $\mathbf{v}$ by a new dependent variables $\mathbf{w} = (w_1, w_2, \ldots, w_n)^T$ through the transformation (see (2.1.23))

$$\mathbf{v} = \mathbf{r}^{(k)} w_k = R\,\mathbf{w}, \text{ say} \tag{4.3.19}$$

The equations (4.3.13) become

$$\mathbf{l}^{(L)} B^{(\alpha)} \mathbf{r}^{(k)} \frac{\partial w_k}{\partial x'_\alpha} + \mathbf{l}^{(L)} B^{(\alpha)} \frac{\partial \mathbf{r}^{(k)}}{\partial x'_\alpha} w_k + F^{(L)} = 0, \quad L = 1, 2, \ldots, n \tag{4.3.20}$$

where

$$F^{(L)} = \mathbf{l}^{(L)}(F_0 \mathbf{r}^{(k)}) w_k \tag{4.3.21}$$

and each of $w_1, w_2, \ldots, w_n$ are at most of the order of δ. Using (4.3.11), we write (4.3.20) in the form

$$\frac{dw_L}{d\sigma_L} + \sum_{k \neq L} \mathbf{l}^{(L)} B^{(\alpha)} \mathbf{r}^{(k)} \frac{\partial w_k}{\partial x'_\alpha} + \mathbf{l}^{(L)} B^{(\alpha)} \frac{\partial \mathbf{r}^{(k)}}{\partial x'_\alpha} w_k + F^{(L)} = 0,$$
$$L = 1, 2, \ldots, n. \tag{4.3.22}$$

If we create an arbitrary disturbance on a given steady solution $\mathbf{u}_0(x_\alpha)$, in general, the disturbance will break into n modes propagating with the characteristic velocities $c_1, c_2, \ldots, c_n$. The locus Ω in space-time of the wavefront Ω_t of the part of the disturbance moving with the velocity c_k will be a member of the family of characteristic surfaces $\phi^{(k)}$ = constants. However, we consider here only those disturbances which consist of only a single mode, i.e., in which the disturbance stays in the neighbourhood of the characteristic surface $\phi^{(M)}(x_\alpha, t) = 0$ where M is a fixed integer from $1, 2, \ldots, n$. This means that we make a high frequency or short wave assumption, i.e., that the disturbance is localized in the neighbourhood of the wavefront so that it is non-zero only over a distance of the order of ϵ from the wavefront where ϵ is small compared to the radius of curvature R of the wavefront and compared to the characteristic length H in the steady state over which $\mathbf{u}_0$ varies significantly. We further assume, for simplicity, that the characteristic $\phi^{(M)} = 0$ is simple (we can easily extend the theory when $\phi^{(M)} = 0$ is a multiple characteristic, see Bhatnagar and Prasad, 1971). We shall show now that each $w_L(L \neq M)$ is small compared to w_M.

We consider (4.3.22) for $L \neq M$. Then the bicharacteristic curve along which σ_L varies remains in the disturbed region only over a distance of the order of ϵ and integrating (4.3.22) along σ_L we get

$$w_L = O\left(\frac{\epsilon\delta}{R}\right) + O\left(\frac{\epsilon\delta}{H}\right) + O(\delta\epsilon), \quad L \neq M \tag{4.3.23}$$

Substituting (4.3.23) in (4.3.22) for $L = M$ and neglecting all terms of orders $\epsilon\delta/R, \epsilon\delta/H$ and $\delta\epsilon$ we get

$$\frac{dw_M}{d\sigma_M} + \left\{\mathbf{l}_0^{(M)} B_0^{(\alpha)} \frac{\partial \mathbf{r}_0^{(M)}}{\partial x'_\alpha}\right\} w_M + \left\{\mathbf{l}_0^{(M)}(F_0 \mathbf{r}_0^{(M)})\right\} w_M = 0$$

(there is no sum over M)(4.3.24)

and the relation (4.3.19), to the same approximation, becomes

$$\mathbf{v} = \mathbf{r}_0^{(M)} w_M \tag{4.3.25}$$

The approximate transport equation (4.3.24) is valid whenever δ and ϵ are two small quantities. In applications it is usually taken that $\delta = \epsilon$ which we shall also choose in discussion hence forth.

We may use the operator $\frac{d}{dt} = \frac{\partial}{\partial t} + \chi^{(\alpha)} \frac{\partial}{\partial x_\alpha}$, where $\boldsymbol{\chi}$ is the bicharacteristic velocity, instead of $\frac{d}{d\sigma_M}$ and also drop the subscript M from all quantities including w, then the perturbation $\mathbf{v} = \mathbf{u} - \mathbf{u}_0$ due to waves in a simple characteristic field in short wave approximation is given by

$$\mathbf{v} = w\mathbf{r}_0 \tag{4.3.26}$$

Note that $\tilde{w}$ in the previous section is of order one but w here is of order δ. The transport equation (4.3.24) can be written as

$$(\mathbf{l}_0 A_0 \mathbf{r}_0)\frac{dw}{dt} + \left(\mathbf{l}_0 B_0^{(\alpha)} \frac{\partial \mathbf{r}_0}{\partial x'_\alpha}\right) w + (\mathbf{l}_0(F_0 \mathbf{r}_0))w = 0 \qquad (4.3.27)$$

where we have again used the result $\mathbf{l}A\mathbf{r} = \mathbf{l}_0 A_0 r_0 + O(\epsilon)$. The relation (4.3.26) shows that the perturbation $\mathbf{v}$ is proportional to the right eigenvector. Comparing this result with the result (3.1.30) we can easily see the relation between a simple wave and high frequency approximation (see also the comment after the equation (3.1.75)).

When we choose $\delta = \epsilon$, the transport equation (4.3.27) has an error of the order ϵ^2. Since we approximated the system in a neighbourhood of the exact characteristic surface or the exact nonlinear wavefront, this equation contains the derivative $\frac{d}{dt}$ along the *exact* rays and the tangential derivative $\frac{\partial}{\partial x'_\alpha}$ are along the *exact* characteristic surface Ω. We also notice that we have put a subscript 0 on $\mathbf{l}$ and $\mathbf{r}$ in equations (4.3.24 - 26) so as to represent the value of these quantities for $\mathbf{u} = \mathbf{u}_0$ but we have used the value of $\mathbf{n}$ to be that of the exact wavefront. In a nonlinear problem we have three types of wavefronts: exact wavefront $\Omega_t|_{\text{exact}}$ obtained from the equation $\det(A\phi_t + B^{(\alpha)}\phi_{x_\alpha}) = 0$ with $\mathbf{u} = \mathbf{u}(\mathbf{x}, t)$ the exact solution, approximate wavefront $\Omega_t|_{\text{appr}}$ obtained from that with $u = \mathbf{u}_0 + \mathbf{r}_0 w$ and the linear wavefront $\Omega_t|_{\text{lin}}$ for $u = \mathbf{u}_0(\mathbf{x})$. We postulate that for moderately strong nonlinear waves (i.e., weak but not so weak that $\Omega_t|_{\text{exact}}$ is very close to $\Omega_t|_{\text{lin}}$), the $\Omega_t|_{\text{exact}}$ and $\Omega_t|_{\text{appr}}$ are close but both have positions very far from $\Omega_t|_{\text{lin}}$ and geometrical shapes quite different from it. We shall give examples of calculated $\Omega_t|_{\text{appr}}$ to justify the statement in chapter 6. Experimental results of Sturtevant and Kulkarni (1976) also support this. Thus, we assume that for a moderately strong nonlinear wave, the unit normal $\mathbf{n}$ of the exact nonlinear wavefront and that of the approximate one differ by a quantity of the order of ϵ. In this case we calculate $\mathbf{n}$ appearing in $\mathbf{l}_0$ and $\mathbf{r}_0$ in (4.3.27) from the ray equations (2.4.6 - 7) by substituting $\mathbf{u} = \mathbf{u}_0 + w\mathbf{r}_0$ and retaining terms upto order ϵ:

$$\frac{d\mathbf{x}}{dt} = \chi_0 + \{(\nabla_u \chi)_0 \cdot \mathbf{r}_0\} w \qquad (4.3.28)$$

$$\frac{d\mathbf{n}}{dt} = \Psi_0 + \{(\nabla_u \Psi)_0 \cdot (\mathbf{r}_0 w)\} \qquad (4.3.29)$$

The expression on the right hand side of (4.3.29) has tangential differential operators $\partial/\partial\eta^\alpha_\beta$ operating on $\mathbf{r}_0$ and w. Detailed expressions for the terms on the right hand side of these two equations will be derived and discussed in the next subsection for waves in a polytropic gas.

Equations (4.3.28 - 29) cannot be solved alone since an additional quantity w appears in them. The system of equations (4.3.27 - 29) form a complete system of $2m$ scalar equations (since $|\mathbf{n}| = 1$, only $m-1$ of the m equations in (4.3.29) are independent), which can be uniquely solved for the amplitude w, the position $\mathbf{x}$ of the wavefront and unit normal $\mathbf{n}$, provided initial position $\mathbf{x}_0$ of the wavefront and the initial amplitude distribution on it are known. When we do so, the operators $\frac{d}{dt}$, $\frac{\partial}{\partial x'_\alpha}$, and $\frac{\partial}{\partial \eta^\alpha_\beta}$ are no longer in tangential directions of the exact characteristic surface but that of the approximate characteristic surface $\Omega|_{\text{appr}}$ which gives the weakly nonlinear wavefront $\Omega_t|_{\text{appr}}$. This is another weakly nonlinear ray theory, which we claim to be more accurate than the CPW theory discussed in section 4.2, especially in the linear caustic region. In order to distinguish this theory from CPW theory (which is also a weakly nonlinear ray theory), we denote it as WNLRT.

We are now in a position to point out the reason for failure of the CPW theory in capturing the correct nonlinear rays. Let us first note that CPW theory is not an ordinary perturbation scheme, it uses a perturbation method using multiple scales which incorporates the nonlinearization technique of Whitham. Even then, the perturbation scheme is such that the eikonal equation (4.2.9) obtained from the $O(1)$ terms is the same as the linear eikonal equation. An approximation of the system in the neighbourhood of the exact characteristic surface gives the exact eikonal equation and hence, exact ray equations. Later on, depending on the accuracy of the transport equation, we approximate the exact ray equations appropriately by equations (4.3.28 - 29). In section 4.4, we shall develop a special perturbation scheme in which the leading order terms give again to these ray equations.

4.3.2 Upstream propagating waves in a steady flow of a polytropic gas

We consider the Euler equations (2.3.16 - 18) of a polytropic gas in three-dimensions. The system has five eigenvalues given by (2.3.19). In this and the next section we are interested in upstream propagating waves since, as explained in the next chapter, these are the waves which get trapped near a sonic point and determine the stability of a transonic flow. When the angle between the fluid velocity $\mathbf{q}$ and the normal $\mathbf{n}$ of the wavefront in a pulse in short wave approximation is acute, the upstream propagating waves correspond to the first eigenvalue in (2.3.19). Therefore we take

$$c = \langle \mathbf{n}, \mathbf{q} \rangle - a \tag{4.3.30}$$

for which the ray velocity is

$$\boldsymbol{\chi} = \mathbf{q} - \mathbf{n}a \tag{4.3.31}$$

The time rate of change of $\mathbf{n}$ along a ray given by (2.4.7) reduces to

$$\frac{d\mathbf{n}}{dt} = \mathbf{L}a - n_\beta \mathbf{L} q_\beta \equiv \Psi \tag{4.3.32}$$

which may be compared with the result (2.4.25) for the eigenvalue $\langle \mathbf{n}, \mathbf{q} \rangle + a$.

The left and right eigenvectors corresponding to the eigenvalue c in (4.3.30) are

$$\mathbf{l} = \left(0, n_1, n_2, n_3, -\frac{1}{\rho a}\right), \ \mathbf{r} = \left(-\frac{\rho}{a}, n_1, n_2, n_3, -\rho a\right)^T \tag{4.3.33}$$

Following the notation of the last section, we denote the basic steady flow (on which perturbation is created) by a subscript 0. Then, the perturbation $\mathbf{v}$ given by (4.3.19) is

$$v_1 = \rho - \rho_0 = -\frac{\rho_0}{a_0} w; \ v_{\alpha+1} = q_\alpha - q_{\alpha 0} = n_\alpha w, \ \alpha = 1, 2, 3;$$
$$v_5 = p - p_0 = -\rho_0 a_0 w \tag{4.3.34}$$

Further,

$$\mathbf{l}_0 A_0 \mathbf{r}_0 = 2 \tag{4.3.35}$$

and

$$\mathbf{l}_0 B_0^{(\alpha)} \frac{\partial \mathbf{r}_0}{\partial x'_\alpha} = -a_0 \frac{\partial n_\alpha}{\partial x'_\alpha} + \frac{1}{\rho_0 a_0}(q_{\alpha 0} - a_0 n_\alpha) \frac{\partial(\rho_0 a_0)}{\partial x'_\alpha}$$

Since $\rho_0 a_0$ are independent of t, $\partial(\rho_0 a_0)/\partial x'_\alpha = \partial(\rho_0 a_0)/\partial x_\alpha$. Further,

$$\frac{\partial n_\alpha}{\partial x'_\alpha} = \left(\frac{\partial}{\partial x_\alpha} + \frac{n_\alpha}{c}\frac{\partial}{\partial t}\right) n_\alpha = \frac{\partial n_\alpha}{\partial x_\alpha} + \frac{1}{c}\frac{\partial}{\partial t}(|\mathbf{n}|^2)$$

The second term vanishes since $|n|^2 = 1$. Hence $\partial n_\alpha / \partial x'_\alpha = \partial n_\alpha / \partial x_\alpha$. Therefore,

$$\mathbf{l}_0 B_0^{(\alpha)} \frac{\partial \mathbf{r}_0}{\partial x'_\alpha} = -a_0 \langle \nabla, \mathbf{n} \rangle + \frac{1}{\rho_0 a_0}(q_{\alpha 0} - a_0 n_\alpha) \frac{\partial(\rho_0 a_0)}{\partial x_\alpha} \tag{4.3.36}$$

When we evaluate the term $\mathbf{l}_0(F_0 \mathbf{r}_0)$ for Euler equations, we get

$$\mathbf{l}_0(F_0 \mathbf{r}_0) = \gamma \langle \nabla, \mathbf{q}_0 \rangle + \langle \mathbf{n}, \langle \mathbf{n}, \nabla \rangle \mathbf{q}_0 \rangle \tag{4.3.37}$$

Using (4.3.33), the expression $a^2 = \frac{\gamma p}{\rho}$ and (4.3.34), we find that perturbation in the sound velocity a is given by

$$a - a_0 = -\frac{\gamma - 1}{2} w + O(\epsilon^2) \tag{4.3.38}$$

so that

$$\mathbf{q} - \mathbf{n} a = \mathbf{q}_0 - \mathbf{n} a_0 + \frac{\gamma + 1}{2} \mathbf{n} w + O(\epsilon^2) \tag{4.3.39}$$

The expansion for $\boldsymbol{\Psi} = \mathbf{L} a - n_\beta \mathbf{L} q_\beta$ now becomes

$$\boldsymbol{\Psi} = \mathbf{L} a_0 - n_\beta \mathbf{L} q_{\beta 0} - \frac{\gamma + 1}{2} \mathbf{L} w \tag{4.3.40}$$

Collecting all of these results we finally get the following ray equations correct up to $O(\epsilon)$ terms

$$\frac{d\mathbf{x}}{dt} = \mathbf{q}_0 - \mathbf{n} a_0 + \frac{\gamma + 1}{2} \mathbf{n} w \tag{4.3.41}$$

$$\frac{d\mathbf{n}}{dt} = \mathbf{L} a_0 - n_\beta \mathbf{L} q_{\beta 0} - \frac{\gamma + 1}{2} \mathbf{L} w \tag{4.3.42}$$

The approximate transport equation along these approximate nonlinear rays is

$$\frac{dw}{dt} = (K - a_0 \Omega) w \tag{4.3.43}$$

where

$$K = -\frac{1}{2\rho_0 a_0}(q_{\alpha 0} - a_0 n_\alpha)\frac{\partial(\rho_0 a_0)}{\partial x_\alpha} - \frac{1}{2}\{\gamma\langle\nabla, \mathbf{q}_0\rangle + \langle\mathbf{n}, \langle\mathbf{n}, \nabla\rangle\mathbf{q}_0\rangle \tag{4.3.44}$$

and Ω is the mean curvature defined by (2.2.22).

Equations (4.3.41 - 44) are the equations of WNLRT for upstream propagating waves on a steady flow of a polytropic gas.

4.4 An asymptotic derivation of WNLRT

Geometric derivation in the last section as well as application of the Fermat's principle in section 3.2.7 show that the ray equations of the WNLRT contain the amplitude of the wavefront in such a way that the nonlinear rays not only stretch in the longitudinal direction but also deviate from the linear rays due to the rotation of the nonlinear wavefront because of a non-zero gradient of the amplitude along it. To achieve this we construct in this section a formal perturbation scheme which leads for the phase function ϕ an eikonal equation of the form

$$Q(\nabla\phi, \phi_t, \mathbf{x}, t, w(\mathbf{x}, t)) = 0 \tag{4.4.1}$$

such that the leading order amplitude w (or $\tilde{w}$ used in this section) appears in the eikonal equation itself. Then we derive the transport equation for $\tilde{w}$ along the nonlinear rays given by the characteristic curves of the first order partial differential equation (4.4.1).

4.4.1 Derivation of the eikonal and transport equations

We consider a hyperbolic system of first order quasilinear partial differential equations,

$$A(\mathbf{u})\mathbf{u}_t + B^{(\alpha)}(\mathbf{u})\mathbf{u}_{\mathbf{x}_\alpha} = 0 \qquad \alpha = 1, 2, \ldots m \tag{4.4.2}$$

Here $\mathbf{x} \in R^m$ are the space variables, $\mathbf{u}(\mathbf{x}, t) \in R^n$ are the dependent variables and A and $B^{(\alpha)}(\mathbf{u})$ are smooth $n \times n$ matrix-valued functions of $\mathbf{u}$.

We look for a generalized asymptotic expansion of solutions of (4.4.2) of the following form:

$$\mathbf{u}(\mathbf{x}, t, \epsilon) = \epsilon \tilde{\mathbf{v}} \left(\mathbf{x}, t, \frac{\phi(\mathbf{x}, t, \epsilon)}{\epsilon}, \epsilon \right) \tag{4.4.3}$$

$$\tilde{\mathbf{v}}(\mathbf{x}, t, \theta, \epsilon) = \tilde{\mathbf{v}}_0(\mathbf{x}, t, \theta, \epsilon) + \epsilon \tilde{\mathbf{v}}'(\mathbf{x}, t, \theta, \epsilon), \quad \theta = \phi/\epsilon \tag{4.4.4}$$

Here ϵ is a small parameter, so this ansatz represents a small amplitude high-frequency solution. The function $\phi(\mathbf{x}, t, \epsilon) \in R$ and the functions $\tilde{\mathbf{v}}_0(\mathbf{x}, t, \theta, \epsilon)$ and $\tilde{\mathbf{v}}'(\mathbf{x}, t, \theta, \epsilon)$ will be chosen so that (4.4.3) gives an asymptotic solution of (4.4.2) as $\epsilon \to 0$. In particular, ϕ is the phase function associated with the leading order solution $\mathbf{u} = \epsilon \tilde{\mathbf{v}}_0$. When carrying out the expansion, we assume that the derivatives, ϕ_{x_α}, are of order one with respect to ϵ.

This method of expansion is similar to the Chapman-Enskog expansion. For other work in nonlinear hyperbolic waves using Chapman-Enskog expansions see Hunter (1995). The leading order solution $\tilde{\mathbf{v}}_0$ and the correction $\tilde{\mathbf{v}}'$ depend on ϵ explicitly. It is therefore not necessary to include any higher order terms in the expansion (4.4.4), since they can be absorbed into $\tilde{\mathbf{v}}'$. As a result of this explicit ϵ dependence, the solution $\tilde{\mathbf{v}}$ can be decomposed into a leading order approximation, $\tilde{\mathbf{v}}_0$ and a perturbation $\epsilon \tilde{\mathbf{v}}'$ in different ways, since terms in $\tilde{\mathbf{v}}'$ can be absorbed into $\tilde{\mathbf{v}}_0$. One way to specify the decomposition uniquely is to require that

$$\mathbf{l}_0.\tilde{\mathbf{v}}' = 0 \tag{4.4.5}$$

where the left null vector $\mathbf{l}_0$ is defined below. However, other choices are possible. For example in gas dynamics we could require that $\tilde{\mathbf{v}}'$ contains no pressure perturbations.

We now derive the asymptotic equations. We will obtain an asymptotic solution which satisfies (4.4.2) up to terms of the order ϵ^3. Higher order approximations can be derived in a similar way, although the resulting equations rapidly become very complicated. Use of (4.4.3) in (4.4.2) gives

$$\left\{ \phi_t A(\epsilon \tilde{\mathbf{v}}) + \phi_{\mathbf{x}_\alpha} B^{(\alpha)}(\epsilon \tilde{\mathbf{v}}) \right\} \tilde{\mathbf{v}}_\theta + \epsilon \left\{ A(\epsilon \tilde{\mathbf{v}}) \tilde{\mathbf{v}}_t + B^{(\alpha)}(\epsilon \tilde{\mathbf{v}}) \tilde{\mathbf{v}}_{\mathbf{x}_\alpha} \right\} = 0 \tag{4.4.6}$$

Here, $\tilde{\mathbf{v}}_\theta$ is the partial derivative of $\tilde{\mathbf{v}}$ at fixed $\mathbf{x}, t$ and $\tilde{\mathbf{v}}_t$, $\tilde{\mathbf{v}}_{\mathbf{x}_\alpha}$ are the partial derivatives at a fixed θ.

We note that if $\tilde{\mathbf{v}}(\mathbf{x},t,\theta,\epsilon)$ satisfies (4.4.6) when $\theta = \epsilon^{-1}\phi(\mathbf{x},t,\epsilon)$ (rather than for all θ, as is usually assumed in the method of multiple scales), then (4.4.3) gives a solution of the original equation (4.4.2). We are therefore free to regard any coefficient in (4.4.6) which does not contain derivatives as functions of $\mathbf{x}, t$ with θ evaluated at $\epsilon^{-1}\phi(\mathbf{x},t,\epsilon)$. Using (4.4.4) in (4.4.6) and Taylor expanding the coefficient matrices, we obtain

$$\begin{aligned}
&\left\{\phi_t A(\epsilon\tilde{\mathbf{v}}_0) + \phi_{x_\alpha} B^{(\alpha)}(\epsilon\tilde{\mathbf{v}}_0)\right\}\tilde{\mathbf{v}}_{0\theta} \\
+\ &\epsilon\left\{\left[\phi_t A(\epsilon\tilde{\mathbf{v}}_0) + \phi_{x_\alpha} B^{(\alpha)}(\epsilon\tilde{\mathbf{v}}_0)\right]\tilde{\mathbf{v}}'_\theta + A(\epsilon\tilde{\mathbf{v}}_0)\tilde{\mathbf{v}}_{0t} + B^{(\alpha)}(\epsilon\tilde{\mathbf{v}}_0)\tilde{\mathbf{v}}_{0x_\alpha}\right\} \\
+\ &\epsilon^2\left\{\left[\phi_t(\nabla_{\mathbf{u}}A)(\epsilon\tilde{\mathbf{v}}_0)\cdot\tilde{\mathbf{v}}' + \phi_{x_\alpha}(\nabla_{\mathbf{u}}B^{(\alpha)})(\epsilon\tilde{\mathbf{v}}_0)\cdot\tilde{\mathbf{v}}'\right]\tilde{\mathbf{v}}_{0\theta}\right. \\
+\ &\left. A(\epsilon\tilde{\mathbf{v}}_0)\tilde{\mathbf{v}}'_t + B^{(\alpha)}(\epsilon\tilde{\mathbf{v}}_0)\tilde{\mathbf{v}}'_{x_\alpha}\right\} = O(\epsilon^3) \qquad (4.4.7)
\end{aligned}$$

As we remarked above, $\tilde{\mathbf{v}}(\mathbf{x},t,\theta,\epsilon)$ is only required to satisfy this equation when $\theta = \frac{\phi}{\epsilon}$. We can therefore evaluate all the coefficients at this value of θ to obtain the equation

$$\begin{aligned}
&[\phi_t A_0 + \phi_{x_\alpha} B_0^{(\alpha)}]\tilde{\mathbf{v}}_{0\theta} \\
+\ &\epsilon\left\{(\phi_t A_0 + \phi_{x_\alpha} B_0^{(\alpha)})\tilde{\mathbf{v}}'_\theta + A_0\tilde{\mathbf{v}}_{0t} + B_0^{(\alpha)}\tilde{\mathbf{v}}_{0x_\alpha}\right\} \\
+\ &\epsilon^2\left[\left\{\phi_t(\nabla_{\mathbf{u}}A)_0\cdot\tilde{\mathbf{v}}' + \phi_{x_\alpha}(\nabla_{\mathbf{u}}B^{(\alpha)})_0\cdot\tilde{\mathbf{v}}'\right\}\tilde{\mathbf{v}}_{0\theta} + A_0\tilde{\mathbf{v}}'_t + B_0^{(\alpha)}\tilde{\mathbf{v}}'_{x_\alpha}\right] \\
=\ &O(\epsilon^3) \qquad (4.4.8)
\end{aligned}$$

where the subscript 0 indicates that the coefficients are evaluated at $\mathbf{u} = \epsilon\tilde{\mathbf{v}}_0(\mathbf{x},t,\ \epsilon^{-1}\phi(t,\mathbf{x},\epsilon),\epsilon)$ so that they are functions of $\mathbf{x}, t$ and ϵ. For example,

$$B_0^{(\alpha)}(\mathbf{x},t,\epsilon) = B^{(\alpha)}\left(\epsilon\tilde{\mathbf{v}}_0(\mathbf{x},t,\epsilon^{-1}\phi(\mathbf{x},t,\epsilon),\epsilon)\right) \qquad (4.4.9)$$

Note that the meaning of a quantity with subscript 0 in the section 4.3.1 (see 4.3.17) is different from that in this section.

The three terms in (4.4.8) are not completely separated as coefficients of the powers of ϵ^0, ϵ and ϵ^2 are also dependent on ϵ. The first term, which is of order ϵ^0, vanishes up to this order and, therefore, we impose that it is exactly zero i.e.,

$$\left\{\phi_t A_0 + \phi_{x_\alpha} B_0^{(\alpha)}\right\}\tilde{\mathbf{v}}_{0\theta} = 0 \qquad (4.4.10)$$

When we choose the leading term $\tilde{\mathbf{v}}_0$ in the high frequency asymptotic limit $\epsilon \to 0$ to satisfy this equation, the first term in (4.4.8) vanishes and we get a relation

$$\begin{aligned}
& \left\{\phi_t A_0 + \phi_{x_\alpha} B_0^{(\alpha)}\right\} \tilde{\mathbf{v}}'_\theta + A_0 \tilde{\mathbf{v}}_{0t} + B_0^{(\alpha)} \tilde{\mathbf{v}}_{0x_\alpha} \\
+ \quad & \epsilon \left\{ \left[\phi_t (\nabla_{\mathbf{u}} A)_0 \cdot \tilde{\mathbf{v}}' + \phi_{x_\alpha} (\nabla_{\mathbf{u}} B^{(\alpha)})_0 \cdot \tilde{\mathbf{v}}'\right] \tilde{\mathbf{v}}_{0\theta} \right. \\
+ \quad & \left. A_0 \tilde{\mathbf{v}}'_t + B_0^{(\alpha)} \tilde{\mathbf{v}}'_{x_\alpha} \right\} = O(\epsilon^2) \qquad (4.4.11)
\end{aligned}$$

between $\tilde{\mathbf{v}}_0$ and $\tilde{\mathbf{v}}'$ with error of the order ϵ^2. To obtain a nontrivial solution for $\tilde{\mathbf{v}}_0$, we then require that ϕ satisfies the eikonal equation

$$\det[\phi_t(\mathbf{x}, t, \epsilon) A_0(\mathbf{x}, t, \epsilon) + \phi_{x_\alpha}(\mathbf{x}, t, \epsilon) B_0^{(\alpha)}(\mathbf{x}, t, \epsilon)] = 0 \qquad (4.4.12)$$

We note that this eikonal equation is associated with the function $\mathbf{u} = \epsilon \tilde{\mathbf{v}}_0(\mathbf{x}, t, \epsilon^{-1}\ \phi(\mathbf{x}, t, \epsilon), \epsilon)$ and thus we are able to incorporate the leading order wave amplitude correction in the eikonal equation itself. We denote left and right null vectors associated with the phase $\phi(t, \mathbf{x}, \epsilon)$ and the perturbed state $u = \epsilon \tilde{\mathbf{v}}_0$ by $\mathbf{l}_0(\mathbf{x}, t, \epsilon)$ and $\mathbf{r}_0(\mathbf{x}, t, \epsilon)$, respectively, i.e., $\mathbf{l}_0$ and $\mathbf{r}_0$ satisfy

$$\mathbf{l}_0 . \left(\phi_t A_0 + \phi_{x_\alpha} B_0^{(\alpha)}\right) = 0 \qquad (4.4.13)$$

and

$$\left(\phi_t A_0 + \phi_{x_\alpha} B_0^{(\alpha)}\right) \mathbf{r}_0 = 0 \qquad (4.4.14)$$

Here

$$\mathbf{l}_0(\mathbf{x}, t, \epsilon) = \mathbf{l}\left(\mathbf{n}(\mathbf{x}, t, \epsilon), \epsilon \tilde{\mathbf{v}}_0\right) \qquad (4.4.15)$$

$$\mathbf{r}_0(\mathbf{x}, t, \epsilon) = \mathbf{r}\left(\mathbf{n}(\mathbf{x}, t, \epsilon), \epsilon \tilde{\mathbf{v}}_0\right) \qquad (4.4.16)$$

where

$$\mathbf{n}(\mathbf{x}, t, \epsilon) = \frac{\nabla\phi}{|\nabla\phi|}, \qquad \nabla\phi = (\phi_{x_1}, \phi_{x_2}, \dots, \phi_{x_m}) \qquad (4.4.17)$$

Also we normalize $\mathbf{l}_0$ so that

$$\mathbf{l}_0 A_0 \mathbf{r}_0 = 1 \qquad (4.4.18)$$

A solution of (4.4.10) is given by

$$\tilde{\mathbf{v}}_0(\mathbf{x}, t, \theta, \epsilon) = \tilde{w}(\mathbf{x}, t, \theta, \epsilon) \mathbf{r}_0(\mathbf{x}, t, \epsilon) \qquad (4.4.19)$$

where $\tilde{w}$ is an arbitrary scalar valued amplitude function. Taking the scalar product of (4.4.11) with the left null vector $\mathbf{l}_0$ we obtain

$$\begin{aligned} & \mathbf{l}_0\left(A_0\tilde{\mathbf{v}}_{0t} + B_0^{(\alpha)}\tilde{\mathbf{v}}_{0x_\alpha}\right) \\ + \ & \epsilon\mathbf{l}_0\left\{\left[\phi_t(\nabla_{\mathbf{u}}A)_0\cdot\tilde{\mathbf{v}}' + \phi_{x_\alpha}(\nabla_{\mathbf{u}}B^{(\alpha)})_0\cdot\tilde{\mathbf{v}}'\right]\tilde{\mathbf{v}}_{0\theta} + A_0\tilde{\mathbf{v}}'_t + B_0^{(\alpha)}\tilde{\mathbf{v}}'_{x_\alpha}\right\} \\ = \ & O(\epsilon^2) \end{aligned} \tag{4.4.20}$$

To eliminate $\tilde{\mathbf{v}}'$ from this equation, we solve (4.4.11) iteratively for $\tilde{\mathbf{v}}'$ in terms of $\tilde{\mathbf{v}}_0$. In order that the eliminant has an error of order ϵ^2 consistent with (4.4.20), we note that we need to solve $\tilde{\mathbf{v}}'$ with error of order ϵ i.e., we consider only the leading order terms in (4.4.11)

$$\left\{\phi_t A_0 + \phi_{x_\alpha}B_0^{(\alpha)}\right\}\tilde{\mathbf{v}}'_\theta + A_0\tilde{\mathbf{v}}_{0t} + B_0^{(\alpha)}\tilde{\mathbf{v}}_{0x_\alpha} = O(\epsilon) \tag{4.4.21}$$

We use (4.4.19) in (4.4.21). A solution of the resulting equation for $\tilde{\mathbf{v}}'$ is then

$$\tilde{\mathbf{v}}'(\mathbf{x},t,\theta,\epsilon) = b_t(\mathbf{x},t,\theta,\epsilon)\mathbf{s}'_0 + b_{x_\beta}(\mathbf{x},t,\theta,\epsilon)\mathbf{s}_0^{(\beta)} + b(\mathbf{x},t,\theta,\epsilon)\mathbf{s}_0 + O(\epsilon), \tag{4.4.22}$$

where b is the scalar amplitude such that

$$b_\theta = \tilde{w} \tag{4.4.23}$$

and the vectors $\mathbf{s}_0(\mathbf{x},t,\epsilon)$, $\mathbf{s}'_0(t,\mathbf{x},\epsilon)$ and $\mathbf{s}_0{}^{(\beta)}(\mathbf{x},t,\epsilon)$ satisfy

$$\begin{aligned} \left(\phi_t A_0 + \phi_{x_\alpha}B_0^{(\alpha)}\right)\mathbf{s}_0 = & -\left(A_0\mathbf{r}_{0t} + B_0^{(\alpha)}\mathbf{r}_{0x_\alpha}\right) \\ & + \left(\mathbf{l}_0(A_0\mathbf{r}_{0t} + B_0^{(\alpha)}\mathbf{r}_{0x_\alpha})\right)A_0\mathbf{r}_0, \end{aligned} \tag{4.4.24}$$

$$\left(\phi_t A_0 + \phi_{x_\alpha}B_0^{(\alpha)}\right)\mathbf{s}'_0 = -\left(A_0\mathbf{r}_0\right) + \left(\mathbf{l}_0A_0\mathbf{r}_0\right)A_0\mathbf{r}_0, \tag{4.4.25}$$

$$\left(\phi_t A_0 + \phi_{x_\alpha}B_0^{(\alpha)}\right)\mathbf{s}_0^\beta = -\left(B_0^{(\beta)}\mathbf{r}_0\right) + \left(\mathbf{l}_0B_0^{(\beta)}\mathbf{r}_0\right)A_0\mathbf{r}_0, \tag{4.4.26}$$

We again remind the reader that $\tilde{w}$ used in section 4.2 and the present section is of order 1 but w in the section 4.3 is of order $\epsilon(=\delta)$.

These equations do not have a unique solution. This is because there is some arbitrariness in how $\tilde{\mathbf{v}}$ is decomposed into $\tilde{\mathbf{v}}_0$ and $\tilde{\mathbf{v}}'$. But if we impose the condition (4.4.5) on $\tilde{\mathbf{v}}'$, then we can choose the unique solutions of (4.4.24 - 26) satisfying

$$\mathbf{l}_0\mathbf{s}_0 = \mathbf{l}_0\mathbf{s}'_0 = \mathbf{l}_0\mathbf{s}_0{}^\beta = 0 \tag{4.4.27}$$

Finally, use of (4.4.19) and (4.4.22) in (4.4.20) gives the following transport equation for $\tilde{w}$,

$$\tilde{w}_t + \chi_{\alpha 0}\tilde{w}_{x_\alpha} - \Omega\tilde{w} + \epsilon\left[\left(\Gamma^t b_t + \Gamma^\alpha b_{x_\alpha} + \Gamma b\right)\tilde{w}_\theta + W b_t + V^\alpha b_{x_\alpha} + D^{\alpha\beta} + Eb\right] = O(\epsilon^2) \quad (4.4.28)$$

Note that $D^{\alpha\beta}$ contains linear terms in the second order derivatives of b as seen below. The coefficients are functions of $(\mathbf{x}, t, \epsilon)$ given by

$$\left.\begin{aligned}
\chi_{\alpha 0} &= \mathbf{l}_0\, B_0^{(\alpha)}\mathbf{r}_0 \\
\Omega &= -\left(\mathbf{l}_0\, A_0\mathbf{r}_{0t} + \mathbf{l}_0 B_0^{(\alpha)}\mathbf{r}_{0x_\alpha}\right) \\
\Gamma &= \mathbf{l}_0\,\{\left(\phi_t(\nabla_\mathbf{u}A)_0 + \phi_{x_\alpha}(\nabla_\mathbf{u}B^{(\alpha)})_0\right)\cdot \mathbf{s}_0\}\mathbf{r}_0 \\
\Gamma^t &= \mathbf{l}_0\,\{\left(\phi_t(\nabla_\mathbf{u}A)_0 + \phi_{x_\beta}(\nabla_\mathbf{u}B^{(\beta)})_0\right)\cdot \mathbf{s}'_0\}\mathbf{r}_0 \\
\Gamma^\alpha &= \mathbf{l}_0\,\{\left(\phi_t(\nabla_\mathbf{u}A)_0 + \phi_{x_\beta}(\nabla_\mathbf{u}B^{(\beta)})_0\right)\cdot \mathbf{s}_0^{(\alpha)}\}\mathbf{r}_0 \\
W &= \mathbf{l}_0\,\left(A_0\mathbf{s}_0 + A_0\mathbf{s}'_{0t} + B_0^{(\beta)}\mathbf{s}'_{0x_\beta}\right) \\
V^\alpha &= \mathbf{l}_0\,\left(B_0^{(\alpha)}\mathbf{s}_0 + A_0\mathbf{s}_{0t}^{(\alpha)} + B_0^{(\beta)}\mathbf{s}_{0x_\beta}^{(\alpha)}\right) \\
D^{\alpha\beta} &= \mathbf{l}_0\left\{A_0\mathbf{s}'_0 b_{tt} + A_0\mathbf{s}_0^{(\beta)} b_{x_\beta t} + B_0^{(\alpha)}\mathbf{s}'_0 b_{t x_\alpha} + \right. \\
&\qquad \left. B_0^{(\alpha)}\mathbf{s}_0^{(\beta)} b_{x_\alpha x_\beta}\right\} \\
E &= \mathbf{l}_0\left(A_0\mathbf{s}_{0t} + B_0^{(\alpha)}\mathbf{s}_{0x_\alpha}\right)
\end{aligned}\right\} \quad (4.4.29)$$

The material presented in this section is the result of collaboration between J.K.Hunter and the author during the summer of 1995.

4.4.2 Ray formulation of the asymptotic equations

The eikonal equation (4.4.10) can be equivalently written in the form (see equation (2.4.10)),

$$Q \equiv \phi_t(\mathbf{l}_0 A_0\mathbf{r}_0) + \phi_{x_\alpha}(\mathbf{l}_0 B_0^{(\alpha)}\mathbf{r}_0) = 0, \quad \alpha = 1, \ldots, m \quad (4.4.30)$$

From the characteristic equations of (4.4.30) we obtain (see equations (2.4.6-7))

$$\frac{dx_\alpha}{ds} = \frac{\partial Q}{\partial \phi_{x^\alpha}} = \mathbf{l}_0 B_0^{(\alpha)} \mathbf{r}_0 = \chi_{\alpha_0} \tag{4.4.31}$$

and a differential equation for $\mathbf{n} = \nabla\phi/|\nabla\phi|$

$$\frac{dn_\alpha}{ds} = -n_\beta \mathbf{l}_0 \left(-c_0 \frac{\partial A_0}{\partial \eta_\beta^\alpha} + n_\gamma \frac{\partial B_0^\gamma}{\partial \eta_\beta^\alpha} \right) \mathbf{r}_0 \equiv \Psi_{\alpha_0}, \quad \text{say} \tag{4.4.32}$$

where $\partial/\partial\eta_\beta^\alpha$ is defined by (2.4.8) and $c_0 = \phi_t/|\nabla\phi|$. The operator

$$\frac{d}{ds} = \frac{\partial}{\partial t} + \chi_{\alpha_0} \frac{\partial}{\partial x_\alpha} \tag{4.4.33}$$

and $\partial/\partial\eta_\beta^\alpha$ appearing in (4.4.31-32), are differentiations in directions tangential to an approximate characteristic surface $\phi(\mathbf{x}, t)$ = constant in $(\mathbf{x}, t)$ space. In addition, the derivatives $\partial/\partial\eta_\beta^\alpha$ are tangential to a wavefront Ω_t : $\phi(\mathbf{x}, t)$ = constant with t = constant in $(x_1, \ldots, x_m)$- space. Because of the choice (4.4.18), d/ds represents the time-rate of change along a ray and may be denoted by the symbol d/dt. The transport equation (4.4.28) can then be written as

$$\frac{d\tilde{w}}{ds} = \Omega\tilde{w} - \epsilon \Big[\Big(\Gamma^t b_t + \Gamma^\alpha b_{x_\alpha} + \Gamma b \Big) \tilde{w}_\theta + W b_t + V^\alpha b_{x_\alpha} + D^{\alpha\beta} + Eb \Big] + O(\epsilon^2) \tag{4.4.34}$$

Equations (4.4.31 - 34) form a complete set of equations of the nonlinear ray theory with error $O(\epsilon^2)$. The amplitude $\mathbf{u} = \epsilon\mathbf{v_0} = \epsilon\tilde{w}\mathbf{r}_0$, correct up to first order in ϵ, appears in the bicharacteristic velocity $\chi_{\alpha_0}(\mathbf{x}, t, \epsilon)$ and the rate of turning Ψ_{α_0} of the rays in a complicated way.

To make these equations more tractable, we approximate $\mathbf{l}_0$ and $\mathbf{r}_0$ defined by (4.4.13 - 16) as follows. We now define $\bar{\mathbf{l}}$ and $\bar{\mathbf{r}}$ as

$$\bar{\mathbf{l}} = \mathbf{l}_0(\mathbf{n}, 0) \qquad \bar{\mathbf{r}} = \mathbf{r}_0(\mathbf{n}, 0)$$

Then

$$\mathbf{l}_0 = \bar{\mathbf{l}} + \epsilon\{(\nabla_{\mathbf{u}}\mathbf{l})_0 \cdot \tilde{\mathbf{v}}_0\} + O(\epsilon^2) = \bar{\mathbf{l}} + \epsilon\{(\nabla_{\mathbf{u}}\mathbf{l})_0 \cdot \mathbf{r}_0\}\tilde{w} + O(\epsilon^2) \tag{4.4.35}$$

where $(\nabla_{\mathbf{u}}\mathbf{l})_0$ is the value of $(\nabla_{\mathbf{u}}\mathbf{l})$ evaluated at $\mathbf{u} = 0$ keeping $\mathbf{n}$ fixed and is a notation different from that introduced by equation (4.4.9) for the use of the subscript 0. Similarly,

$$\mathbf{r}_0 = \bar{\mathbf{r}} + \epsilon\{(\nabla_{\mathbf{u}}\mathbf{r})_0 \cdot \mathbf{r}_0\}\tilde{w} + O(\epsilon^2) \qquad (4.4.36)$$

The vectors $\bar{\mathbf{l}}$ and $\bar{\mathbf{r}}$ still depend on the leading order term $\tilde{\mathbf{v}}_0$ in the solution and the nonlinear phase ϕ, through $\mathbf{n}$. Also if

$$A_* = A(\mathbf{u}=0) \ \text{ and } \ B_*^{(\alpha)} = B^{(\alpha)}(\mathbf{u}=0) \ \text{ are constant matrices} \qquad (4.4.37)$$

(see (4.2.8)) then we have

$$A_0 = A_* + \epsilon((\nabla_u A)_* \cdot \mathbf{r}_0)\tilde{w} + O(\epsilon^2) \qquad (4.4.38)$$

and

$$B_0^{(\alpha)} = B_*^{(\alpha)} + \epsilon((\nabla_u B^{(\alpha)})_* \cdot \mathbf{r}_0)\tilde{w} + O(\epsilon^2) \qquad (4.4.39)$$

where $(\nabla_{\mathbf{u}}B^{(\alpha)})_*$ is the value of $(\nabla_{\mathbf{u}}B^{(\alpha)})$ evaluated at $\mathbf{u}=0$. The important point in simplifying the equations now is to realize that a nonlinear wavefront given by the phase function $\phi(\mathbf{x},t,\epsilon)$ may differ significantly from the corresponding linear wavefront given by the linear phase function $\phi^*(\mathbf{x},t)$ introduced in section 4.2. This has been emphasized in section 4.3.1 just above the equations (4.3.28 - 29) and can be seen from the large number of results we shall present in Chapter 6. The partial derivatives ϕ_{x_α} of the nonlinear phase and $\phi^*_{x_\alpha}$ of the linear phase (i.e., the unit normal $\mathbf{n}$ of a nonlinear wavefront and $\mathbf{n}_*$ of the corresponding linear wavefront) also differ significantly. One may think that the nonlinear ray theory which is being considered here may be valid only on the length scale over which the linear theory or CPW nonlinear theory are valid. But this is not so. In the derivation of this theory we have made no reference to the length scales associated with the linear theory. The numerical results in chapter 6 show that this theory is valid even in a caustic region of the linear theory where the normal $\mathbf{n}$ of a nonlinear wavefront and $\mathbf{n}_*$ of the corresponding linear wavefront differ a great deal. In fact, the theory is valid on a much larger length scale than the radii of curvature of the initial wavefront (see comments in section 6.4 on wavefronts with initially sinusoidal shape and the corresponding results in chapter 10). Therefore, while trying to make further approximation in some of the terms in (4.4.31 - 34), we keep $\mathbf{n}$ and the operators $\partial/\partial\eta^\alpha_\beta$ (tangential derivatives on the nonlinear wavefront) unchanged and use Taylor's expansion with respect to $\epsilon \mathbf{v}_0$ at 0. Following this we can approximate some of the terms as follows:

$$\begin{aligned} \mathbf{l}_0 B_0^{(\alpha)} \mathbf{r}_0 &= \bar{\mathbf{l}} B_*^{(\alpha)} \bar{\mathbf{r}} + \epsilon \Big[((\nabla_{\mathbf{u}} \mathbf{l})_0 \cdot \bar{\mathbf{r}}) B_*^{(\alpha)} \bar{\mathbf{r}} + \bar{\mathbf{l}}((\nabla_{\mathbf{u}} B^{(\alpha)})_* \cdot \bar{\mathbf{r}}) \bar{\mathbf{r}} \\ &+ \bar{\mathbf{l}} B_*^{(\alpha)} (\nabla_{\mathbf{u}} \mathbf{r})_0 \cdot \bar{\mathbf{r}} \Big] \tilde{w} + O(\epsilon^2) \end{aligned} \tag{4.4.40}$$

$$\begin{aligned} \mathbf{l}_0 A_0 \mathbf{r}_{0t} &= \bar{\mathbf{l}} A_* \bar{\mathbf{r}}_t + \epsilon \left[((\nabla_{\mathbf{u}} \mathbf{l})_0 \cdot \bar{\mathbf{r}}) A_* \bar{\mathbf{r}}_t + \bar{\mathbf{l}}((\nabla_{\mathbf{u}} A)_* \cdot \bar{\mathbf{r}}) \bar{\mathbf{r}}_t \right. \\ &+ \left. \bar{\mathbf{l}} A_* (\nabla_{\mathbf{u}} \mathbf{r})_{0t} \cdot \bar{\mathbf{r}} + \bar{\mathbf{l}} A_* (\nabla_{\mathbf{u}} \mathbf{r})_0 \cdot \bar{\mathbf{r}}_t \right] \tilde{w} \\ &+ \epsilon \bar{\mathbf{l}} A_* ((\nabla_{\mathbf{u}} \mathbf{r})_0 \cdot \bar{\mathbf{r}}) \tilde{w}_t + O(\epsilon^2) \end{aligned} \tag{4.4.41}$$

and

$$\begin{aligned} \mathbf{l}_0 B_0 \mathbf{r}_{0_{x_\alpha}} &= \bar{\mathbf{l}} B_*^{(\alpha)} \bar{\mathbf{r}} + \epsilon \Big[((\nabla_{\mathbf{u}} \mathbf{l})_0 \cdot \bar{\mathbf{r}}) B_*^{(\alpha)} \bar{\mathbf{r}}_{x_\alpha} + \bar{\mathbf{l}}((\nabla_{\mathbf{u}} B^{(\alpha)})_* \cdot \bar{\mathbf{r}}) \bar{\mathbf{r}}_{x_\alpha} \\ &+ \bar{\mathbf{l}} A_* (\nabla_{\mathbf{u}} \bar{\mathbf{r}})_{0x_\alpha} \cdot \bar{\mathbf{r}} + \bar{\mathbf{l}} B_*^{(\alpha)} (\nabla_{\mathbf{u}} \mathbf{r})_0 \cdot \bar{\mathbf{r}}_{x_\alpha} \Big] \tilde{w} \\ &+ \epsilon \bar{\mathbf{l}} B_*^{(\alpha)} ((\nabla_{\mathbf{u}} \mathbf{r})_0 \cdot \bar{\mathbf{r}}) \tilde{w}_{x_\alpha} + O(\epsilon^2) \end{aligned} \tag{4.4.42}$$

Therefore

$$\mathbf{l}_0 A_0 \mathbf{r}_{0t} + \mathbf{l}_0 B_0^{(\alpha)} \mathbf{r}_{0x_\alpha} = \bar{\mathbf{l}} A_* \bar{\mathbf{r}}_t + \bar{\mathbf{l}} B_*^{(\alpha)} \bar{\mathbf{r}}_{x_\alpha} + O(\epsilon) = -\bar{\Omega} + O(\epsilon) \tag{4.4.43}$$

where

$$\bar{\Omega} = -(\bar{\mathbf{l}} A_* \bar{\mathbf{r}}_t + \bar{\mathbf{l}} B_*^{(\alpha)} \bar{\mathbf{r}}_{x_\alpha}) \tag{4.4.44}$$

Substituting (4.4.40 - 44) in (4.4.31 - 34) and retaining terms only up to order ϵ, we get the full set of equations of WNLRT (note $d/ds = d/dt$)

$$\begin{aligned} \frac{dx_\alpha}{ds} &= \bar{\mathbf{l}} B_*^{\alpha} \bar{\mathbf{r}} + \epsilon \Big[((\nabla_{\mathbf{u}} \mathbf{l})_0 \cdot \bar{\mathbf{r}}) B_*^{(\alpha)} \bar{\mathbf{r}} + \bar{\mathbf{l}}((\nabla_{\mathbf{u}} B^{(\alpha)})_* \cdot \bar{\mathbf{r}}) \bar{\mathbf{r}} + \\ &+ \bar{\mathbf{l}} B_*^{(\alpha)} (\nabla_{\mathbf{u}} \mathbf{r})_0 \cdot \bar{\mathbf{r}} \Big] \tilde{w} + O(\epsilon^2) \end{aligned} \tag{4.4.45}$$

$$\begin{aligned} \frac{dn_0^\alpha}{ds} &= -\epsilon \mathbf{n}^\beta \bar{\mathbf{l}} \left[\left\{ -\bar{c}(\nabla_{\mathbf{u}} A)_* \cdot \frac{\partial \bar{\mathbf{r}}}{\partial \eta_\beta^\alpha} + n_\gamma (\nabla_{\mathbf{u}} B^\gamma)_* \cdot \frac{\partial \bar{\mathbf{r}}}{\partial \eta_\beta^\alpha} \right\} \tilde{w} \right. \\ &+ \left. \{ -\bar{c}(\nabla_{\mathbf{u}} A)_* \cdot \bar{\mathbf{r}} + n_\gamma (\nabla_{\mathbf{u}} B^\gamma)_* \cdot \bar{\mathbf{r}} \} \frac{\partial \tilde{w}}{\partial \eta_\beta^\alpha} \right] \bar{\mathbf{r}} \\ &+ O(\epsilon^2) \quad \beta = 1, 2 \ldots, m \end{aligned} \tag{4.4.46}$$

where

$$\bar{c} = c_0(\mathbf{n}, \mathbf{u} = 0)$$

and we note that Ψ_* is zero because A_* and $B_*^{(\alpha)}$ are constants, and

$$\begin{aligned}
\frac{d\tilde{w}}{dt} &= \bar{\Omega}\tilde{w} + \epsilon\left[((\nabla_{\mathbf{u}}\mathbf{l})_0 \cdot \bar{\mathbf{r}})A_*\bar{\mathbf{r}}_t + \bar{\mathbf{l}}((\nabla_{\mathbf{u}}A)_* \cdot \bar{\mathbf{r}})\bar{\mathbf{r}}_t\right.\\
&+ \left.\bar{\mathbf{l}}A_*(\nabla_{\mathbf{u}}\mathbf{r})_0 \cdot \bar{\mathbf{r}}_t\right]\tilde{w} + \epsilon\left[((\nabla_{\mathbf{u}}\mathbf{l})_0 \cdot \bar{\mathbf{r}})B_*^{(\alpha)}\bar{\mathbf{r}}_{x_\alpha}\right.\\
&+ \left.\bar{\mathbf{l}}((\nabla_{\mathbf{u}}B^{(\alpha)})_* \cdot \bar{\mathbf{r}})\bar{\mathbf{r}}_{x_\alpha} + \bar{\mathbf{l}}B_*^{(\alpha)}(\nabla_{\mathbf{u}}\mathbf{r})_0 \cdot \bar{\mathbf{r}}_{x_\alpha}\right]\tilde{w}\\
&+ \epsilon\left\{\bar{\mathbf{l}}A_*((\nabla_{\mathbf{u}}\mathbf{r})_0 \cdot \bar{\mathbf{r}})\tilde{w}_t + \bar{\mathbf{l}}B_*^{(\alpha)}((\nabla_{\mathbf{u}}\mathbf{r})_0 \cdot \bar{\mathbf{r}})\tilde{w}_{x_\alpha}\right\}\\
&- \epsilon\left[\left(\Gamma^t b_t + \Gamma^\alpha b_{x_\alpha} + \Gamma b\right)\tilde{w}_\theta + W b_t + V^\alpha b_{x_\alpha} + D^{\alpha\beta} + Eb\right]\\
&+ O(\epsilon^2) \qquad (4.4.47)
\end{aligned}$$

If the terms of the order ϵ are also neglected in the ray equations (4.4.45 - 46), these equations decouple from the transport equation (4.4.47) and give the linear rays. In order to retain the nonlinear effects it is necessary to retain in the ray equations, terms at least up to order ϵ. The situation for the transport equation (4.4.47) is different. The exact solution presented in Chapter 6 and numerical results (Prasad and Sangeeta (1999)) show that inclusion of the order ϵ in terms of (4.4.45 - 46) changes $\bar{\Omega}$ by order 1 in the caustic region leading to an order 1 change in the value of $\tilde{w}$ in finite time. This is in contrast to what we expect in a perturbation method. But it is not surprising when we note that the neglect of $O(\epsilon)$ terms (4.4.45 - 46) (i.e., linear theory) changes $\bar{\Omega}$ from a finite curvature to as infinite curvature in the caustic region which is reached in finite time. It is different with the transport equation (4.4.47), which, with only the first term on the right hand side, always leads to a finite value of $\tilde{w}$ everywhere whenever the terms of order ϵ is included in (4.4.45 - 46) and $\epsilon\tilde{w}$ (i.e., w of the section 4.3) is not chosen to be too small. During the competition of convergence of linear rays and the opposing effect of nonlinearity, a balance is reached which leads to a finite change in $\bar{\Omega}$ and hence a finite change in $\tilde{w}$. There is no mathematical proof so far for the amplitude to be finite due to nonlinearity but the extensive numerical computation with small (but not very small) values of amplitude $\epsilon\,\tilde{w}$ leads to this conjecture. In all these cases the effect of inclusion of the terms of order ϵ in (4.4.47) will remain small in finite time. Thus, to get only the leading order correction to the amplitude, it is not necessary to retain the last four terms in (4.4.47) which are multiplied by ϵ and then we get

$$\frac{d\tilde{w}}{dt} = \bar{\Omega}\tilde{w} \qquad (4.4.48)$$

These arguments convice us that we have indeed deduced a weakly nonlinear theory (i.e., equations (4.4.45 - 47)) in which $\tilde{w}$ has error $O(\epsilon^2)$ (i.e., the solution $\mathbf{u}$ has an error $O(\epsilon^3)$). However, in the solution of the simpler WNLRT (i.e., equations (4.4.45 - 46) and (4.4.48)), the amplitude $\tilde{w}$ has an error $O(\epsilon)$.

The transport equation looks exactly the same as the linear transport equation but it now contains all the leading order nonlinear effects since in it $d\tilde{w}/dt$ represents the time-rate of change along the nonlinear rays and $\mathbf{n}$ appearing in $\bar{\Omega}$ is the normal of the nonlinear wavefront. The equation (4.4.48) along with the equations (4.4.45 - 46) is equivalent to the transport equation

$$\tilde{w}_t + \Big\{\bar{\mathbf{l}}B_*^{(\alpha)}\bar{\mathbf{r}} + \epsilon\Big[(\nabla_{\mathbf{u}}\mathbf{l})_0 \cdot \bar{\mathbf{r}}B_*^{(\alpha)}\bar{\mathbf{r}} + \bar{\mathbf{l}}((\nabla_{\mathbf{u}}B^{(\alpha)})_*.\bar{\mathbf{r}})\bar{\mathbf{r}} \\ + \bar{\mathbf{l}}B_*^{(\alpha)}(\nabla_{\mathbf{u}} \cdot \mathbf{r})_0\bar{\mathbf{r}}\Big]\tilde{w}\Big\}\tilde{w}_{x_\alpha} = \bar{\Omega}\tilde{w} \quad (4.4.49)$$

and $\bar{\Omega}$, which contains derivatives of $\mathbf{n}$ and which is calculated using the coupled system (4.4.45 - 46) and (4.4.48), remains finite everywhere including the points on the caustic, where the corresponding value Ω^* by linear theory tends to infinity. The equations (4.4.45 - 46) and (4.4.48) form the final coupled system of equations of our nonlinear ray theory. Retaining the other terms of order ϵ in (4.4.47) will modify the results only by effects of order ϵ^2 in the solution $\mathbf{u}$ since these terms in (4.4.47) are actually of order ϵ^2 in the original equation (4.4.7). Equations (4.4.45 - 46) and (4.4.48) are the same as the equations obtained for a weakly nonlinear ray theory (WNLRT) in section 4.3.

4.4.3 Comparison with other theories

The WNLRT developed in the last two sections is valid over a length scale L over which the assumptions involved in the derivation of the equations are valid. This length L can be determined only from the solution of this approximate theory. One exact solution, called the composite simple wave solution in chapter 6, and the extensive numerical solution by Prasad and Sangeeta (1999), show that this L is large compared to the length scale R of the order of principal radii of curvature of the initial wavefront. The CPW ray theory is valid over a length scale L_c which is small compared to R. On this scale L_c, the linear and nonlinear wavefronts are not only close but have the same shape and the amplitude given by the linear theory remains small. Thus, $\frac{L_c}{R} << 1 << \frac{R}{L}$. We shall show that over the length scale L_c, the equation (4.4.48) reduces to the leading order equation obtained from CPW theory in addition to some extra terms which can be neglected. We now examine the equation (4.4.48) over a length scale L_c. On this length scale, the linear wavefront and the corresponding nonlinear wavefront originating from a same initial wavefront are close to one another and their unit normals denoted respectively by $\mathbf{n}_*$ and $\mathbf{n}$ differ by a quantity of order ϵ. We denote the rate of change along the linear ray by $\frac{d^*}{ds^*}$ i.e.,

$$\frac{d^*}{ds^*} = \mathbf{l}_* \, A_*^\alpha \, \mathbf{r}_* \, \frac{\partial}{\partial x^\alpha} \, , \; (x_\alpha) = (x_0 = t, x_1 = x, \cdots x_m = x_m) \quad (4.4.50)$$

where we have not set $\mathbf{l}A\mathbf{r} = 1$, and used $A^0 = A$, $A^\alpha = B^{(\alpha)}$ and

$$\mathbf{l}_* = \bar{\mathbf{l}}(\mathbf{n}_*, 0), \quad \mathbf{r}_* = \bar{\mathbf{r}}(\mathbf{n}_*, 0) \quad (4.4.51)$$

The quantities with subscript and superscript $*$ represent the same quantities as in section 4.2.

The summation convention in this section extends to the range $0, 1, 2, \cdots, m$. The rate of change $\frac{d}{ds}$ along the nonlinear ray (see equations (4.4.45 - 46)) for $|\mathbf{n} - \mathbf{n}_*| = O(\epsilon)$ can be written as

$$\begin{aligned}
\frac{d}{ds} &= \mathbf{l}_* A_*^\alpha \mathbf{r}_* \frac{\partial}{\partial x_\alpha} + \epsilon \left\{ \left((\nabla_{\mathbf{n}} \bar{\mathbf{l}})_* \cdot \left(\frac{\mathbf{n} - \mathbf{n}_*}{\epsilon} \right) \right) A_*^\alpha \mathbf{r}_* \right. \\
&+ \left. \mathbf{l}_* A_*^\alpha \left((\nabla_{\mathbf{n}} \bar{\mathbf{r}})_* \cdot \left(\frac{\mathbf{n} - \mathbf{n}_*}{\epsilon} \right) \right) \right\} \frac{\partial}{\partial x_\alpha} \\
&+ \epsilon \tilde{w} \left[((\nabla_{\mathbf{u}} \mathbf{l})_* \cdot \mathbf{r}_*) A_*^\alpha \mathbf{r}_* + \mathbf{l}_* \left((\nabla_{\mathbf{u}} A^\alpha)_* \cdot \mathbf{r}_* \right) \mathbf{r}_* \right. \\
&+ \left. \mathbf{l}_* A_*^\alpha \left((\nabla_{\mathbf{u}} \mathbf{r})_* \cdot \mathbf{r}_* \right) \right] \frac{\partial}{\partial x_\alpha} + 0(\epsilon^2)
\end{aligned} \quad (4.4.52)$$

where $\nabla_{\mathbf{n}} = \left(\frac{\partial}{\partial n_1}, \cdots, \frac{\partial}{\partial n_n} \right)$. The middle term in the square bracket is important and we write it along with the first term on the right hand side of (4.4.53). Thus,

$$\frac{d}{ds} = \frac{d^*}{ds^*} + \epsilon \left\{ \mathbf{l}_* \left(\nabla_{\mathbf{u}} A^\alpha \right)_* \cdot \mathbf{r}_* \right) \mathbf{r}_* \} \, \tilde{w} \frac{\partial}{\partial x_\alpha} + \epsilon \tilde{w} S^\alpha \frac{\partial}{\partial x_\alpha} + \epsilon T^\alpha \frac{\partial}{\partial x_\alpha} + 0(\epsilon^2) \quad (4.4.53)$$

where

$$S^\alpha = ((\nabla_{\mathbf{u}} \mathbf{l})_* \cdot \mathbf{r}_*) A_*^\alpha \mathbf{r}_* + \mathbf{l}_* A_*^\alpha ((\nabla_{\mathbf{u}} \mathbf{r})_* \cdot \mathbf{r}_*) \quad (4.4.54)$$

$$T^\alpha = \left(\left(\nabla_{\mathbf{n}}\bar{\mathbf{l}}\right)_* \cdot \left(\frac{\mathbf{n}-\mathbf{n}_*}{\epsilon}\right)\right) A_*^\alpha \mathbf{r}_* + \mathbf{l}_* A_*^\alpha \left(\left(\nabla_{\mathbf{n}}\bar{\mathbf{r}}\right)_* \cdot \left(\frac{\mathbf{n}-\mathbf{n}_*}{\epsilon}\right)\right) \tag{4.4.55}$$

The second term in (4.4.53) contains in it the nonlinear stretching of the rays as given in CPW theory. In fact, if we make a transformation from (x_α)-coordinates to $(\phi^*, y^1, ..., y^n)$-coordinates (where ϕ^* is the linear phase function)

$$\phi^* = \phi^*(x_0, x_1, \ldots, x_m), \quad y_\alpha = x_\alpha, \quad \alpha = 1, 2, \ldots n \tag{4.4.56}$$

then

$$\frac{\partial}{\partial x_0} = \phi^*_{x_0}\frac{\partial}{\partial x_0}, \quad \frac{\partial}{\partial x_\alpha} = \phi^*_{x_\alpha}\frac{\partial}{\partial \phi^*} + \frac{\partial}{\partial y_\alpha}, \ \alpha = 1, 2, \ldots, n \tag{4.4.57}$$

Now with $\theta^* = \frac{\phi^*}{\epsilon}$

$$\epsilon\{\mathbf{l}_*((\nabla_u A^\alpha)_* \cdot \mathbf{r}_*)\mathbf{r}_*\} w \frac{\partial}{\partial x_\alpha} = \mathcal{G}_* w \frac{\partial}{\partial \theta^*} + 0(\epsilon)$$

where

$$\mathcal{G}_* = \{\mathbf{l}_*(\phi^*_{x_\alpha}(\nabla_u A^\alpha)_* \cdot \mathbf{r}_*)\mathbf{r}_*\} \tag{4.4.58}$$

since $\frac{\partial}{\partial\theta^*} = \frac{1}{\epsilon}\frac{\partial}{\partial\phi^*}$. Further,

$$\epsilon S^\alpha \frac{\partial}{\partial x_\alpha} = \{((\nabla_u \mathbf{l})_* \cdot \mathbf{r}_*)(A_*^\alpha \phi^*_{x_\alpha} \mathbf{r}_*) + (\mathbf{l}_* A_*^\alpha \phi^*_{x_\alpha})((\nabla_u \mathbf{r})_* \mathbf{r}_*)\} \frac{\partial}{\partial \theta^*} + 0(\epsilon) \tag{4.4.59}$$

in which all terms of order one vanish because $A_*^\alpha\ \phi^*_{x_\alpha}\mathbf{r}_* = 0$, and $\mathbf{l}_* A_*^\alpha \phi^*_{x_\alpha} = 0$.

On the length scale L_c, $\mathbf{n} - \mathbf{n}_* = 0(\epsilon)$, so that

$$\epsilon T^\alpha \frac{\partial}{\partial x_\alpha} = T^\alpha \phi^*_{x_\alpha} \frac{\partial}{\partial \theta^*} + 0(\epsilon) \tag{4.4.60}$$

and here, too all the terms of order 1 vanish due to the same reason i.e., $A_*^\alpha \phi^*_{x_\alpha} \mathbf{r}_* = 0$ and $\mathbf{l}_* A_*^\alpha \phi^*_{x_\alpha} = 0$. Thus, to the leading order, the transport equation (4.4.48 or 49) reduces to the transport equation of the CPW theory

$$\frac{d^*}{dt^*} w + \mathcal{G} w w_{\theta^*} + \Omega_* w = 0 \tag{4.4.61}$$

Note that the assumption $|\mathbf{n} - \mathbf{n}^*| = O(\epsilon)$ breaks down as soon the nonlinear wavefront starts approaching a caustic region of the linear theory.

One of the most interesting outcomes of this theory is a derivation of the weak shock ray theory (Prasad (1993), p.95), from the WNLRT consisting of the equations (4.4.45 - 46) and (4.4.49). Shock ray theory consists of the shock ray (see chapter 9) equations and an infinite system of compatibility conditions. Unlike the WNLRT, shock ray theory is exact because ϵ is of the order of the shock thickness which is zero in the inviscid theory and hence the high frequency approximation is exactly satisfied. But the shock ray theory is as difficult as the original problem, in fact, more difficult due to horrendously long expressions present even in the first few (say the second itself) of the infinite number of compatibility conditions involved in it. An infinite number of equations remain involved even if a weak shock assumption is made. As mentioned here, the weak shock ray theory can be derived from the WNLRT of this paper. This derivation is not only simple but also much more transparent for the Euler's equations of gas dynamics, which we shall see in Chapter 9. In passing, we mention that an attempt has been made in showing such a relation between WNLRT and shock ray theory by Anile et al. (1993), (pp. 85-87) without making any distinction between a linear ray, nonlinear ray or shock ray.

Chapter 5

Stability of solutions near a singularity of sonic type

5.1 Introduction

One of the most intriguing phenomena in fluid dynamics is the acceleration and deceleration of the fluid flow through the velocity of sound. This is mainly due to the presence of sonic singularity of the steady flow and the genuine nonlinearity of gas dynamic equations. Small amplitude waves are trapped in a neighbourhood of these singularities where they grow and decay for a long period during which time the genuine nonlinearity of the governing equations start to dominate.

Both the exact and approximate solutions of one- or two - dimensional steady equations of motion of a polytropic gas have established beyond doubt the theoretical existence of continuously accelerating and decelerating isentropic, mixed transonic flows through the converging or diverging nozzles or past solid boundaries. Theoretically then, a fluid element should be able to accelerate continuously from a subsonic state to a supersonic state and vice versa. Continuous flows accelerating steadily through the speed of sound could always be obtained but most of the early experiments showed that a shock wave necessarily appeared where a continuously decelerating flow should have existed according to theoretical predictions. Kantrowitz (1947) studied the stability of quasi-one-dimensional steady transonic flows by superposing unsteady disturbances and came to a definite conclusion that a flow in a de Laval nozzle continuously accelerating through the speed of sound is stable. The reversed flow is unstable and the downstream part of the latter would be replaced by an accelerating flow terminated by a shock wave. A study of the stability of two-dimensional plane flows past an aerofoil surface was done by Kuo (1951) and his conclusion that two-dimensional flows decelerating continuously through the speed of sound are unstable was also not in contradiction with the early experimental results of his time. The faith in the instability of a continuously decelerating transonic flow grew stronger when other arguments (Spee (1971)) such as the non-existence of neighbouring flows (Morawetz 1964) poured in.

Later, in the late 1960's and early 1970's, theoretical and experimental investigations at National Aerospace Laboratory (NLR), the Netherlands (Nieuwland (1966), Nieuwland and Spee (1968) and Spee (1971)) showed that two-dimensional continuously decelerating transonic flows were not unstable in the strict sense. It would be possible to obtain these flows experimentally as closely as possible if we could reduce the model imperfections and boundary layer effects.

The reason for this transonic controversy was explained (Prasad (1973)) by showing that the upstream propagating waves are trapped near a sonic point due to the vanishing of all components of the ray velocity (4.3.31) at the sonic point in the steady solution when the wavefront becomes normal to the flow direction. Such a pulse stays in an ϵ neighbourhood of the sonic point (where ϵ is a small quantity of the order of the amplitude of the pulse) for a long time interval during which period the nonlinear evolution becomes very important and the perturbation remains bounded as seen in the Example 1.5.6.

The transition of a steady flow from a subsonic flow to a supersonic flow as we move with the fluid or the reverse transition takes place across a sonic surface at every point of which $|\mathbf{q}|^2 = a^2$. We note that a sonic point i.e., a point on the sonic surface (sonic type of singular point for a general hyperbolic system) can be defined only in a flow which can reduce to a steady flow in an appropriate Galilean frame (or a self-similar flow in a suitable similarity coordinate system - see later in this chapter). For such flows, there exists a frame with the help of which $|\mathbf{q}|^2 = a^2$ defines sonic surface uniquely. We choose a sonic point P^* and denote all quantities in the steady flow at P^* by a superscript $*$ for example $\mathbf{q}^* = \mathbf{q}_0(x^*)$, then

$$|\mathbf{q}^*| = a^* \tag{5.1.1}$$

If we consider a wavefront such that its normal $\mathbf{n} = \mathbf{q}^*/|\mathbf{q}^*|$, then

$$\boldsymbol{\chi}^* = \mathbf{q}^* - \mathbf{n}^* a^* = 0 \tag{5.1.2}$$

i.e., *all components of the ray velocity in the steady flow vanish at a sonic point when the wavefront is orthogonal to the stream line.* In such a flow, small amplitude upstream propagating waves with wavefronts orthogonal to the streamline at P^* are temporarily trapped at P^* since $c^* = \langle \mathbf{n}^*, \boldsymbol{\chi}^* \rangle = 0$.

For a system of first order equations (4.3.1) with at least one real eigenvalue $c = c(\mathbf{x}, \mathbf{u}, \mathbf{n})$, *a point $P^*(\mathbf{x}^*)$ in a steady solution $\mathbf{u}_0(\mathbf{x})$ is defined to be a sonic type of singularity if all components* $\chi^{(\alpha)} = \mathbf{l}B^{(\alpha)}\mathbf{r}/(\mathbf{l}A\mathbf{r})$, *of the ray velocity* $\boldsymbol{\chi}$ (see equation (2.4.6)) *vanish at the point P^* for a particular value* of $\mathbf{n} = \mathbf{n}^*$ i.e., $\chi_{\alpha 0}(\mathbf{x}^*, \mathbf{u}_0(\mathbf{x}^*), \mathbf{n}^*) = 0$, $\alpha = 1, 2, \ldots, m$.

Note that in the case of the Euler equations of motion of a gas in one space dimension, the normal $\mathbf{n}^* = \mathbf{q}^*/|\mathbf{q}^*|$ for a sonic singularity automatically reduces to (1,0,0) since $\mathbf{q} = (q, 0, 0)$. Therefore, in order for the above definition of a sonic type of singularity to be consistent with that for a hyperbolic system in two independent variables x and t, some restriction will be required on the hyperbolic system (4.3.1).

In order to gain some understanding of the sonic type of singularity, let us consider one-dimensional steady transonic flow in a de Laval nozzle.

Example 5.1.1 Transonic flow in de Laval nozzle

A *de Laval nozzle* is a duct with cylindrical symmetry, say about the x-axis and consists of a converging *entry* section and a diverging *exhaust* section. Consider a steady one-dimensional isentropic flow of a gas through the nozzle of a cross-sectional area $A(x)$. The fluid velocity q_0 at a section x of the nozzle and the area A satisfy the relation (Courant and Friedrichs (1948))

$$(q_0 - a_0)\frac{dq_0}{dx} = \frac{q_0 a_0^2}{A(q_0 + a_0)}\frac{dA}{dx} \tag{5.1.3}$$

where the subscript 0 represents the value in the steady solution.

We assume that at the point x^* the steady flow is sonic i.e., the characteristic velocity $c_0^* = q_0^* - a_0^* = 0$. Assuming that there is a smooth transition of the flow at the sonic point i.e., $(dq/dx)^*$ is finite and non-zero, from (5.1.3) we get $(dA/dx)^* = 0$ i.e., the cross-sectional area of the nozzle must be stationary at x^*. Since the eigenvalue c_0, representing the velocity of propagation of upstream propagating waves changes sign at which point the flow either accelerates or decelerates through the speed of sound, the equation (5.1.3) also shows that A must have either maximum or minimum at x^*.

Bernaulli's equation for a steady flow is $q_0^2 + \frac{2}{\gamma-1}a_0^2 =$ constant throughout the flow. With the help of this equation we can express the differential of $c_0 = q_0 - a_0$ in terms of dq_0 as

$$dc_0 = \left(1 + \frac{(\gamma - 1)q_0}{2a_0}\right) dq_0 \tag{5.1.4}$$

so that in a neighbourhood of the sonic point $dc_0 \sim \frac{\gamma+1}{2}dq_0$.

Therefore, in a neighbourhood of a sonic point, an approximate form of (5.1.3) is

$$\frac{dc_0}{dx} = \beta\frac{x - x^*}{c_0}\ , \quad \beta = \frac{(\gamma + 1)a^{*2}}{4A^*}\left(\frac{d^2A}{dx^2}\right)^* \tag{5.1.5}$$

Equation (5.1.5) shows that *the sonic point where $c_0 = 0$ is a singular point of the steady equation* (5.1.3).

The cross-sectional area of the nozzle in the neighbourhood of the sonic point is given by

$$\begin{aligned} A &= A^* + \frac{1}{2}\left(\frac{d^2A}{dx^2}\right)^* (x - x^*)^2 + \dots \\ &= A^*\left\{1 + \frac{2\beta}{(\gamma + 1)a^{*2}}(x - x^*)^2 + \dots\right\} \end{aligned} \tag{5.1.6}$$

If $\beta < 0$, the cross-section A is maximum at $\xi = 0$. The singular point is a center or focus and there does not exist any continuous steady solution through the nozzle with a sonic point at the throat. When $\beta > 0$, the nozzle area A is minimum at $\xi = 0$, the singular point is a node (Fig. 5.1.1). The figure shows that there are only four continuous steady flows represented by lof, lob, aob and aof with a sonic state at $x = 0$. Apart from this, there are an infinity of subsonic flows such as ghi, an infinity of supersonic flows such as pqr and an infinity of flows with weak shock waves such as pqmni and lo$m'n'$f

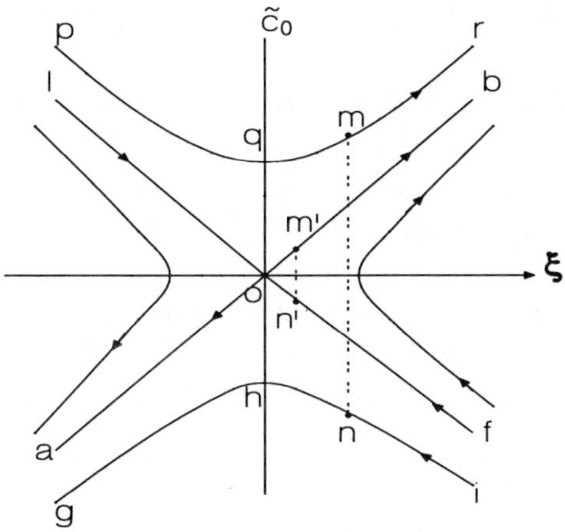

Fig. 5.1.1: Phase-plane of 5.1.5 with $\beta > 0$.

Kulikovskii and Slobodkina (1967) developed a very elegant theory to discuss the stability of steady solutions passing through the sonic type of singular points of a hyperbolic system in two independent variables. Their special transformation of a perturbation equation in a neighbourhood of such singular points leads to an equation which is independent of the steady solution on which perturbation is created. Hence, the transformed equation not only governs the perturbation of the steady solution but also the steady solution itself. The theory was extended by Bhatnagar and Prasad (1971) in the case of a multiple eigenvalue of a system of first order equations when the eigenspace of the eigenvalue under consideration is complete. As far as the author knows, this is the only example of a theory where the perturbation equation governs also the basic solution which is perturbed. We call this theory the BKPS theory. Since the case of multiple eigenvalue involves quite difficult algebraic calculations, we consider only the case of a simple eigenvalue. We first present general results on the approximation of a hyperbolic system in the neighbourhood of a characteristic curve and then present the BKPS theory.

5.2 One-dimensional weakly nonlinear wave propagation

Consider a system of n first order partial differential equations in two independent variables

$$A(x, \mathbf{u})\mathbf{u}_t + B(x, \mathbf{u})u_x + \mathbf{C}(x, \mathbf{u}) = 0 \tag{5.2.1}$$

where the symbols have their usual meaning. Let us assume that (5.2.1) has a real and simple eigenvalue $c(x, \mathbf{u})$ satisfying $\det[B - cA] = 0$. The corresponding left and right eigenvectors are denoted by $\mathbf{l}(x, \mathbf{u})$ and $\mathbf{r}(x, \mathbf{u})$ respectively. Consider a steady solution $\mathbf{u}_0(x)$ of (5.2.1) satisfying

$$B_0 \frac{d\mathbf{u}_0}{dx} + C_0 = 0\,, \;\; B_0 = B(x, \mathbf{u}_0(x)),\; C_0 = C(\mathbf{x}, \mathbf{u}_0) \tag{5.2.2}$$

and a small amplitude perturbation of this steady flow: $\mathbf{u} = \mathbf{u} - \mathbf{u}_0$ as introduced in section 4.3.

As mentioned at the end of the section 4.1, the difference between the CPW theory and WNLRT disappears when we consider a system of equations in two independent variables. Since, in this case, the unit normal $\mathbf{n}$ of the wavefront does not appear either in $c, \mathbf{l}$ or $\mathbf{r}$, the quantities with a subscript or superscript $*$ in section 4.2, those with a subscript 0 in section 4.3 and those with a bar above in section 4.4 (see the definition just before the expression (4.4.35)) are the same. In this section we follow the notations of the section 4.3.

Let $\phi(x, t) = 0$ be a characteristic (embedded in a one parameter family of characteristic curves) of (5.2.1) corresponding to the solution $\mathbf{u} = \mathbf{u}_0 + \mathbf{v}$, where $|\mathbf{v}|$ is of the order of ϵ. We now use a coordinate system (ϕ, x')

$$\phi = \phi(x, t)\,, \;\; x' = x \tag{5.2.3}$$

and approximate the system (5.2.1), using (5.2.2) and introducing $\theta = \phi/\epsilon$, in an ϵ neighbourhood of (yet unknown) characteristic curves $\phi = 0$. Following the procedure either of the section 4.2 or the section 4.3, we get

$$\mathbf{v} = w\mathbf{r}_0, \quad \mathbf{r}_0 = \mathbf{r}(x, \mathbf{u}_0(x)) \tag{5.2.4}$$

where w is the amplitude of the perturbation and is $O(\epsilon)$. The transport for w is

$$\frac{\partial w}{\partial t} + c(x, w)\frac{\partial w}{\partial x} = K\, w \tag{5.2.5}$$

where

$$c(x, w) = c(x, \mathbf{u}_0(x) + w\mathbf{r}_0) \tag{5.2.6}$$

and

$$K(x) = -\frac{1}{(\mathbf{l}A\mathbf{r})_0}\mathbf{l}_0\left\{(\nabla_u \mathbf{C})_0 \cdot \mathbf{r}_0 + (\nabla_u B)_0 \cdot \mathbf{r}_0 + \mathbf{l}_0 B_0 \frac{d\mathbf{r}_0}{dx}\right\} \tag{5.2.7}$$

(5.2.5) is an approximate transport equation giving the amplitude of the nonlinear wave propagating with the exact eigenvalue c corresponding to $\mathbf{u} = \mathbf{u}_0 + w\mathbf{r}_0$. Since w is a small quantity of order ϵ, we can approximate (5.2.5) in the neighbourhood of the approximate nonlinear characteristic curve. To do this, we note

$$c = c_0(x) + c_w w + 0(\epsilon^2) \tag{5.2.8}$$

where

$$c_0(x) = c(x, u_0(x)), \; c_w(x) = (\nabla_u c)_0 \cdot \mathbf{r}_0 \tag{5.2.9}$$

Therefore, retaining terms up to order ϵ in (5.2.5), we get the final approximate transport equation

$$w_t + (c_0(x) + c_w(x)w)w_x = K(x)w \tag{5.2.10}$$

The expression for c in terms of $A, B, \mathbf{l}$ and $\mathbf{r}$ is $c = (\mathbf{l}B\mathbf{r})/(\mathbf{l}A\mathbf{r})$, hence we get

$$c_0(x) = (\mathbf{l}_0 B_0 \mathbf{r}_0)/\mathbf{l}_0 A_0 \mathbf{r}_0) \tag{5.2.11}$$

and

$$c_w(x) = \frac{1}{(\mathbf{l}B\mathbf{r})_0}\mathbf{l}_0 \left\{(\nabla_u B \cdot \mathbf{r})_0 - c_0(\nabla_u A \cdot \mathbf{r})_0\right\} \mathbf{r}_0 \tag{5.2.12}$$

Example 5.2.1 Consider a reducible hyperbolic system (3.1.26) in two independent variables x and t. Take the basic state $\mathbf{u}_0(x) = 0$ so that c_0 and c_w are constant and $K = 0$. Then the equation (5.2.10) reduces to the Burgers' equation

$$u_t + uu_\xi = 0 \tag{5.2.13}$$

where

$$u = c_w w \text{ and } \xi = x - c_0 t \tag{5.2.14}$$

Unlike the cases in multi-dimensions, the one-dimensional model equation (5.2.10) for a weakly nonlinear wave propagation is very simple. Given the initial shape of a pulse or boundary condition along a moving boundary $x = X_p(t)$ (such as in the piston problem, see section 3.1.1), the wave profile can be easily computed by solving the characteristic equation along with the compatibility condition (see equations (1.2.3 - 4) for Burgers' equation). For certain initial or boundary values, a shock (or a number of shocks) would appear in a solution of (5.2.10). It is simple to deal with the shock provided we know the correct conservation form of (5.2.10). For this, it would be necessary to start with the original equations of the physical system under consideration in conservation form from which the equations (5.2.1) were derived, for example, the equations of motion of a gas: (2.1.8) with (2.1.20). The procedure given in this section (or those in chapter 4) may be modified (see Hunter (1995), section 3.2.2.1 equation (3.2.31)) to get the correct conservation form, say for simplicity, of (5.2.13):

$$u_t + \left(\frac{1}{2}u^2\right)_\xi = 0 \tag{5.2.15}$$

With an appropriate conservation form of (5.2.10), shocks can be fitted in the solution. The conservation form (5.2.15) with the flux function proportional to the square of the amplitude of the wave is a general result for a weakly nonlinear ray theory. This is consistent with the result of shock fitting in multi-space dimensions as stated in the theorem 9.2.1 in chapter 9.

5.2.1 BKPS theory

In this section we shall study one-dimensional steady solutions and their stability in a neighbourhood of a sonic type of singularity. A complete study of this type (i.e., study of steady solutions and their stability) is possible with the help of a single partial differential equation, which is also very simple.

A steady solution $\mathbf{u}_0(x)$ of (5.2.1) is governed by (5.2.2). We assume that the eigenvalue c vanishes at a point x^* in the steady solution i.e.,

$$c^* \equiv c_0(x^*, U_0(x^*)) = 0 \tag{5.2.16}$$

Multiplying (5.2.2) on the left by l_0 and using $\mathbf{l}_0 B_0 = c_0 \mathbf{l} A_0$ we get

$$c_0 \left(l_0 A_0 \frac{du_0}{dx} \right) + l_0 C_0 = 0 \tag{5.2.17}$$

(5.2.17) can be satisfied only if $l_0 C_0$ also vanishes at x^* showing that *the point where a characteristic velocity vanishes in a steady solution is a sonic type of singular point of the steady equations* (5.2.2).

The approximate partial differential equation governing the motion of weak unimodal waves on the steady solution $\mathbf{u}_0(x)$ is (5.2.10) where c_0, c_w and K are functions of x only. This approximation is valid for perturbations of the order of ϵ, in an ϵ neighbourhood of a characteristic curve given by $\frac{dx}{dt} = c_0$ and over a time-scale of the order of unity. Let us select a characteristic curve to be the one passing through the singular point x^* at time $t = 0$. Since $c_0 = 0$ at $x^* = 0$, this characteristic (for the solution $\mathbf{u} = \mathbf{u}_0(x)$) is a straight line in the (x, t)-plane parallel to the t-axis and passing through the point $(x^*, 0)$. Therefore, the approximate equation is valid in an ϵ neighbourhood of the singular point x^* and over a time-scale of the order of unity i.e., it governs the motion of those small amplitude waves which stay in the neighbourhood of the singular point for the time interval of order of unity. Therefore, it determines the complete history of an isolated, unsteady transonic-type of pulse. In the neighbourhood of the singular point, we expand the quantities c_0, c_w and K in the powers of $\xi = x - x^*$ and retain only the most dominant terms consistent with our approximation. Then we get

$$w_t + (c_x^* \xi + c_w^* w) w_\xi = K^* w \ , \quad \xi = x - x^* \tag{5.2.18}$$

where from (5.2.11 - 12) and (5.2.16) we get

$$c_x^* = \frac{1}{(\mathbf{l}A\mathbf{r})^*} \mathbf{l}^* \left\{ (B_x)^* + \left((\nabla_u B)^* \cdot \left(\frac{d\mathbf{u}}{dx} \right)^* \right) \right\} \mathbf{r}^* \tag{5.2.19}$$

$$c_w^* = \frac{1}{(\mathbf{l}A\mathbf{r})^*}\mathbf{l}^*((\nabla_u B)^* \cdot \mathbf{r}^*)\mathbf{r}^* \tag{5.2.20}$$

and using (5.2.7) and noting that $(\mathbf{l}B)^* = c^*(\mathbf{l}A)^* = 0$

$$K^* = -\frac{1}{(\mathbf{l}A\mathbf{r})^*}\mathbf{l}^*\left\{(\nabla_u C)^* \cdot \mathbf{r}^* + ((\nabla_u B)^* \cdot \mathbf{r}^*)\left(\frac{d\mathbf{u}}{dx}\right)^*\right\} \tag{5.2.21}$$

where Q^* for any quantity Q is defined by

$$Q^* = Q(x^*, \mathbf{u}_0(x^*)) \tag{5.2.22}$$

Kulikovskii and Slobodkina (1967) have shown that equation (5.2.18) is valid even if the effect of the perturbation away from the sonic point is non-zero. They achieved this by a suitable transformation of dependent and independent variables. Therefore, there is no loss of generality if we base our discussion of the stability of the steady solutions on the equation (5.2.19).

We assume that the characteristic field under consideration is genuinely nonlinear which implies

$$c_w \neq 0 \tag{5.2.23}$$

The values of the constants c_x^*, c_w^* and K^* depend on $\mathbf{u}^*$ and also on the derivatives $\left(\frac{d\mathbf{u}}{dx}\right)^*$ i.e., on the particular steady solution on which the perturbation is considered. To remove this dependence we replace w by a new unknown $\tilde{c} = c_x^*\xi + c_w^* w$. Then

$$\frac{\partial \tilde{c}}{\partial t} + \tilde{c}\frac{\partial \tilde{c}}{\partial \xi} = \alpha\tilde{c} + \beta\xi \tag{5.2.24}$$

where the constants α and β are defined by

$$\alpha = K^* + c_x^* \text{ and } \beta = -K^* c_x^* \tag{5.2.25}$$

We note that the quantity $\tilde{c} = c_x^*(x - x^*) + c_w^* w$ represents an approximate value of the eigenvalue c in the ϵ-neighbourhood of the singular point i.e.,

$$c = \tilde{c} + 0(\epsilon)^2 \tag{5.2.26}$$

In fact, $c_0 - c^* = c_x^*\xi + 0(\epsilon^2)$ and $c - c_0 = c_w^*\xi + 0(\epsilon^2)$ so that $c - c^* = \tilde{c}$. Based on this Kulikovskii and Slobodkina (1967) reach a remarkable result that *the equation* (5.2.24) *governs not only the perturbation* *of the steady solution* $\mathbf{u}_0$ *but also the steady solution itself in the* ϵ *neighbourhood of the singular point.* An explicit proof of this important result that these constants are indeed independent of $\left(\frac{d\mathbf{u}}{dx}\right)^*$ was given by Bhatnagar and Prasad (1971) in a more general case: the case when the eigenvalue c is multiple. Thus, the coefficient α and β depend only on the value $\mathbf{u}_*$ i.e., the sonic state of the physical system governed by the equations (5.2.1).

Therefore, the equation (5.2.24) describes also the steady solution in the neighbourhood of the singular point. To evaluate the constants α and β, we do not have to work out the complicated algebra of the general theory. They can be determined from an approximate form of the ordinary differential equations describing the steady solution. For example, in the de Laval nozzle problem, equation (5.1.5) gives $\alpha = 0$ and β.

It is a simple matter to write the general unsteady solution of (5.2.24) with the help of the characteristic equations

$$\frac{d\tilde{c}}{dt} = \alpha\,\tilde{c} + b\xi \tag{5.2.27}$$

and

$$\frac{d\xi}{dt} = \tilde{c} \tag{5.2.28}$$

However, any steady solution $\tilde{c}_0 = \tilde{c}_0(\xi)$ of (5.2.24) can also be obtained from a solution of (52.27 - 28) in the form $c_0 = c_0(t)$, $\xi = \xi_0(t)$.

The general solution of equations (5.2.27 - 28) is

$$\tilde{c} = \mu\lambda_1 e^{\lambda_1 t} + \nu\lambda_2 e^{\lambda_2 t}\ ,\ \xi = \mu e^{\lambda_1 t} + \nu e^{\lambda_2 t} \tag{5.2.29}$$

where μ and ν are arbitrary constants and λ_1, λ_2 are the two roots of the equation

$$\lambda^2 - \alpha\lambda - \beta = 0 \tag{5.2.30}$$

When the two roots of this algebraic equation are complex (as in the case of Example 5.1.1 describing flow through a nozzle of maximum area where $\alpha = 0$ and $\beta < 0$), the singular point $(0, 0)$ in $(\tilde{c}, \xi)$-plane is a center and no continuous steady solution can exist. Hence *a necessary condition for the existence of a continuous steady solution through the critical point is that the two roots* λ_1 *and* λ_2 *are real. We assume that* $\lambda_1 > \lambda_2$.

The behaviour of the integral curves of (5.2.27 - 28) in the $(\xi, \tilde{c})$-plane for various real values of λ_1 and λ_2 has been shown in Figures 5.2.1 - 4 where arrows indicate the direction of the increasing t. When λ_1 and λ_2 are real and distinct, the singular point $\tilde{c} = 0, \xi = 0$ being either a node or a saddle point, there are two important continuous steady solutions $\tilde{c} = \lambda_1 \xi$ and $\tilde{c} = \lambda_2 \xi$ with a sonic transition. If it is a node, there are an infinity of such steady solutions. However, we can construct an infinity of discontinuous solutions also, such as AA_1B_1B in Figs. 5.2.1 - 3, where a shock A_1B_1 has been fitted by finding points A_1 and B_1 such that $c(A_1) + c(B_1) = 0$. Multiplicity of all such steady solutions with shocks have been discussed by Prasad (1969) and their stability has been discussed by Bhatnagar and Prasad (1971).

We establish an important relation between the equation (5.2.24) and its characteristic equations (5.2.27 - 28). From equations (5.2.27 - 28) we derive

$$\frac{d\tilde{c}}{d\xi} = \frac{\alpha \tilde{c} + \beta \xi}{\tilde{c}} \tag{5.2.31}$$

which governs the steady solutions of (5.2.24). The equation (5.2.24) interpreted as a directional derivative in (ξ, t)-plane, means that the space rate of change of $\tilde{c}$ as we move along a characteristic curve is $\frac{\alpha \tilde{c} + \beta \xi}{\tilde{c}}$. From the equation (5.2.31), it follows that the space-rate of change of $\tilde{c}$ in a steady solution has the same value $\frac{\alpha \tilde{c} + \beta \xi}{\tilde{c}}$. Any steady solution $\tilde{c}_0(x)$ consisting of a segment of an integral curve (single valued in x) of (5.2.27 - 28) can be taken as an unperturbed steady solution. At any instant, a disturbance of the steady solution will be represented by a curve $\tilde{c} = c_0(x) + \Delta \tilde{c}(x)$ as shown in Fig. 5.2.1. We shall consider only those disturbances which are bounded in space and, therefore, any disturbance will be represented geometrically by an area bounded by a closed curve in the $(\xi, \tilde{c})$-plane, a part of the boundary being a segment of the curve representing the steady solution. During the propagation, different points of the boundary curve of the disturbance will move along the integral curves of the characteristic equations. The direction of propagation, shown by the arrows in the figures, can be easily obtained.

Consider on the $(\xi, \tilde{c})$-plane, an arbitrary part of an area S bounded by a closed curve whose points move in accordance with equations (5.2.27 - 28). Since the vector field given by the right-hand side of

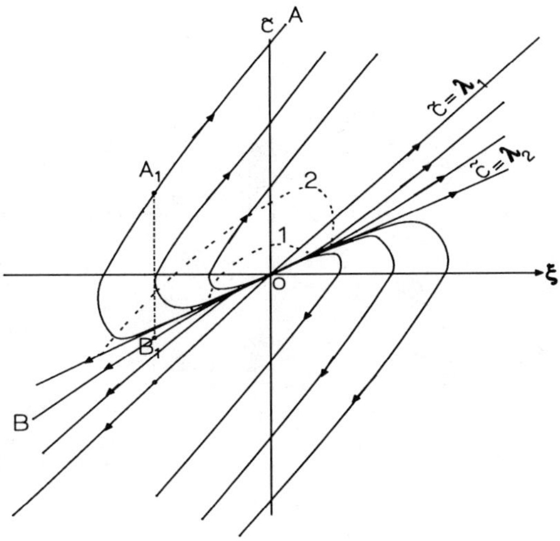

Fig. 5.2.1: Phase-plane for $0 < \lambda_2 < \lambda_1,\ \alpha > 0$.

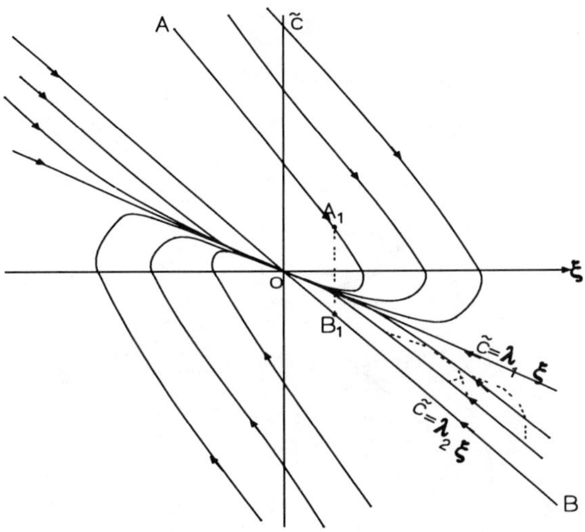

Fig. 5.2.2: Phase-plane for $\lambda_2 < \lambda_1 < 0,\ \alpha < 0$.

these equations has a constant divergence

$$\frac{\partial}{\partial \xi}\left(\frac{d\xi}{dt}\right) + \frac{\partial}{\partial \tilde{c}}\left(\frac{d\tilde{c}}{dt}\right) = \alpha = \lambda_1 + \lambda_2 = \frac{1}{S}\frac{dS}{dt} \tag{5.2.32}$$

it follows that $S = S_0 e^{\alpha t}$ where S_0 is the value of S at $t = 0$.

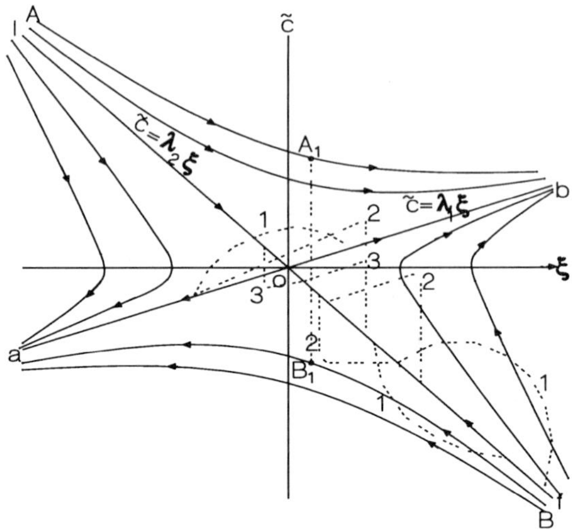

Fig. 5.2.3: Phase-plane for $\lambda_2 < 0 < \lambda_1$, $\alpha < 0$.

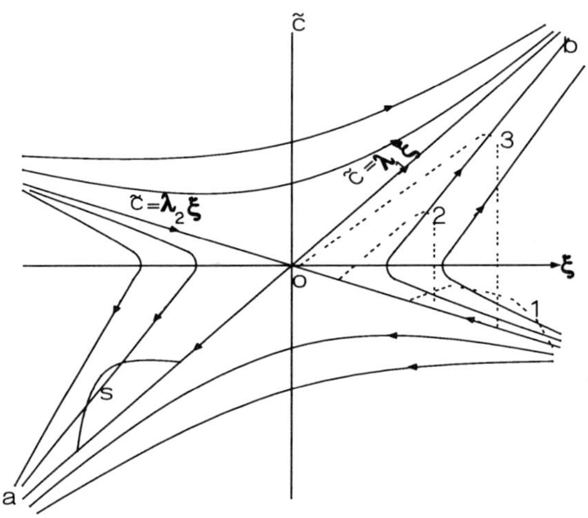

Fig. 5.2.4: Phase-plane for $\lambda_2 < 0 < \lambda_1$, $\alpha > 0$.

With the help of the above results we are now in a position to follow the motion, the growth or decay and the change in the shape of a disturbance as it moves in the phase-plane of the equations (5.2.27 - 28). We discuss briefly the various particular cases.

When $\lambda_1 > \lambda_2 > 0$, we have a singularity in the form a node with positive characteristic directions as shown in Fig. 5.2.1. There are an infinity of steady solutions through the critical point. Since $\alpha = \lambda_1 + \lambda_2 > 0$, any perturbation grows with time without bounds. The leading and trailing fronts always move away from the singular point with increasing velocity of departure from the origin. In the Fig. 5.2.1 we have shown by the broken lines a typical disturbance with positive values of ∇c on an initial steady solution $\tilde{c} = \lambda_2 x$. 1 and 2 represent the position at two instances and we can easily see that at the second instance a shock wave has developed in the perturbation on the right hand side front. This shock will move towards right with increasing velocity and increasing strength.

Thus, *any steady state solution* (Fig. 5.2.1) *passing through the singular point is unstable when* $\lambda_1 > \lambda_2 > 0$.

In all other particular cases, namely $\lambda_2 < \lambda_1 < 0$ (Fig. 5.2.2) and $\lambda_2 < 0 < \lambda_1$ (Figs. 5.2.3 - 4), it is not difficult to discuss the stability of a steady solution. We just state the final result, the details are given in Kulikovskii and Slobodkina's original paper.

When $\alpha = \lambda_1 + \lambda_2 < 0$, *all steady solutions are stable near the singular point. When* $\alpha > 0$, *all steady solutions are unstable with the exception of the solution represented by an integral curve passing through a saddle point singularity with* $\frac{d\tilde{c}}{dx} > 0$.

Example 5.2.2 Neutral stability of an accelerating transonic flow

We consider the steady transonic flow in the de Laval nozzle near its throat where the cross sectional area is minimum. The steady solution is given by (5.1.5). Hence, the equation (5.2.24) reduces to $\tilde{c}_t + \tilde{c}\tilde{c}_x = \beta x$ where β is given by (5.1.5) and $\alpha = 0$. In this case $\lambda_2 < 0 < \lambda_1 = \sqrt{\beta}$ and the phase-plane is given in Fig. 5.1.1. This is a special case in which the area S of a perturbation of a steady solution remains constant (see equation (5.2.32)). The continuously accelerating flow aob is stable. The continuously decelerating flow lof is neutrally stable. A perturbation of it will get trapped near the sonic point O and the flow lof will be replaced by lo$m'n'$f. The position of the shock $m'n'$ will depend on the initial area of the perturbation which will be equal to the area of the triangle $Om'n'$. This explains the difficulty in obtaining a continuously decelerating transonic flow experimentally. The stability of all other solutions including those with shocks can also be discussed (see Bhatnagar and Prasad (1970)).

It is now a simple matter to use the above theory to discuss the stability in the case of particular problems as shown in the above example. Kulikovskii and Slobodkina (1967) have used it to study the stability of magnetohydrodynamic flow of a fluid through a channel, where they have met all types of singularities depending on the values of the flow variable in the critical state. Bhatnagar and Prasad (1970) have shown that all one-dimensional steady flows of a polytropic gas when the thermal effect of radiation is considered are stable at the sonic point.

5.2.2 Sonic type of singularity in self-similar solutions

Self-similar solutions and one-dimensional steady solutions in fluid dynamics share a common interesting feature in that, sometimes, a singular point or a singular surface appears in the phase-space of the flow variables. The one-dimensional steady solutions, where the flow variables depend on just one spatial coordinate, form a particular case of self-similar solutions. The singularities of the system of ordinary differential equations describing these solutions can be classified into two groups. The singularities of the first group represent equilibrium states or *uniform flows* and they generally form the starting points or end points of the steady flows at infinite distance. Examples of such singularities are the Rankine-Hugoniot points in the steady flows (Ludford (1951); Prasad (1969) and von-Mises (1950)) describing shock structure. We can call them *natural singularities* since their appearance is independent of the dissipative terms such as viscosity or heat conduction terms included in the equations of motion. The appearance of the other group of singularities depends on the dissipative terms included in the equations of motion and hence we call them *pseudo-singularities.* A singularity of the second group in the case of steady solutions corresponds to those points where an eigenvalue of the original equations of motion becomes zero; in the case of a self-similar solution it corresponds to the states on one of the characteristics of the equations. For example, in the steady solution through a Laval nozzle, a singularity appears at the critical point where the particle velocity equals isentropic sound velocity. If we consider any steady solution with viscosity as the only dissipative mechanism, no singularity of the second group will appear, because the equations of motion with the viscous terms included are parabolic in nature with infinite speed of propagation for disturbances. Thus, the singularities in the second group depend on our physical assumptions. However, they play a very important role in determining the existence and uniqueness of a possible steady or self-similar solution. Zel'dovich and Raizer (1967) offer an illuminating discussion of the self-similar solutions of the second kind and also of the role played by pseudo-singularities in determining the unknown but unique exponent δ of the similarity variable $\xi = r/t^{\delta}$. In these self-similar solutions of the second kind, the equations of motion and the initial and boundary conditions contain only one independent dimensional constant so that they are not sufficient to determine the exponent; however, it can be uniquely determined from a necessary condition that the integral curve representing the solution must pass through the singular point.

In this section we shall discuss the extension of the method of the last section to study the stability of self-similar solutions near pseudo-singularities as developed by Bhatnagar and Prasad (1970).

Consider the equations of motion of a polytropic gas in the form

$$\frac{\partial \rho}{\partial t} + u\frac{\partial \rho}{\partial r} + \rho\frac{\partial u}{\partial r} + \frac{(\nu-1)\rho u}{r} = 0 \tag{5.2.33}$$

$$\frac{\partial u}{\partial t} + u\frac{\partial u}{\partial r} + \frac{1}{\rho}\frac{\partial p}{\partial r} = 0 \tag{5.2.34}$$

and

$$\frac{\partial p}{\partial t} + u\frac{\partial p}{\partial r} - \frac{\gamma p}{\rho}\left(\frac{\partial \rho}{\partial t} + u\frac{\partial \rho}{\partial r}\right) = 0 \tag{5.2.35}$$

where $\nu = 1, 2$ and 3 correspond respectively to one-dimensional, axi-symmetric and spherically symmetric motions. We introduce the nondimensional variables π, g, v, φ and τ defined by equations

$$\left.\begin{aligned} &p(x,t) = m(t)\dot{R}^2(t)\pi(\varphi,t), \;\; \rho(x,t) = m(t)g(\varphi,\tau) \\ &u(x,t) = \dot{R}(t)v(\varphi,\tau), \;\; \phi = \tfrac{r}{R(t)}, \tau = \delta\, ln\, R(t), \end{aligned}\right\} \tag{5.2.36}$$

where $m(t)$ and $R(t)$ are two positive functions of time with dimensions of density and length, respectively and δ is a dimensionless constant. In terms of these new variables, equations (5.2.27 - 5.2.29) become

$$\frac{1}{\delta}\frac{\partial g}{\partial \tau} + (v-\varphi)\frac{\partial g}{\partial \varphi} + g\frac{\partial v}{\partial \varphi} + \frac{(\nu-1)vg}{\varphi} + \left(\frac{\dot{m}R}{m\dot{R}}\right)g = 0 \qquad (5.2.37)$$

$$\frac{1}{\delta}\frac{\partial v}{\partial \tau} + (v-\varphi)\frac{\partial v}{\partial \varphi} + \frac{1}{g}\frac{\partial \pi}{\partial \varphi} + \frac{R\ddot{R}}{\dot{R}^2}v = 0 \qquad (5.2.38)$$

and

$$\frac{1}{\delta}\frac{\partial \pi}{\partial \tau} + (v-\varphi)\frac{\partial \pi}{\partial \varphi} - \frac{\gamma\pi}{g}\left\{\frac{1}{\delta}\frac{\partial g}{\partial \tau} + (v-\varphi)\frac{\partial g}{\partial \varphi}\right\} + \left\{2\frac{R\ddot{R}}{\dot{R}^2} - (\gamma-1)\frac{\dot{m}R}{m\dot{R}}\right\}\pi = 0 \qquad (5.2.39)$$

It is well known that similarity solutions are possible for the equations (5.2.33 - 5.2.35) and, in such cases, π, g, v are functions of only one variable, called the similarity variable ϕ, so that $g_\tau = v_\tau = \pi_\tau = 0$. This implies that the functions $m(t)$ and $R(t)$ satisfy

$$\frac{R\ddot{R}}{\dot{R}^2} = \text{constant} = A, \text{ say } ; \quad \frac{\dot{m}R}{m\dot{R}} = \text{constant} = B, \text{ say} \qquad (5.2.40)$$

Therefore, if $\pi_0(\phi)$, $g_0(\phi)$, $v_0(\phi)$ give a self-similar solution, they satisfy

$$\frac{v_0-\varphi}{g_0}\frac{dg_0}{d\varphi} + \frac{dv_0}{d\varphi} + \frac{(\nu-1)v_0}{\varphi} + B = 0 \qquad (5.2.41)$$

$$(v_0-\varphi)\frac{dv_0}{d\varphi} + \frac{1}{g_0}\frac{d\pi_0}{d\varphi} + Av_0 = 0 \qquad (5.2.42)$$

and

$$(v_0-\varphi)\left(\frac{d\pi_0}{d\varphi} - \frac{\gamma\pi_0}{g_0}\frac{dg_0}{d\varphi}\right) + \{2A - (\gamma-1)B\} = 0 \qquad (5.2.43)$$

As discussed by Sedov (1959), the set of equations (5.2.41 - 5.2.43) has many pseudo-singular points and there are a large number of physically realistic solutions (Zel'dvoich and Raizer (1967)) for which the integral curves pass through a singular point, where the value ϕ^* of ϕ is, in general, not zero. We can immediately apply Kulikovskii and Slobodkina's method to study the stability of these self-similar solutions in the neighbourhood of the singular points. Here $\phi - \phi^*$ take the role of the spatial coordinate ξ and τ that of time t. Finally the self-similar solutions, being independent of the new time variable τ, take the role of the steady flow. We also find that when $m(t)$ and $R(t)$ satisfy (5.2.40) the coefficients in the equations do not contain the new variable τ.

Example 5.2.3 Stability of a flow behind an imploding shock

Let us now take a simple example and apply the above method to it in order to study the stability in the neighbourhood of the singular point. We imagine a spherically symmetric flow ($\nu = 3$) in which a strong shock wave travels to the center of the symmetry through a gas of uniform initial density ρ_0 and zero pressure. Whatever may be the origin of the wave, the above limiting motion (i.e., when the shock radius is very small) must be self-similar (Zel'dovich and Raizer (1967)). This problem was first studied independently by Landau and Stanyukovich (see Stanyukovich (1960) and Guderley (1942)), and has been discussed in detail as one of the self-similar solutions of the second kind by Zel′dovich and Raizer. The origin of time is taken to be the instant of collapse when $R(t)$, the radius of shock, is zero. Thus, the time t up to the instant of collapse is negative and we can take

$$m = \text{ constant } = \rho_0, \ \ R(t) = A(-t)^\delta \tag{5.2.44}$$

Instead of working with variables π, g and v we use a new system of dependent variables V, G and Z and a new spatial coordinate η defined by

$$\left.\begin{aligned} &\eta = \text{In}\,\phi, \ \ G(\eta,\tau) = g(\phi,\tau), \\ &V(\eta,\tau) = \delta\tfrac{v(\phi,\tau)}{\phi}, \ \ Z(\eta,\tau) = \tau\delta^2\tfrac{\pi(\phi,\tau)}{g\phi^2}, \end{aligned}\right\} \tag{5.2.45}$$

The equations (5.2.41 - 44) transform to

$$\frac{dV_0}{d\eta} + \frac{V_0 - \delta}{G_0}\frac{dG_0}{d\eta} + 3V_0 = 0, \tag{5.2.46}$$

$$(V_0 - \delta)\frac{dV_0}{d\eta} + \frac{Z_0}{\gamma G_0}\frac{dG_0}{d\eta} + \frac{1}{\gamma}\frac{dZ_0}{d\eta} + \frac{2Z_0}{\gamma} + V_0(V_0 - 1) = 0 \tag{5.2.47}$$

and

$$\frac{(\gamma - 1)Z_0}{G_0}\frac{dG_0}{d\eta} - \frac{dZ_0}{d\eta} - 2\left[\frac{\delta - 1}{V_0 - \delta} + 1\right] Z_0 = 0 \tag{5.2.48}$$

We notice that when equations (5.2.37 - 39) are expressed in terms of G, V, Z, η and τ, the independent variable η does not appear explicitly in the coefficients and hence the value η^* of η at the singular point will not appear explicitly in our results.

The characteristic of the equations (5.2.37 - 39) in (η, τ)-plane are

$$\frac{d\eta}{d\tau} = V - \delta, \quad \frac{d\eta}{d\tau} = (V - \delta) \pm \sqrt{Z} \tag{5.2.49}$$

We define the Mach number μ by

$$\mu = \frac{V - \delta}{\sqrt{Z}} \tag{5.2.50}$$

Solving equations (5.2.46 - 48) for $dV_0/d\eta$, $1/G_0\, dG_0/d\eta$ and $dZ_0/d\eta$ and using the relation (5.2.50) we get

$$\frac{d\mu_0}{d\eta} = -\frac{[(\delta - 1)/\gamma] + 2V_0 + \delta}{\sqrt{Z_0}} + \frac{(\gamma + 1)(V_0 - \delta)}{2Z_0^{\frac{3}{2}}} \frac{f(V_0)}{\mu_0^2 - 1} \tag{5.2.51}$$

and

$$\frac{dV_0}{d\eta} = \frac{2\sqrt{Z_0}}{\gamma + 1} \frac{d\mu_0}{d\eta} - \left\{ \frac{3\gamma - 1}{\gamma + 1} V_0 - \frac{2}{\gamma + 1} \right\} \tag{5.2.52}$$

where

$$f(V_0) = \frac{2(\delta - 1)}{\gamma}(V_0 - \delta) + V_0(2V_0 + 1 - 3\delta) \tag{5.2.53}$$

In order that the solution of equations (5.2.46 - 5.2.47) satisfies correct boundary conditions at the shock and at infinity it is necessary (Zel'dovich and Raizer (1967)) that the integral curve in (Z_0, V_0)-plane must pass through the singular point (Z^*, V^*) determined by the equations

$$Z^* = (\delta - V^*)^2 \text{ and } f(V^*) = 0 \tag{5.2.54}$$

and this determines a unique value of the exponent δ. The equation $f(V^*) = 0$ has two roots and V^* is the larger of the two roots. At the critical point $\mu_0 = \mu^*$ ($= -1$ in this particular problem) and $f(V_0)/(\mu_0^2 - 1)$ is of the form $0 \div 0$. Therefore, we differentiate the numerator and denominator of $f(V_0)/(\mu_0^2 - 1)$ and use the relation

(5.2.52). This gives us the following equation for $d\mu_0/d\eta$ at the critical point

$$\left(\frac{d\mu_0}{d\eta}\right)^2 - \alpha_1\left(\frac{d\mu_0}{d\eta}\right) - \beta_1 = 0 \tag{5.2.55}$$

where

$$\alpha_1 = \frac{[(\delta-1)/\gamma] + 2V^* + \delta}{Z^*}\left\{\frac{V^*-\delta}{\mu^*} - \sqrt{Z}^*\right\} + \frac{(V^*-\delta)(1-5\delta)}{2Z^*\mu^*} \tag{5.2.56}$$

and

$$\beta_1 = -\frac{(V^*-\delta)}{4\mu^* Z^{*\frac{3}{2}}}\left\{\frac{2(\delta-1)}{\gamma} + 4V^* + (1-3\delta)\right\}\{(3\gamma-1)V^*-2\} \tag{5.2.57}$$

Therefore, the eigenvalue (in (τ, η)-plane)

$$C = V - \delta + \sqrt{Z},$$

which vanishes at the critical point in the case of the self-similar solution, satisfies the approximate equation

$$\frac{\partial C}{\partial \tau} + C\frac{\partial C}{\partial \eta} = \alpha C + \beta(\eta - \eta^*) \tag{5.2.58}$$

in the neighbourhood of the critical point, where the constants α and β are given by

$$\alpha = \alpha_1\sqrt{Z}^* = -\frac{1}{2}(5\delta - 1) \tag{5.2.59}$$

and

$$\beta - 1 = \beta_1 Z^* = -\left\{\frac{\delta-1}{\gamma} + 2V^* + \frac{1-3\delta}{2}\right\}\left\{\frac{3\gamma-1}{2}V^* - 1\right\} \tag{5.2.60}$$

since $\mu^* = -1$ and $\sqrt{Z}^* = \delta - V^*$.

In our particular problem we have $\delta = 0.638$ for $\gamma = 3$ and tends to 1 as γ tends to 1. Thus, for all physically realistic values of γ we have $\alpha < 0$. The time τ decreases from ∞ to $-\infty$ continuously as t increases from $-\infty$ to 0 up to the instant of collapse.

For $\gamma = 1.4$, we have $\delta = 0.717$, $V^* = 0.469$, $\alpha = -1.29$ and $\beta = 0.0397$, $\lambda_1 = 0.030$, $\lambda_2 = -1.322$ and the singular point $\eta = \eta^*, C = 0$ of the characteristic equations $d\eta/d\tau = C$, $dC/d\tau = \alpha C + \beta(\eta - \eta^*)$ is a saddle point. The phase-plane will be similar to that in the Fig. 5.2.3 with the only exception being that the origin O will correspond to the point $(\eta^*, 0)$. If we integrate equations (5.2.46 - 48) numerically from the shock boundary, we can easily verify that in the neighbourhood of the critical point, the self-similar flow is represented by the line lof. In order to study the growth of perturbations with time we must reverse the direction of all arrows on the integral curves since, as time increases to zero, τ decreases to $-\infty$. After reversing the direction of arrows we get the direction of propagation of waves as t increases. Owing to this change in the direction of arrows, our case now corresponds to that of $\alpha > 0$ of the previous section where only one of the four steady flows passing through the saddle point is stable. Since $\alpha < 0$, the area of a perturbation $S = S_0 e^{\alpha\tau}$ increases without limit as τ tends to $-\infty$, i.e., as t tends to zero from negative side. The leading and trailing fronts of a disturbance of the solution represented by log moves away from the critical point and even though the area of disturbance in (η, C)-plane increases, its boundary tends to coincide with lof as t tends to -0 or τ tends to $-\infty$. Therefore, our self-similar flow is stable in the neighbourhood of the critical point for radially symmetric disturbances bounded in space.

Thus, we have proved that *the spherically symmetric flow behind an imploding strong shock wave moving into a gas of uniform density* ρ_0 *and zero pressure* (Guderley (1942)) *is locally stable at the singular point.*

Some other results have also been proved using the above method. One of them is:

All similarity solutions by Hunter (1963) for the flow into a cavity in a liquid, with the variable speed of the cavity boundary are unstable (Tagare and Prasad (1970)).

The BKPS theory discussed in these sections, gives a local nonlinear stability of steady and self-similar solutions. If a steady or self-similar solution is unstable in the neighbourhood of a singular point, we come to a definite conclusion that the solution is globally unstable. However, we can not make a definite conclusion about the solution when it is stable in the neighbourhood of a singular point because the source of instability may be at points away from the singular point.

5.3 Waves in a multi-dimensional steady transonic flow

Consider a two- or three-dimensional steady transonic flow with density $\rho_0(\mathbf{x})$, fluid velocity $\mathbf{q}_0(\mathbf{x})$ and pressure $p_0(\mathbf{x})$ and let $P^*(\mathbf{x}^*)$ be a point on the sonic surface. We wish to consider an upstream propagating small amplitude trapped pulse in short wave approximation in an ϵ neighbourhood of the point P^*. Equations of such upstream propagating waves in a general steady solution (i.e., not necessarily in a transonic region) have been derived in section 4.3.2, which we use now.

We use a local coordinate system $(t', \xi, \eta_1, \eta_2)$ defined by

$$t' = t\,,\ \ \xi = \langle \mathbf{n}^*, \mathbf{x} - \mathbf{x}^* \rangle\,,\ \ \eta_p = \langle \mathbf{a}^{(p)}, \mathbf{x} - \mathbf{x}^* \rangle\,,\ \ p = 1, 2 \qquad (5.3.1)$$

where

$$\mathbf{n}^* = \mathbf{q}^*/|\mathbf{q}^*| \qquad (5.3.2)$$

and the matrix

$$N = \begin{bmatrix} n_1^* & n_2^* & n_3^* \\ a_1^{(1)} & a_2^{(1)} & a_3^{(1)} \\ a_1^{(2)} & a_2^{(2)} & a_3^{(2)} \end{bmatrix} \qquad (5.3.3)$$

is orthogonal. Short wave assumption for the pulse implies the derivative of the amplitude w with respect to ξ to be $O(\epsilon^{-1})$ but those with respect to η_1 and η_2 need to be of order 1. Therefore, we need to introduce a variable ξ' by

$$\xi' = \xi/\epsilon \qquad (5.3.4)$$

We note that ξ' is the local value of θ^* defined by (4.2.5).

The fluid velocity $\mathbf{q}_0$ in the steady flow near P^* is given by

$$\mathbf{q}_0 = (a^* + \epsilon a^* q_{N0}\,,\ \ \epsilon a^* q_{T10},\ \ \epsilon a^* q_{T20}) \qquad (5.3.5)$$

where the first, second and third components are the components of the fluid velocity in steady flow in ξ, η_1 and η_2 directions, respectively. Using the Bernoulli's equation, we get the following expression for the sound velocity

$$a_0 = a^*\{1 - \frac{1}{2}(\gamma - 1)\epsilon q_{N0}\} + O(\epsilon^2) \qquad (5.3.6)$$

Now we use the equations (4.3.41 - 43) of the WNLRT derived in the section 4.3.2. The time-rate of change of w along the nonlinear rays is given by

$$\frac{dw}{dt} = \frac{\partial w}{\partial t} + \langle \mathbf{q}_0 - \mathbf{n}a_0 + \frac{\gamma+1}{2}\mathbf{n}\tilde{w}, \nabla \rangle w$$

i.e.,

$$\frac{dw}{dt} = \frac{\partial w}{\partial t} + \langle \mathbf{q}_0, \nabla \rangle w - \left(a_0 - \frac{\gamma+1}{2}\epsilon\tilde{w}\right) \langle \mathbf{n}, \nabla \rangle w \tag{5.3.7}$$

where we have used $w = \epsilon\tilde{w}$ in some terms. $\tilde{w}$ is of the order unity. Using the transformation (5.3.3), we get

$$\frac{dw}{dt} = \frac{\partial w}{\partial t} + \left\{ (\mathbf{q}_0, \mathbf{n}^*) - \left(a_0 - \frac{\gamma+1}{2}\epsilon\tilde{w}\right) \langle \mathbf{n}, \mathbf{n}^* \rangle \right\} \frac{\partial w}{\partial \xi} + \sum_{p=1}^{2} \left\{ \langle \mathbf{q}_0, \mathbf{a}^{(p)} \rangle - \left(a_0 - \frac{\gamma+1}{2}\epsilon\tilde{w}\right) \langle \mathbf{n}, \mathbf{a}^{(p)} \rangle \right\} \frac{\partial w}{\partial \eta_p} \tag{5.3.8}$$

As the trapped nonlinear wavefront moves slowly near the sonic point P^*, its normal $\mathbf{n}$ will differ from $\mathbf{n}^* = \mathbf{q}^*/|\mathbf{q}^*|$ by a small quantity. We assume that $|\mathbf{n} - \mathbf{n}^*| = O(\epsilon)$.

Since the expression for $\mathbf{q}_0$ is given by (5.3.5), the terms of order 1 in the curly brackets in the coefficients of $\frac{\partial}{\partial \xi}$ and $\frac{\partial}{\partial \eta_p}$ vanish leaving only terms of the order ϵ. But, in short wave approximation $\frac{\partial w}{\partial \xi} = O(\frac{1}{\epsilon})$ and $\frac{\partial w}{\partial \eta^{(p)}} = O(1)$, so that

$$\frac{dw}{dt} = \frac{\partial w}{\partial t} + \left\{ \langle \mathbf{q}_0, \mathbf{n}^* \rangle - \left(a_0 - \frac{\gamma+1}{2}\epsilon\tilde{w}\right) \langle \mathbf{n}, \mathbf{n}^* \rangle \right\} \frac{\partial w}{\partial \xi} + O(\epsilon)$$

Using (5.3.5-6) and $\mathbf{n} - \mathbf{n}^* = O(\epsilon)$, we get

$$\frac{dw}{dt} = \frac{\partial w}{\partial t} + a^* \left\{ \frac{\gamma+1}{2}\epsilon(q_{N0} + \tilde{w}) - \langle \mathbf{n} - \mathbf{n}^*,\ \mathbf{n}^* \rangle \right\} \frac{\partial w}{\partial \xi} \tag{5.3.9}$$

For future reference, we approximate q_{N0} also in the neighbourhood of the $P^*(\mathbf{x}^*)$

$$q_{N0} = \frac{1}{a^*\epsilon} \left\{ \left(\frac{\partial q_0}{\partial \xi}\right)^* \xi + \sum_{p=1}^{2} \left(\frac{\partial q_0}{\partial \eta_p}\right)^* \eta_p \right\} + O(\epsilon)$$

where q_0 is the fluid speed in the steady flow along a stream line (so that $q^* = \langle \mathbf{n}^*, \mathbf{q}^* \rangle = a^*$) and ξ, η_p are of the order of ϵ. Thus, we finally get

$$\frac{dw}{dt} = \frac{\partial w}{\partial t} + (c_\xi \xi + c_{\eta_1}\eta_1 + c_{\eta_2}\eta_2 + c_w w)\frac{\partial w}{\partial \xi} + O(\epsilon) \qquad (5.3.10)$$

where

$$c_\xi = \frac{\gamma+1}{2}\left(\frac{\partial q_0}{\partial \xi}\right)^*, \quad c_{\eta_p} = \frac{\gamma+1}{2}\left(\frac{\partial q_0}{\partial \eta_p}\right)^*, \quad c_w = \frac{\gamma+1}{2}a^* \quad (5.3.11)$$

and we note that all quantities ξ, η_p and w are of the order ϵ in a neighbourhood of the point P^*.

We shall now approximate the term $K - a_0\Omega$ on the right hand side of (4.3.43). In the neighbourhood of the sonic point for a wavefront with normal $\mathbf{n}$ such that $|\mathbf{n}-\mathbf{n}^*| = O(\epsilon)$ and $|\mathbf{q}_0/q_0-\mathbf{n}^*| = O(\epsilon)$, the first term of K is of the order of ϵ, so that

$$\begin{aligned} K &= -\frac{\gamma}{2}\frac{\partial q_{\alpha 0}}{\partial x_\alpha} - \frac{1}{2}n_\alpha n_\beta \frac{\partial_{\beta 0}}{\partial x_\alpha} + O(\epsilon) \\ &= -\frac{\gamma}{2}\frac{\partial q_{\alpha 0}}{\partial x_\alpha} - \frac{1}{2}\frac{q_{\alpha 0}q_{\beta 0}}{q_0^2}\frac{\partial q_{\beta 0}}{\partial x_\alpha} + O(\epsilon) \end{aligned} \qquad (5.3.12)$$

where we are reminded that $q_0 = |\mathbf{q}_0|$.

From the equation of continuity for a steady flow, we get

$$\text{div } \mathbf{q}_0 = -\frac{1}{\rho_0}\left(q_{\beta 0}\frac{\partial \rho_0}{\partial x_\beta}\right) = -\frac{1}{\rho_0}q_0\frac{\partial \rho_0}{\partial s}$$

where $\frac{\partial}{\partial s} = \langle (\mathbf{q}_0/q_0), \nabla \rangle$ is the spatial rate of change along a stream line. Hence

$$\text{div } \mathbf{q}_0 = -\frac{q_0}{\rho_0 a_0^2}\frac{\partial p_0}{\partial s}$$

Further

$$\frac{q_{\alpha 0}q_{\beta 0}}{q_0^2}\frac{\partial q_{\beta 0}}{\partial x_\alpha} = \frac{1}{2q_0}\frac{\partial |\mathbf{q}_0|^2}{\partial s} = \frac{\partial q_0}{\partial s}$$

Taking the inner product of the momentum equation with $\mathbf{q}_0$ for a steady flow, we get

$$q_{\alpha 0}q_{\beta 0}\frac{\partial q_{\alpha 0}}{\partial x_\beta} + \frac{q_0}{\rho}\frac{\partial p_0}{\partial s} = 0$$

Hence,

$$\text{div } \mathbf{q}_0 = \frac{1}{a_0^2}\left(q_{\alpha 0} q_{\beta 0} \frac{\partial q_{\alpha 0}}{\partial x_\beta}\right) = \frac{q_0^2}{a_0^2}\frac{\partial q_0}{\partial s} \tag{5.3.13}$$

Therefore, the expression for K in a neighbourhood of P^* becomes

$$K = -\frac{\gamma}{2}\frac{q_0^2}{a_0^2}\frac{\partial q_0}{\partial s} - \frac{1}{2}\frac{\partial q_0}{\partial s} + O(\epsilon)$$

Using the result that in the transonic region $q_0^2 = a_0^2 + O(\epsilon)$, we get an approximate expression for K

$$K = -\frac{\gamma + 1}{2}\frac{\partial q_0}{\partial s} \tag{5.3.14}$$

For future reference we give the value of K at the sonic point

$$K^* = -\frac{\gamma + 1}{2}\left(\frac{\partial q}{\partial \xi}\right)^* \tag{5.3.15}$$

Now we pass on to an approximation of the equation (4.3.42) in a neighbourhood of the sonic point P^* assuming $|\mathbf{n} - \mathbf{n}^*|$ to be of the order of ϵ. We denote by $\boldsymbol{\Psi}_0$ the expression consisting of the first two terms on the right hand side of (4.3.42) i.e.,

$$\boldsymbol{\Psi}_0 = \mathbf{L}a_0 - n_\beta \mathbf{L}q_{\beta 0} \tag{5.3.16}$$

Then

$$\frac{d\mathbf{n}}{dt} = \boldsymbol{\Psi} = \boldsymbol{\Psi}_0 - \frac{\gamma + 1}{2}\mathbf{L}w \tag{5.3.17}$$

We write

$$\begin{aligned}\boldsymbol{\Psi}_0 &= \mathbf{L}a_0 - \frac{q_{\beta 0}}{q_0}\mathbf{L}q_{\beta 0} - \left(\frac{q_{\beta 0}}{q_0} - n_\beta\right)\mathbf{L}q_{\beta 0} \\ &= \mathbf{L}(a_0 - q_0) - \left(\frac{q_{\beta 0}}{q_0} - n_\beta^*\right)\mathbf{L}q_{\beta 0} - (n_\beta - n_\beta^*)\mathbf{L}q_{\beta 0}\end{aligned} \tag{5.3.18}$$

From (5.3.5) we get

$$q_0 = a^*(1 + \epsilon q_{N0}) + O(\epsilon^2) \tag{5.3.19}$$

which, when combined with (5.3.6), gives

$$a_0 - q_0 = -\frac{\gamma + 1}{2}\epsilon a^* q_{N0} + O(\epsilon^2) \tag{5.3.20}$$

To evaluate the second term in (5.3.18) we choose an instantaneous coordinate system such that x_1, x_2 and x_3 axes are in the directions of ξ_1, η_1 and η_2. Then $\mathbf{n}^* = (1,0,0)$ and $\frac{q_{10}}{q_0} = \{a^*(1+\epsilon q_{N0})\}/\{a^*(1+\epsilon q_{N0})\} + O(\epsilon^2) = 1 + O(\epsilon^2)$ so that $\frac{q_{10}}{q_0} - n_1^* = O(\epsilon^2)$ and $\left(\frac{q_{10}}{q_0} - n_1^*\right)\mathbf{L}q_{10} = O(\epsilon^3)$, Similarly $\left(\frac{q_{20}}{q_0} - n_2^*\right)\mathbf{L}q_{20} = O(\epsilon^2)$ and $\left(\frac{q_{30}}{q_0} - n_3^*\right)\mathbf{L}q_{30} = O(\epsilon^2)$. According to our assumption $|\mathbf{n} - \mathbf{n}^*| = O(\epsilon)$, so that the third term $(n_\beta - n_\beta^*)\mathbf{L}q_{\beta_0} = O(\epsilon^2)$. Therefore, the most dominant term in (5.3.18) is the first term and is of the order of ϵ which is also the order of the term $\mathbf{L}w$ in (5.3.17). To the leading order terms, the equation (5.3.17) reduces to

$$\frac{d\mathbf{n}}{dt} = -\frac{\gamma+1}{2}\mathbf{L}(\epsilon a^* q_{N0} + w)\,, \quad w = O(\epsilon) \tag{5.3.21}$$

The first term in the bracket on the right hand side of (5.3.21) looks deceptive. $\mathbf{L}$ is the differentiation over the length scale of order 1 and the approximation (5.3.5) is valid in a spherical neighbourhood of P^* of radius of the order of ϵ. Hence,

$$\mathbf{L}(\epsilon a^* q_{N0}) = \mathbf{L}(q_0) \tag{5.3.22}$$

the order of which needs to be carefully examined, it may be of the order of 1. However, if the approximation (5.3.5) is valid over a distance of order 1 in a direction transverse to the streamlines, then this term is indeed of order ϵ as in the case of a thin aerofoil discussed below in Case 2.

When all components of the ray velocity χ given by (2.4.6) vanish at some point P^* with a special choice of $\mathbf{n} = \mathbf{n}^*$, it is not necessary that the time-rate of rotation $\boldsymbol{\Psi}$ of the wavefront given by (2.4.7) will also vanish simultaneously. To obtain a deeper understanding of the influence of non-zero turning on a trapped wave, we consider a model which consists of a pair of ordinary differential equations

$$\frac{dx}{dt} = x + n\,, \quad \frac{dn}{dt} = \psi_* = \text{ constant } \neq 0 \tag{5.3.23}$$

with the initial condition

$$x(0) = 0, \quad n(0) = 0 \tag{5.3.24}$$

The solution of this problem is $x = \psi_*(e^t - (t+1))$, $n = \psi_* t$. For small times $x = \frac{1}{2}\psi_* t^2 + O(t^3)$. Thus, we notice that the *displacement*

x in time of the order $\epsilon^{1/2}$ is ϵ, where ϵ is small. This implies that it takes considerably a longer time (say 0.5) in order to make a small displacement (say 0.25). Thus, there is a considerable *trapping* of a quantity whose *displacement* is x due to the nonvanishing *rate of rotation* ψ_*.

We consider now three cases:

Case 1: Pulse in an arbitrary three-dimensional transonic flow

We note that, in general, the value $\mathbf{\Psi}^*$ of $\mathbf{\Psi}_0$ at P^* with $\mathbf{n} = \mathbf{n}^*$ is not zero and is of the order 1. Since $w = O(\epsilon)$ and also $\mathbf{L}w = O(\epsilon)$, we may neglect the term containing w in the equation (5.3.17). Trapped transonic waves are possible only when $|\mathbf{n} - \mathbf{n}^*|$ is small i.e., for small time t. Hence,

$$\mathbf{n} = \mathbf{n}_0 + \mathbf{\Psi}^* t \quad , \quad t = O(\epsilon) \tag{5.3.25}$$

where $\mathbf{n}|_{t=0} = \mathbf{n}_0$ such that $|\mathbf{n}_0 - \mathbf{n}^*| = O(\epsilon)$. Compared to one-dimensional transonic waves discussed in the previous section, there is an additional complication. The question of rotation of a wavefront did not arise in one space variable problem. When $c_0(x^*, \mathbf{u}(x^*)) = 0$ at the sonic point and $|w| = O(\epsilon)$, the wave was trapped at the sonic point. Therefore, even a small amplitude pulse is not strictly trapped at a sonic point in multi-dimensions but as seen from the model equations (5.3.23 - 24) it stays there for a considerable time.

Using (5.3.10, 15 and 25) and setting $a_0 = a^*$, we get an equation for the amplitude

$$w_t + (c_\xi \xi + c_{\eta_1} \eta_1 + c_{\eta_2} \eta_2 + c_w w) w_\xi = (K^* - a^* \Omega) w \tag{5.3.26}$$

where Ω is the value of the initial mean curvature, calculated using $\mathbf{n} = \mathbf{n}_0$. We note an important relation

$$c_\xi = -K^* \tag{5.3.27}$$

The equation (5.3.26) needs to be solved only for a small time $t = O(\epsilon)$. When the initial geometry of the wavefront is given, $\mathbf{n}_0$ is known and Ω appearing in (5.3.26) can be calculated. Then (5.3.26) can be solved by solving both the ray equations

$$\frac{d\xi}{dt} = c_\xi\, \xi + c_{\eta_1} \eta_1 + c_{\eta_2} \eta_2 + c_w w, \quad \frac{d\eta_1}{dt} = 0, \quad \frac{d\eta_2}{dt} = 0 \tag{5.3.28}$$

and the amplitude equation

$$\frac{dw}{dt} = (K^* - a^*\Omega)w \tag{5.3.29}$$

Equation (5.3.29) shows that the wavefront propagates in ξ direction. Since ξ, η_p, and w are of order ϵ, during the time $O(\epsilon)$ the displacement in ξ direction is $O(\epsilon^2)$. Since $\mathbf{n}$ and $\mathbf{n}_0$ (the initial unit normal) are close to $\mathbf{n}^*$, the curvature Ω remains small compared to K and hence the equation (5.3.28) finally becomes

$$\frac{dw}{dt} = K^* w \tag{5.3.30}$$

Ravindran (1979) analyzed this case in detail for small time and investigated propagation of a transonic pulse with a slightly different mathematical model which included the effect of rotation of the wavefront on a trapped plane pulse in a transonic flow.

Case 2: Waves in a two-dimensional transonic flow produced by a thin aerofoil

Consider a two-dimensional steady transonic flow produced by a thin aerofoil with τ as the camber i.e., the ratio of its maximum thickness to its length l. Then τ is a small nondimensional quantity. Assume a free undisturbed flow with high subsonic speed in the direction of the aerofoil, which is chosen as the x-direction. The y direction is perpendicular to the aerofoil. The steady transonic flow with an embedded supersonic flow around the thin aerofoil can be represented by (Guderley (1962))

$$q_{10} = a^* + \tau a^* \bar{q}_{10}\ , \quad q_{20} = \tau^{3/2} a^* \bar{q}_{20}\ , \quad q_{30} = 0 \tag{5.3.31}$$

where $\bar{q}_{10}$ and $\bar{q}_{20}$ are quantities of order 1 and are functions of nondimensional variables

$$\bar{x} = x/l\ , \quad \bar{y} = \tau^{1/2} y/l\ , \quad \bar{t} = t a^* \tau / l, \tag{5.3.32}$$

From Bernoulli's equation, we get the following approximate expression for the sound speed in terms of $\bar{q}_{10}$,

$$a_0 = a^*\{1 - \frac{1}{2}(\gamma - 1)\tau\bar{q}_{10} - \frac{1}{8}(\gamma - 1)(\gamma + 1)\tau^2\bar{q}_{10}^2\} + O(\tau^3) \tag{5.3.33}$$

Using the above scaling it is possible to derive a simpler form of the equations for WNLRT for the upstream propagating small amplitude waves assuming the amplitude parameter ϵ to be same as τ (see Prasad and Krishnan (1977) for an introduction). It is easier to numerically integrate these equations, which we shall write below. What is more important theoretically is that the scaling (5.3.31 - 32) gives a flow field in which (i) waves, which are almost perpendicular to the x-axis, are trapped so that the angle θ, which the normal to the wavefront makes with the x-axis, then is small and will have to be appropriately scaled; (ii) the first two terms in (4.3.42) i.e., $\mathbf{L}a_0$ and $\eta_\beta \mathbf{L} q_{\beta 0}$ become small and of the same order as the third term $\mathbf{L}w$ so that unlike Case 1 discussed above, the wavefront turns slowly and remains trapped in the transonic region for a time interval of the order 1; and (iii) we are able to follow the complete history of the nonlinear wavefront as it transverses the transonic region which extends over a length scale of the order of l. Due to the last consideration, we need to ignore in this case some of the results such as (5.3.10) which are valid only locally. In this case, K and Ω on the right hand side of (4.3.43) are also of the same order as that of w. We introduce a scaled angle θ and scaled amplitude $\bar{w}$ by

$$\bar{\theta} = \theta \tau^{1/2}, \quad \bar{w} = w/(\tau a^*) \tag{5.3.34}$$

We note that $n_1 = \cos\theta$ and $n_2 = \sin\theta$.

The approximate equations which can be derived, following the procedure in Prasad and Krishnan (1977), are

$$\frac{d\bar{x}}{dt} = \frac{1}{2}(\gamma+1)\bar{q}_{10} + \frac{1}{2}\bar{\theta}^2 + \frac{1}{2}(\gamma+1)\bar{w} \tag{5.3.35}$$

$$\frac{d\bar{y}}{dt} = -\theta \tag{5.3.36}$$

$$\frac{d\bar{\theta}}{dt} = \frac{\gamma+1}{2}\left\{\left(\bar{\theta}\frac{\partial}{\partial \bar{x}} - \frac{\partial}{\partial \bar{y}}\right)\bar{q}_{10} + \bar{w}\right\} \tag{5.3.37}$$

and

$$\frac{d\bar{w}}{dt} = (\bar{K} - \bar{\Omega})\bar{w} \tag{5.3.38}$$

where

$$\bar{\Omega} = \frac{1}{2}\left\{\bar{\theta}\frac{\partial\bar{\theta}}{\partial\bar{x}} - \frac{\partial\bar{\theta}}{\partial\bar{y}}\right\}, \quad \bar{K} = -\frac{\gamma+1}{2}\frac{\partial \bar{q}_{10}}{\partial \bar{x}} \tag{5.3.39}$$

Some results based on the local analysis of the approximate equations have been discussed by Prasad and Krishnan (1977), neglecting the term containing $\bar{w}$ in (5.3.37) but they are too cumbersome to be discussed here. Now that the numerical solution of partial differential equations has become routine work due the availability of faster computers and better computational algorithms, it may be worthwhile to compute the full history of transonic pulses using the equations (5.3.35 - 39).

Case 3: Quasi-one-dimensional transonic waves

In Case 1, the gradients of the steady flow in directions perpendicular to the stream lines were of order 1, which resulted in a finite rate of turning of the wavefront. In Case 2, the gradient in y-direction was of the order of ϵ, which resulted in the trapping of waves in the transonic region for a time interval of the order 1. The approximation of the transport equation by expanding the quantities in the steady flow at a sonic point P^* (see (5.3.10) and (5.3.15)) results in (5.3.30).

Now we make an additional assumption that the wavefront remains plane and perpendicular to the stream lines passing through a sonic point P^*. Then $\Omega = 0$. The equation (4.3.42) for $\mathbf{n}$ is no longer relevant. With this assumption, the rate of turning of the wavefront is absent. Hence, the wavefront now stays near the sonic point for a time interval of the order 1. The approximate equation with this additional assumption for a general hyperbolic system in four independent variables was derived by Prasad (1973) in the form

$$\frac{\partial w}{\partial t} + \left(c_\xi \, \xi + \sum_{p=1}^{2} c_{\eta_p} \eta_p + c_w \, w \right) \frac{\partial w}{\partial \xi} = K^* w \tag{5.3.40}$$

where c_ξ, c_{η_p}, c_w and K^* are constants and depend on the steady flow. In the particular case of gas dynamic equations, from (5.3.11) and (5.3.15), we note that

$$c_\xi = -K^* = \frac{\gamma+1}{2} \left(\frac{\partial q_0}{\partial \xi} \right)^* \tag{5.3.41}$$

where $(\partial q_0/\partial \xi)^*$ represents the space rate of change of the fluid speed at the point x^* as we move with the fluid in the steady solution and hence is equal to the acceleration of the fluid elements at the sonic point P^* divided by a^*.

The relation (5.3.41) is very simple to deal with in (ξ_1, t)-plane, where $\xi_1 = \xi + (\sum_{p=1}^{2} c_{\eta_p}\eta_p)/c_\xi$. The transformation from (ξ, η_1, η_2, t) to $(\xi_1, \eta_1' = \eta_1, \eta_2' = \eta_2, t' = t)$ reduces the equation to

$$\frac{\partial w}{\partial t} + (c_\xi\, \xi_1 + c_w\, w)\frac{\partial w}{\partial \xi_1} = K^* w \tag{5.3.42}$$

The relation (5.3.41) is crucial and was responsible for the simplicity of Example 5.2.2 and led to $\alpha = 0$ in the case of the de Laval nozzle problem. It is this relation which makes a pulse bounded in ξ_1 direction to attain a stationary state near the sonic point as seen in Example 1.5.6. By assuming the wavefront to be plane and perpendicular to the flow direction at P^*, we have neglected the rotation of the wavefront, which we hope will not affect qualitatively the signature of the pulse along a ray. The effect of rotation is basically to make the trapped pulse free to move away from the sonic point but in a duration $O(\epsilon^{1/2})$. However, the simplification involved in critically examining the results with the help of (5.3.42) is so great that it is worth considering this model equation. For a complete analysis of this equation and description of the full history of the trapped waves, reference may be made to the work of Prasad (1973).

Chapter 6

WNLRT in a polytropic gas

6.1 Basic equations

Consider Euler equations (2.3.16 - 18) governing the motion of a polytropic gas. In section 4.3.2 we derived the equations of WNLRT for an upstream propagating wave (i.e., the wave corresponding to the eigenvalue $c_1 = \langle \mathbf{n}, \mathbf{q} \rangle - a$) on a given study solution $(\rho_0(\mathbf{x}), \mathbf{q}_0(\mathbf{x}), p_0(\mathbf{x}))$. In this section we shall derive the equations of WNLRT for a downstream propagating wave Ω_t (corresponding to the eigenvalue $c_5 = \langle \mathbf{n}, \mathbf{q} \rangle + a$) running into a uniform state at rest ($\rho_0 =$ constant, $\mathbf{q} = 0$ and $p_0 =$ constant). These equations can be deduced from those in the section 4.3.2 by taking $\rho_0 =$ constant, $\mathbf{q}_0 = 0$ and $p_0 =$ constant and then changing the sign of a_0 everywhere. This gives the ray equations

$$\rho - \rho_0 = (\rho_0/a_0)w, \quad \mathbf{q} = \mathbf{n}w, \quad p - p_0 = \rho_0 a_0 w \tag{6.1.1}$$

$$\frac{d\mathbf{x}}{dt} = \left(a_0 + \frac{\gamma+1}{2}w\right)\mathbf{n}, \quad \frac{d\mathbf{n}}{dt} = -\frac{\gamma+1}{2}\mathbf{L}w \tag{6.1.2}$$

and the transport equation for the amplitude w (which is assumed to be small)

$$\frac{dw}{dt} = \Omega a_0 w \tag{6.1.3}$$

where

$$\frac{d}{dt} = \frac{\partial}{\partial t} + \left(a_0 + \frac{\gamma+1}{2}w\right)\langle \mathbf{n}, \nabla\rangle \tag{6.1.4}$$

$$\Omega = -\frac{1}{2}\langle \nabla, \mathbf{n}\rangle = \text{ mean curvature of } \Omega_t \tag{6.1.5}$$

and

$$\mathbf{L} = \nabla - \mathbf{n}\langle \mathbf{n}, \nabla\rangle \tag{6.1.6}$$

We note the expression (2.4.20) for the components of $\mathbf{L}$ in terms of the tangential derivatives $\frac{\partial}{\partial \eta_\beta^\alpha}$ defined by (2.4.8). Another expression for the mean curvature of Ω_t is

$$\Omega = -\left(\frac{\partial n_1}{\partial \eta_3^1} + \frac{\partial n_2}{\partial \eta_3^2}\right) \tag{6.1.7}$$

We would like to emphasize once more (see derivation of 5.2.13) that the system (6.1.2 - 4) is a true generalization of Burgers' equation (1.1.6) to multi-dimensions for the propagation of a multi-dimensional nonlinear wavefront. Some other equations (Hunter (1995) or section 10.4.2) have been called two-dimensional Burgers' equations but they are valid neither for an arbitrarily curved wavefront nor do they account for nonlinearity in directions transversal to the direction of propagation (Prasad and Ravindran (1977)).

Since $|\mathbf{n}| = 1$, only two of the three equations in (6.1.2) are independent. Therefore, equations (6.1.2 - 3) from a system of six coupled equations for determination for successive positions $\mathbf{x}$ of the nonlinear wavefront, the unit normal $\mathbf{n}$ and the wavefront intensity w. The linear theory of Sommerfeld and Runge (1911)

$$\frac{d\mathbf{x}}{dt} = a_0\mathbf{n}, \quad \frac{d\mathbf{n}}{dt} = 0, \quad \frac{dw}{dt} = \Omega a_0 w \tag{6.1.8}$$

follows from the equations of WNLRT if we drop w from equations (6.1.2) so that the ray equations decouple from the amplitude equation. In this case, the rays and the successive positions of the wavefront can be constructed without any reference to the amplitude of the wave. This corresponds to the original statement of Huygens' wavefront construction, a generalization of which to a linear hyperbolic system was presented in section 3.2.2. In our weakly nonlinear theory, the amplitude is related to the curvature of the wavefront (or the ray tube area) by the same equation (i.e., the last equation in

(6.1.8)) but the nonlinear rays stretch due to the presence of w in the equation for $\mathbf{x}$ in (6.1.2) and the wavefront turns due to the non-uniform distribution of the amplitude on the wavefront (represented by $\mathbf{L}w$ in (6.1.2)). Thus, the amplitude of the wave modifies the rays and the wavefront geometry which, in turn, affects the growth and decay of the amplitude. Further, we note that only the tangential derivatives on Ω_t of w and $\mathbf{n}$ appear on the right hand sides of the equations (6.1.2 - 3). Therefore, given the initial position Ω of the wavefront and the distribution of the amplitude on it, all quantities on the right hand sides of (6.1.2 - 3) can be completely determined at $t = 0$. Hence, as in the case of a non-characteristic Cauchy problem, the evolution of the nonlinear wavefront and the distribution of the amplitude on it at later times can be determined from these equations. This implies that, in the short wave approximation, the nonlinear wavefront is self-propagating as explained in section 3.3.1. The Huygens' method of wavefront construction (its ray formulation is shown in 3.2.3 in free space) has now been very elegantly extended to the construction of a nonlinear wavefront in the short-wave limit, in this extension the amplitude also affects the position of the wavefront (Ramanathan (1985)). Finally, we mention that this theory can be easily extended to waves of arbitrary amplitude in the short-wave limit (Srinivasan (1987)). However, in this case, the equations can not be put in the elegant form of (5.1.2 - 3) and they will be valid for a time interval of the order of ϵ.

6.1.1 Non-dimensional form of equations of WNLRT in two-space-dimensions

In two dimensions, the components n_1, n_2 of the unit normal can be expressed in terms of θ, the angle which the normal to the wavefront makes with the x-axis (using x for x_1 and y for x_2): $n_1 = \cos\theta, n_2 = \sin\theta$. The mean curvature is now given by the expression $\Omega = \frac{1}{2}\left(\sin\theta\frac{\partial\theta}{\partial x} - \cos\theta\frac{\partial\theta}{\partial y}\right)$. We non-dimensionalize the x and y coordinates by a typical length L in the problem (say the radius of curvature of Ω_0 at a particular point), w by the sound velocity a_0 in the undisturbed state and time by L/a_0 and denote the non-dimensional quantities also by x, y, t and w. The equations (5.1.2 - 3) reduce to

$$\frac{dx}{dt} = \left(1 + \frac{\gamma+1}{2}w\right)\cos\theta \tag{6.1.9}$$

$$\frac{dy}{dt} = \left(1 + \frac{\gamma+1}{2}w\right)\sin\theta \tag{6.1.10}$$

$$\frac{d\theta}{dt} = -\frac{\gamma-1}{2}\frac{\partial w}{\partial \lambda} \tag{6.1.11}$$

and

$$\frac{dw}{dt} = -\frac{1}{2}w\frac{\partial \theta}{\partial \lambda} \tag{6.1.12}$$

where

$$\frac{\partial}{\partial \lambda} = -\frac{1}{\sin\theta}L_1 = \cos\theta\frac{\partial}{\partial y} - \sin\theta\frac{\partial}{\partial x} \tag{6.1.13}$$

The symbol $\frac{\partial}{\partial \lambda}$ stands for an operator defined by (6.1.13). There is no well-defined variable λ. In order to overcome this difficulty, we used the ray coordinate system which will also allow us to use the two conservation laws (3.3.13 - 14) with $T = 0$. We define a variable ξ', such that $g'd\xi'$ is an element of length (non-dimensional) along the wavefront at time t. Then

$$g'^2 = x_{\xi'}^2 + y_{\xi'}^2, \quad \frac{\partial}{\partial \xi'} = g'\frac{\partial}{\partial \lambda} \tag{6.1.14}$$

$t =$ constant gives the successive positions of the wavefront and $\xi' =$ constant represents the family of associated rays in the (x, y)-plane. t and ξ' can be treated as new independent variables. $\frac{d}{dt}$ in (x, y, t)-space becomes partial derivative $\frac{\partial}{\partial t}$ in the ray coordinate system. Then $\frac{\partial}{\partial t}$ and $\frac{\partial}{\partial \xi'}$ commute. Using

$$(y_{\xi'})_t = (y_t)_{\xi'} = -\left\{\left(1 + \frac{\gamma+1}{2}w\right)x_{\xi'}/g'\right\}_{\xi'}$$

$$(x_{\xi'})_t = (x_t)_{\xi'} = \left\{\left(1 + \frac{\gamma+1}{2}w\right)y_{\xi'}/g'\right\}_{\xi'}$$

we deduce from (5.1.14)

$$\begin{aligned} g'_t &= -\frac{1}{g'^2}\left(1 + \frac{\gamma+1}{2}w\right)(y_{\xi'}x_{\xi'\xi'} - x_{\xi'}y_{\xi'\xi'}) \\ &= -\left(1 + \frac{\gamma+1}{2}w\right)\theta_{\xi'} \end{aligned}$$

$$= \frac{2g\{1+(\gamma+1)w/2\}}{w} w_t, \text{ from (6.1.12)} \qquad (6.1.15)$$

Let m denote the Mach number of the wave intensity relative to the sound speed a_0 in the undisturbed state, then

$$m = 1 + \frac{\gamma+1}{2} w \qquad (6.1.16)$$

Integrating (6.1.15), we get g' in terms of m as

$$g' = \frac{f(\xi')}{(m-1)^2 e^2(m-1)} \qquad (6.1.17)$$

where f is determined from the distribution of the intensity m on the initial wavefront. Using (6.1.14) and (6.1.17) we can transform the pair of equations (6.1.11 - 12) with t and ξ' as independent variables. However, the coefficients in the equations would depend not only on m but also on ξ'. To get rid of the dependence of the coefficients on ξ', we introduce a new variable ξ by $\xi = \int^{\xi'} f(\xi')d\xi'$, so that

$$\frac{\partial}{\partial \xi} = g(m)\frac{\partial}{\partial \lambda} \qquad (6.1.18)$$

where

$$g(m) = (m-1)^{-2} e^{-2(m-1)} \qquad (6.1.19)$$

The equations (6.1.9 - 12) finally reduce to

$$x_t = m\cos\theta, \quad y_t = m\sin\theta \qquad (6.1.20)$$

$$\theta_t + \frac{1}{g} m_\xi = 0 \qquad (6.1.21)$$

and

$$m_t + \frac{m-1}{2g}\theta_\xi = 0 \qquad (6.1.22)$$

The system of four equations (6.1.20 - 22) is hyperbolic provided $m > 1$. One eigenvalues of the system is double and equal to zero so that the corresponding characteristic curves are the rays: $\xi =$ constant. The other two eigenvalues, denoted by c_1 and c_2 are

$$c_1 = -\sqrt{\frac{m-1}{2g^2}}, \quad c_2 = \sqrt{\frac{m-1}{2g^2}} \qquad (6.1.23)$$

The condition $m > 1$ or $w > 0$ for c_1, c_2 to be real implies (with the help of (6.1.1)) that the pressure p due to the wave is greater than the constant pressure p_0 in the undisturbed region, i.e., the wave under consideration is a compression wave. By *compression wave* we mean a curved pulse with nonlinear wavefronts on which the value of $w > 0$. The pulse, which is of finite extent in the direction of the normal to the wave, may be in the shape of a single hump and will deform in the direction of the normal. Therefore, a part of the pulse will have a compression phase and a part an expansion phase interpreted in the usual sense. In the case of an expansion wave in a pulse with pressure p less than p_0 i.e., $m < 1$, these two characteristics are imaginary.

The equations (6.1.20 - 22) of WNLRT give successive positions of a nonlinear wavefront in a curved pulse. Consider now a curved pulse in short wave approximation constituted of a one-parameter family of initial wavefronts at $t = 0$

$$x = x_0(\xi, s), \quad y = y_0(\xi, s) \tag{6.1.24}$$

where s is a parameter identifying a self-propagating wavefront and ξ varies along the wavefront. Let the distribution of the wave intensity at $t = 0$ on these wavefronts be prescribed as

$$m = m(\xi, s) > 1 \tag{6.1.25}$$

The initial value $\theta_0(\xi, s)$ of θ can be obtained from (6.1.24)

Solving the hyperbolic system (6.1.20-22) with initial data (6.1.24-25) we get

$$x = x(\xi, t, s), \quad y = y(\xi, t, s) \tag{6.1.26}$$

and

$$\theta = \theta(\xi, t, s), \quad m = m(\xi, t, s) \tag{6.1.27}$$

These four expressions give the position of a nonlinear wavefront (designated by a value of s) at time t and the distribution of the amplitude m on it. In the case of an expansion wave, i.e., $m < 1$, the system of four equations is not hyperbolic. However, the initial value problem could still be solved numerically to get a complete description of a nonlinear expansion wave (Ramanathan, Prasad and Ravindran (1984)).

It is interesting to note that on a given wavefront $s =$ constant, the two equations (6.1.21 - 22) decouple in the (ξ, t)-plane from the other two equations (6.1.20). The characteristic form of (6.1.21 - 22), with t and ξ as independent variables, is

$$Q \equiv \theta - \sqrt{8(m-1)} = \text{ constant along } \frac{d\xi}{dt} = -\sqrt{\frac{m-1}{2g^2}} \tag{6.1.28}$$

and

$$R \equiv \theta + \sqrt{8(m-1)} = \text{ constant along } \frac{d\xi}{dt} = \sqrt{\frac{m-1}{2g^2}} \tag{6.1.29}$$

It follows that changes in the value m on the compression wavefront and the slope θ of the normal to the wavefront propagate on the wavefront with velocities $\pm\sqrt{\frac{m-1}{2g^2}}$ in the (ξ, t)-plane. As $m \to 1+0$, both characteristic velocities tend to zero and for an expansion wave ($w < 0$) they become complex showing that no waves propagate on the wavefront. However, care is necessary in interpreting the results in the physical (x, y)-plane since $g \to \infty$ as $m \to 1$ so that a small arc of the wavefront corresponding to fixed $d\xi$ is $gd\xi$ which $\to \infty$.

6.1.2 A simple wave solution

We now discuss an interesting exact solution of the weakly nonlinear wave equations (6.1.20 - 22). This solution represents a simple wave moving on the wavefront. Consider a simple wave solution of (6.1.21 - 22) in which the Riemann invariant Q is constant i.e.,

$$\theta - \sqrt{8(m-1)} = \theta - 2\sqrt{(\gamma+1)w} = \text{ constant } = Q_0, \text{ say} \tag{6.1.30}$$

in a domain in (ξ, t)-plane. In such a simple wave, θ can be eliminated from the compatibility condition (6.1.29), so that m or w satisfies

$$m_t + \sqrt{\frac{m-1}{2g^2}} m_\xi = 0 \tag{6.1.31}$$

or

$$\frac{\partial w}{\partial t} + \left(\frac{\gamma+1}{4}w\right)^{\frac{1}{2}} \frac{\partial w}{\partial \lambda} = 0 \tag{6.1.32}$$

or in terms of t, x and y as independent variables it satisfies

$$\frac{\partial w}{\partial t} + \left\{\left(1 + \frac{\gamma+1}{2}w\right)\cos\theta - \left(\frac{\gamma+1}{4}w\right)^{\frac{1}{2}} \sin\theta\right\} \frac{\partial w}{\partial x}$$

$$+\left\{\left(1+\frac{\gamma+1}{2}w\right)\sin\theta+\left(\frac{\gamma+1}{4}w\right)^{\frac{1}{2}}\cos\theta\right\}\frac{\partial w}{\partial y}=0 \quad (6.1.33)$$

where w and θ are related by (6.1.30)

We call a characteristic curve of this simple wave, either in the (ξ,t)-plane or in (z,y,t)-space, a "characteristic curve on the wavefront" or briefly CCWF. For the simple wave under consideration, these CCWFs are given by

$$\frac{dx}{dt}=\left(1+\frac{\gamma+1}{2}w\right)\cos\theta-\left(\frac{\gamma+1}{4}w\right)^{\frac{1}{2}}\sin\theta \qquad (6.1.34)$$

$$\frac{dy}{dt}=\left(1+\frac{\gamma+1}{2}w\right)\sin\theta-\left(\frac{\gamma+1}{4}w\right)^{\frac{1}{2}}\cos\theta \qquad (6.1.35)$$

and

$$\frac{dw}{dt}=0,\quad \frac{d\theta}{dt}=0 \qquad (6.1.36)$$

where w and θ satisfy (6.1.30). We shall get another family of CCWFs by taking the Riemann invariant $R=R_0$ in (6.1.29) to be constant on the moving wavefront. Thus, in the case of a compression wavefront, a linear wave moving along the ray: $\theta=$ constant, $x=x_0+t\cos\theta$, $y=y_0+t\sin\theta$, breaks due to nonlinearity into two waves, one moves along the CCWF (6.1.34 - 36) and the other, given by taking the second Riemann invariant to be constant, moves along the second family of CCWFs.

Consider now a solution of (6.1.31) in the (ξ,t)-plane. Since the eigenvalue $\sqrt{(m-1)/(2g^2)}=\{\frac{1}{\sqrt{2}}(m-1)^{5/2}e^{2(m-1)}\}$ is an increasing function of m, any initial distribution $m_0(\xi)$ with $m_0'(\xi)<0$ will end up in a multi-valued solution $m(\xi,t)$ after a critical time t_c. It is simple to find the time t_c from the implicit form of the solution

$$m=m_0(\xi_0),\quad \xi_0=\xi-\sqrt{\frac{m-1}{2g^2}}t \qquad (6.1.37)$$

which relates the derivative m_ξ at (ξ,t) to the derivative $m_0'(\xi_0)$ of the initial data connected by the characteristic curve $\xi-\sqrt{\frac{m-1}{2g^2}}t=\xi_0$. This relation is

$$m_\xi=\frac{m_0'(\xi_0)}{1+\frac{t}{2\sqrt{2}}(m_0-1)^{3/2}e^{2(m_0-1)}m_0'(\xi_0)(5+4(m_0-1))} \qquad (6.1.38)$$

It shows that for $m > 1$, if $m_0' < 0$ on a point set S (which is a union of open intervals) of the initial line, there exists a time t_c where $\lim_{t \to t_c - 0} m_\xi = \infty$ for some value of ξ:

$$t_c = 2\sqrt{2} \min_{\xi_0 \in S} \{m_0 - 1)^{3/2} e^{2(m_0 - 1)} |m_0'(\xi)|(5 + 4(m_0 - 1))\}^{-1} \quad (6.1.39)$$

(6.1.39) was first derived by Ramanathan, Prasad and Ravindran in 1984.

The appearance of a multi-valued solution beyond t_c means that the simple wave solution of (6.1.21 - 22) breaks down and one needs to consider a property defined weak solution containing discontinuities in m and θ. A discontinuity in θ means that a kink on the wavefront would appear in the solution.

6.2 Geometrical features of a nonlinear wavefront

In this section we shall use the equations of WNLRT to study all the possible shapes which a nonlinear wavefront can have. Some of these shapes involve kinks (see section 3.3.3). Hence, we need to consider the kinematical conservation form of the equations (6.1.21 - 22), which can be obtained from (3.3.13 - 14) by choosing $T = 0$ and $C = m$

$$(g \sin \theta)_t + (m \cos \theta)_\xi = 0 \quad (6.2.1)$$

$$(g \cos \theta)_t - (m \sin \theta)_\xi = 0 \quad (6.2.2)$$

where g is a function of m given by (6.1.19):

$$g = (m - 1)^{-2} e^{-2(m-1)} \quad (6.2.3)$$

Let us denote the kink velocity in (ξ, t)-plane by s. Then either directly from (6.2.1 - 2) or from (3.3.25) we get (setting $K = s$, $T = 0$, $C = m$)

$$s^2 = \frac{m_l^2 - m_r^2}{g_r^2 - g_l^2} \quad (6.2.4)$$

where the subscripts l and r on any quantity represent the limiting values of the quantity as we approach the kink from lower and higher values of ξ at a fixed time. Similarly, from (3.3.29) we get

$$\cos(\theta_r - \theta_l) = \frac{m_l g_l + m_r g_r}{m_l g_r + m_r g_l} \tag{6.2.5}$$

6.2.1 Elementary wave solutions and their interpretation as elementary shapes

Elementary wave solutions of (6.2.1 - 3) are solutions of the form $m(\xi, t) = m(\xi/t)$, $\theta(\xi, t) = \theta(\xi/t)$. *These are centered rarefaction wave solutions with the center at the origin of the* (ξ, t)*-plane and shock waves passing through the origin.*

We denote centered rarefaction waves of the first and second characteristic family by 1-R and 2-R, respectively. In a 1-R wave the corresponding Riemann invariant is constant i.e., $\theta + \sqrt{(8(m-1))} =$ constant. Suppose the constant state on the left of the 1-R wave in the (ξ, t)-plane is (m_l, θ_l), then by rotation of the coordinate axes we can always choose $\theta_l = 0$ i.e., for 1-R wave we have the relation (6.1.29)

$$\theta + \sqrt{(8(m-1))} = \sqrt{(8(m_l - 1))} \tag{6.2.6}$$

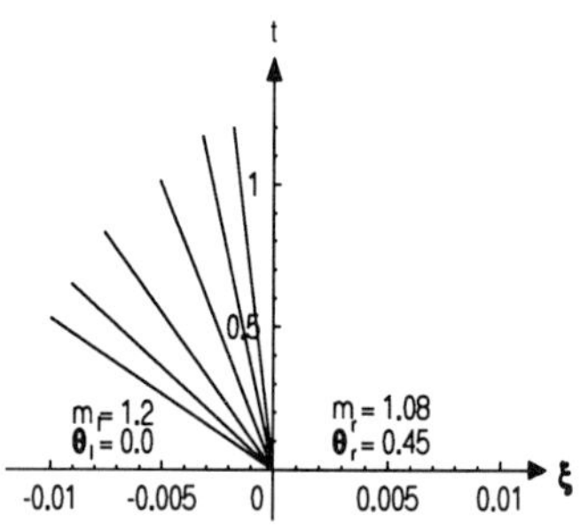

Fig. 6.2.1(a): An example of a 1-R wave i.e., centered simple wave of the first family in (ξ, t)-plane. The fan of characteristic curves are shown in the figure.

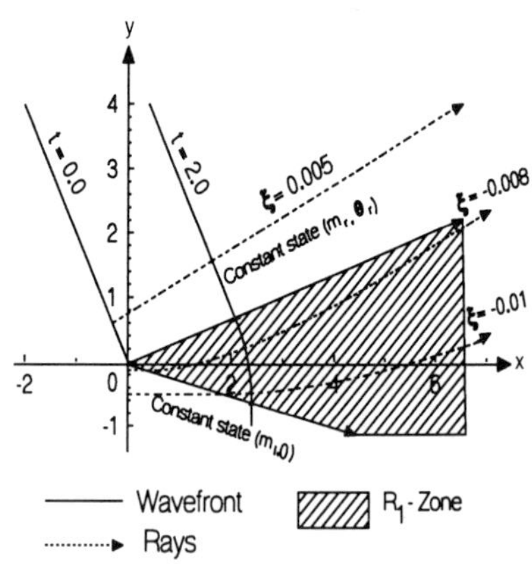

Fig. 6.2.1(b): Geometrical features of the front associated with the solution in Fig. 6.2.1(a). Elementary shape $\mathcal{R}_1$ is a part of the front in R_1-zone.

If the state on any straight characteristic in 1-R wave in (ξ, t)-plane is (m, θ), then $\lambda_1\ (m_l) < \lambda(m)$ which implies $m_l > m$. Then the relation (6.2.6) gives $\theta > 0$. At the trailing end of the 1-R wave in ξ-space, the wave merges into a constant state (m_r, θ_r) and these inequalities remain valid i.e., $m_r < m_l$ and $\theta_r > 0$. Fig. 6.2.1(a) represents a typical 1-R wave solution in (ξ, t)-plane and Fig. 6.2.1(b) represents its image in the (x, y)-plane. Similarly, Fig. 6.2.2(a) represents a typical 2-R wave solution in (ξ, t)-plane and Fig. 6.2.2(b) its image in (x, y)-plane, where we note that $m_r > m_l$ and $\theta_r > 0$. We call a geometrical shape of a front in (x, y)-plane obtained from an elementary wave solution in (ξ, t)-plane an *elementary shape*. We observe that the elementary shapes in Figs 6.2.1(b) (we denote it by $\mathcal{R}_1$) and 6.2.2(b) (we denote it by $\mathcal{R}_2$) are convex smooth wavefronts and look almost the same geometrically but $\mathcal{R}_1$ in Fig. 6.2.1(b) propagates downwards on the wavefront whereas $\mathcal{R}_2$ in Fig. 6.2.2(b) moves upwards. Note that the rays in Fig. 6.2.1(b) cross the $\mathcal{R}_1$ region denoted by R_1-zone from below whereas in Fig. 6.2.2(b) they cross R_2-zone from above.

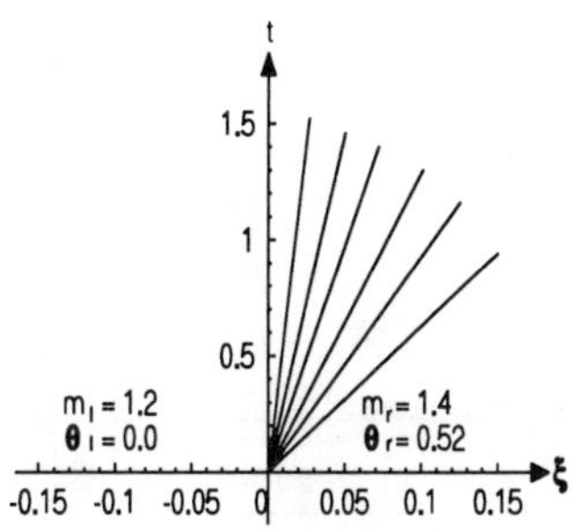

Fig. 6.2.2(a): An example of a 2-R wave i.e., centered simple wave of the second family in (ξ, t)-plane. The fan of characteristic curves is shown in the figure.

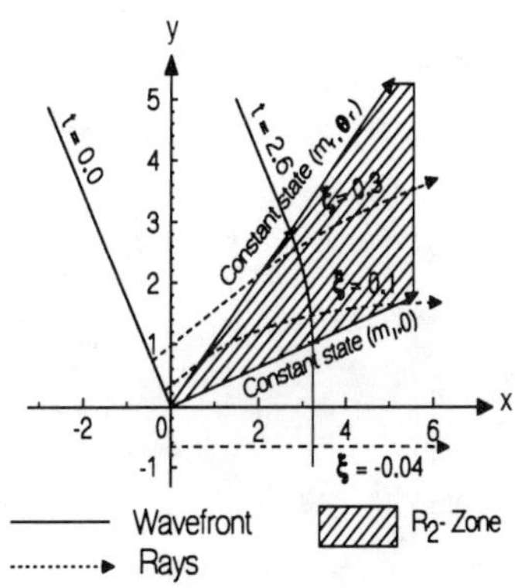

Fig. 6.2.2(b): Geometrical features of the front associated with the solution in Fig.6.2.2(a). Elementary shape $\mathcal{R}_2$ is a part of the wavefront in R_2-zone.

When $(m_l, 0)$ and (m_r, θ_r) satisfy a jump condition, (6.2.5) we get one of the two shocks 1-S and 2-S joining two constant states $(m_l, 0)$ and (m_r, θ_r) and passing through $\xi = 0$ at $t = 0$. The jumps in θ and m across a shock satisfy (since $\theta_l = 0$)

$$\cos\theta_r = \frac{(m_r g_r - m_l g_l)}{(m_l g_r + m_r g_l)} \tag{6.2.7}$$

Since the Lax shock inequality implies $\lambda_1(m_r) < \lambda_1(m_l)$ for 1-S and $\lambda_2(m_r) < \lambda_2(m_l)$, for 2-S, we get $m_r > m_l$ for 1-S and $m_r < m_l$ for 2-S. From the expression for g it follows that when we move with the ray velocity g decreases after crossing the shock ($g_r < g_l$ for 1-S and $g_r > g_l$ for 2-S). The jump relations from (6.2.1 - 2) give

$$sg_r \sin\theta_r = g_r(m_r^2 - m_l^2)(m_l g_r + m_r g_l) \tag{6.2.8}$$

where s is the shock velocity in (ξ, t)-plane which is negative for 1-S and positive for 2-S. This relation shows that for both shocks $\theta_r < 0$. The images of 1-S and 2-S in (ξ, t)-plane to (x, y)-plane are elementary shapes of a front, which are 1-kink (denoted by $\mathcal{K}_1$) and 2-kink (denoted by $\mathcal{K}_2$) as shown in Fig. 6.2.3(a) and Fig. 6.2.3(b), respectively.

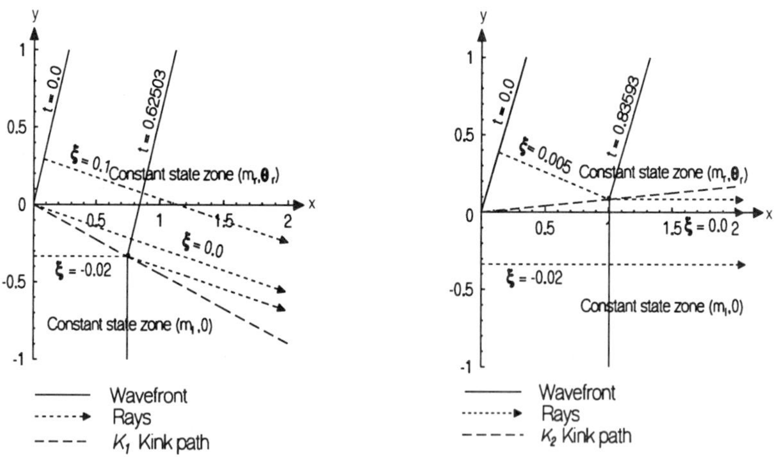

Fig. 6.2.3(a) and (b): Rays are neither created nor lost across a kink but suddenly change their direction and since g decreases after it crosses the kink line, they emerge compressed.

6.2.2 Solution of the Riemann problem and interpretation

In this section we briefly review unpublished works of Baskar, Potdar and Szeftel (1999). A Riemann problem for the system (6.2.1 - 2) consists of solving the system with following initial conditions

$$(m,\theta)|_{t=0} = \begin{cases} (m_l,\theta_l) \ , \ \text{we choose } \theta_l = 0, \xi < 0 \\ \\ (m_r,\theta_r) \ , \ \xi > 0 \end{cases} \tag{6.2.9}$$

where m_l, m_r and θ_r are constant.

We define curves R_α and $S_\alpha (\alpha = 1,2)$ as loci of the points (m_r, θ_r) which can be joined to the point $(m_l, 0)$ by α-R and α-S waves. Fig. 6.2.4 shows these curves for the typical value of $m_l = 1.2$. We note that nonlinear ray theory is valid only for small values of $m-1$, say for $0 < m-1 < 0.25$.

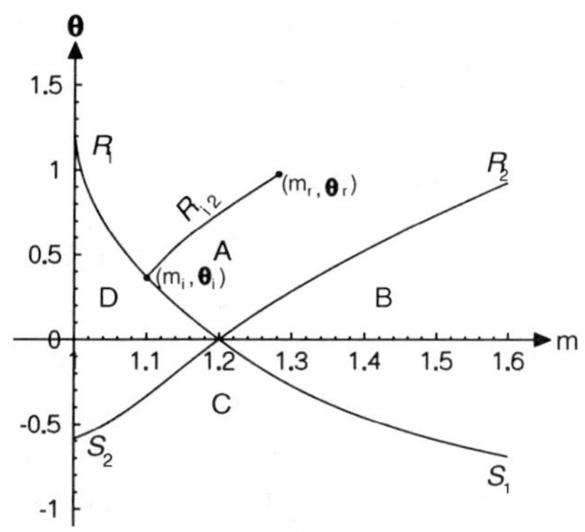

Fig. 6.2.4: R_α and S_α $(\alpha = 1,2)$ curves in (m,θ)-plane for $m = 1.2$.

If we do not go into the question of the existence* of the curves into consideration, the method of solution is simple. Suppose (m_r, θ_r) lies in the domain A bounded by curves R_1 and R_2 as shown in Fig. 6.2.4. We draw a curve R_{i2} which represents the set of points joining (m_r, θ_r) by 2-R wave to an intermediate state (m_i, θ_i), which lies on

*Existence of curves joining appropriate points have been worked out for a system more general than (6.2.1 - 2) in a yet unpublished paper by Baskar and Prasad.

the R_1 curve. Thus, in this case, the solution consists of the state $(m_l, 0)$ on the left of a 1-R wave continuing up to an intermediate constant state (m_i, θ_i), which ends into a 2-R wave to the right of which we have the final state (m_r, θ_r) (see Fig. 6.2.5(a)). The shape of the wavefront at $t = 0$ and $t = t_1 > 0$ is shown in Fig. 6.2.5(b). Since $(m_l, 0)$ is a state on the left, it can be joined to an intermediate state (m_i, θ_i) on its right only if (m_i, θ_i) lies on R_1 and not on R_2.

We describe this result symbolically as

$$(m_r \theta_r) \in A \to \mathcal{R}_1 \mathcal{R}_2 \tag{6.2.10}$$

which means that when (m_r, θ_r) is in A, the resultant wavefront has an elementary shape $\mathcal{R}_1$ propagating below, and $\mathcal{R}_2$ propagating above and these two are separated by a section of plane (or straight) front. Similarly, we get the result

$$(m_r, \theta_r) \in B \to \mathcal{K}_1 \mathcal{R}_2 \tag{6.2.11}$$

as shown in Fig. 6.2.6. Other results are

$$(m_r, \theta_r) \in C \to \mathcal{K}_1 \mathcal{K}_2 \tag{6.2.12}$$

$$(m_r, \theta_r) \in D \to \mathcal{R}_1 \mathcal{K}_2 \tag{6.2.13}$$

An asymptotic shape as t tends to infinity, of a nonlinear wavefront when the initial wavefront is as in Fig. 3.3.1 can be easily obtained. We note that when we observe the wavefront from a very large length scale, the central curved part of the initial wavefront tends to a point and the wavefront takes the shape of a wedge. We also prescribe the same amplitude m_l on the two parts of the wedge. Therefore, we seek a solution of (6.2.1 - 2) with an initial condition

$$m(\xi, 0) = \begin{cases} (m_l, \theta_l) & \xi < 0 \\ (m_l, -\theta_l) & \xi > 0 \end{cases} \tag{6.2.14}$$

Choosing the direction of the x-axis perpendicular to the lower part the wavefront, solving the corresponding Riemann problem and rotating back the x-direction, we get the following solution of (6.2.1 - 2) and (6.2.14)

$$m(\xi, t) = \begin{cases} (m_l, \theta_l) & \xi < -st \\ (m_i, 0) & -st < \xi < st \\ (m_l, -\theta_l) & st < \xi \end{cases} \tag{6.2.15}$$

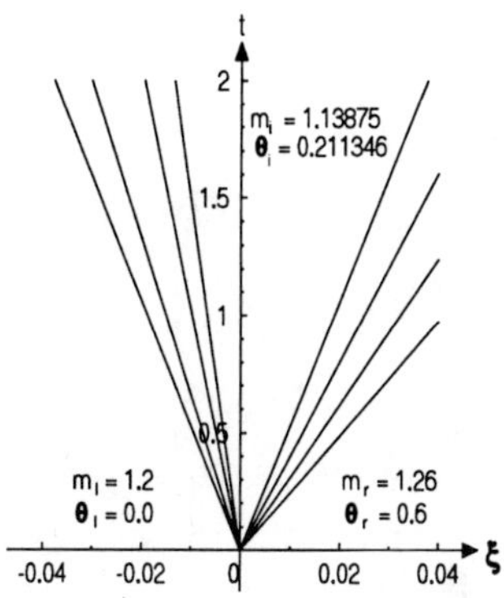

Fig. 6.2.5(a): Solution of the Riemann problem when (m_r, θ_r) is in A.

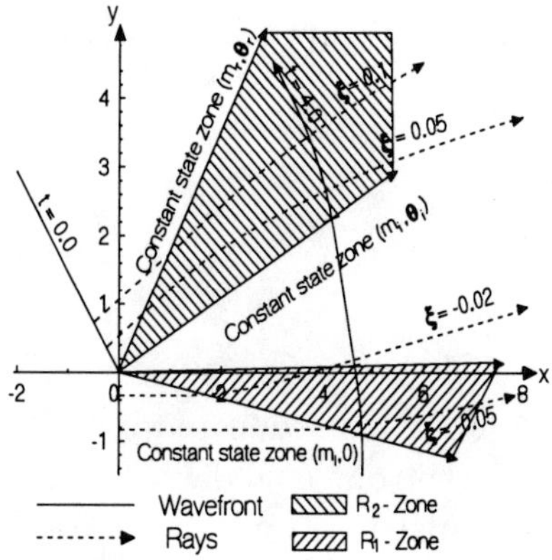

Fig. 6.2.5(b): Shape of the wavefront at $t = 0$ and $t = t > 0$ when (m_r, θ_r) is in A.

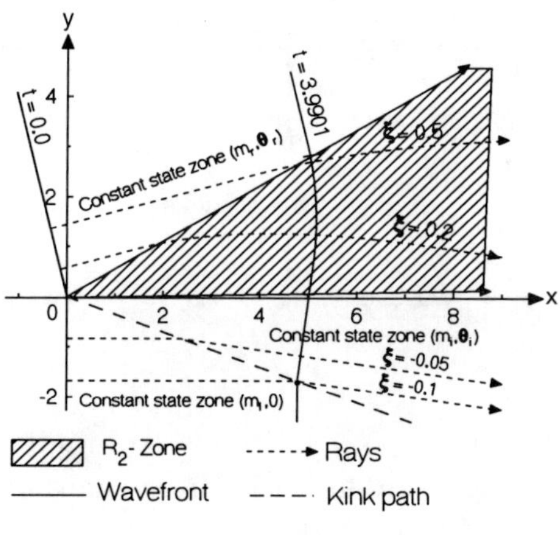

Fig. 6.2.6: When (m_r, θ_r) is in B, the front consist of a $\mathcal{K}_1$ propagating upward and $\mathcal{R}_2$ propagating upward.

where m_i is given by the equation

$$m_i g_{m_i} + m_l g_{m_l} = (m_i g_{m_l} + m_l g_{m_i}) \cos \theta_l \tag{6.2.16}$$

and

$$s = \sqrt{\frac{(m_i^2 - m_l^2)}{(g_{m_l}^2 - g_{m_i}^2)}} \tag{6.2.17}$$

This solution when mapped into the (x, y)-plane gives the shape of the wavefront as shown in Fig. 6.2.7.

Transition from one shape of the wavefront to another shape, e.g. from $\mathcal{R}_1\mathcal{R}_2$ to $\mathcal{K}_1\mathcal{R}_2$, as the point (m_r, θ_r) crosses curves R_1, R_2, S_1 and S_2 have also been discussed. The results of the transition lead to beautiful geometrical patterns.

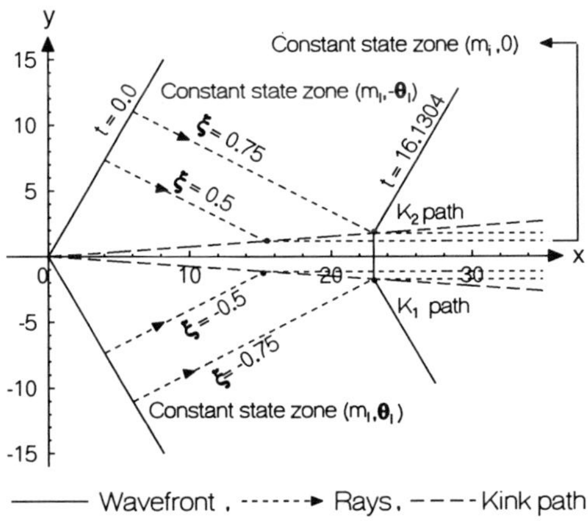

Fig. 6.2.7: Limiting shape, as t tends to infinity, of the nonlinear wavefront originating from an initial front as in Fig.3.3.1.

6.2.3 Interaction of elementary shapes

Elementary shapes on a nonlinear wave propagate on the front. Two elementary shapes, separated by a plane portion of the front, may or may not interact. The process of interaction if it takes place, may take finite or infinite time depending on the strengths of the two elementary shapes. It is not possible to visualize the shape during the process of interaction without a full numerical solution of the conservation laws (6.2.1 - 2). However, when the interaction period is finite we can easily obtain the final results, which will again consist of a pair of elementary shapes. All these geometrically beautiful results can be studied from the corresponding results on the interaction of simple waves and shock waves in (ξ, t)-plane (Courant and Friedrich (1948), Smoller (1983)). We can use Fig. 6.2.4 for this purpose, where we note that the curve R_1, R_2, S_1 and S_2 are meaningful for more general simple waves (not just for centered waves) and shock waves (not necessarily passing through the origin in (ξ, t)-plane). No distinction has been made between the waves, in which characteristics converge (corresponding to compression waves in gas dynamics) and a corresponding shock. This is justified because we are considering only small changes in m. We use the symbols introduced in the previous section with a slight modification. $\mathcal{K}_2\mathcal{K}_1$ would mean a kink of the second family on the lower part of the front (smaller values of ξ) separated by a plane part (m_j, θ_j) of the front from a kink of the first family on the upper part of the front. To reach a state (m_r, θ_r) from $(m_l, 0)$ through $\mathcal{K}_2\mathcal{K}_1$ we need to move along S_2 from $(m_l, 0)$ up to (m_j, θ_j) and then move along S_{j1} from (m_j, θ_j) up to the point (m_r, θ_r). Clearly (m_r, θ_r) is in the region C, which implies

$$\mathcal{K}_2\mathcal{K}_1 \to \mathcal{K}_1\mathcal{K}_2 \tag{6.2.18}$$

with obvious physical interpretation. Such interactions of kinks are clearly seen in the case of propagation of an initially sinusoidal front, for example, see Fig. 10.3.4.

All possible interactions of elementary shapes, namely $\mathcal{K}_1\mathcal{K}_1$, $\mathcal{K}_2\mathcal{K}_2, \mathcal{R}_1\mathcal{K}_1$, $\mathcal{R}_2\mathcal{K}_2, \mathcal{K}_1\mathcal{R}_1$, $\mathcal{K}_2\mathcal{R}_2$, $\mathcal{R}_2\mathcal{R}_1$, $\mathcal{R}_2\mathcal{K}_1$ and $\mathcal{K}_2\mathcal{R}_1$ in addition to $\mathcal{K}_2\mathcal{K}_1$ mentioned above, have been discussed. A geometrical representation of one of these cases, namely

$$\mathcal{R}_1\mathcal{K}_1 \to \mathcal{K}_1\mathcal{R}_2$$

when $\mathcal{K}_1$ is strong compared to $\mathcal{R}_1$ has been shown in Fig. 6.2.8. Note that the scales for x and y used in (a) and (b) are very different.

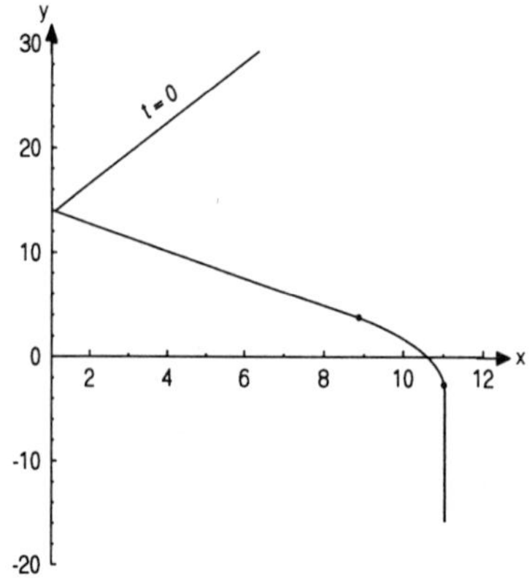

Fig. 6.2.8(a): The front $\mathcal{R}_1\mathcal{K}_1$ before the interaction.

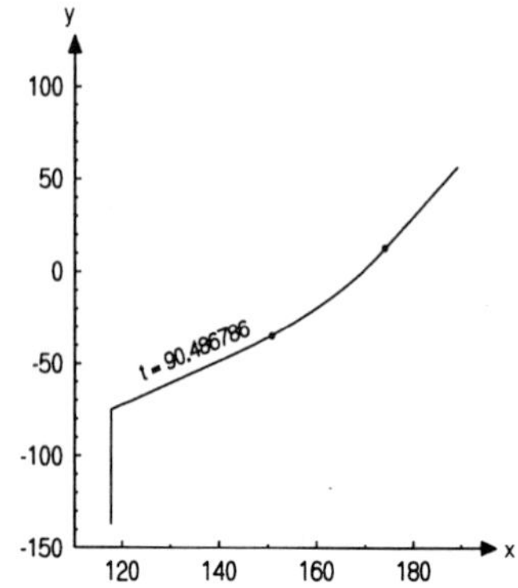

Fig. 6.2.8(b): The front $\mathcal{K}_1\mathcal{R}_2$ after the interaction.

6.3 Exact solution of an initial value problem

Given an initial position Ω_0 of a nonlinear wavefront and amplitude distribution on it, we have already indicated how to use the equations of the WNLRT to find the successive positions of the front Ω_t for $t > 0$ as long as the front remains smooth (see equations (6.1.24 - 27) where $s =$ constant identifies a front. Let us collect all equations and initial conditions here to describe the evolution of a two-dimensional nonlinear wavefront with kinks.

Ray equations:

$$x_t = m\cos\theta \ , \ y_t = m\sin\theta \tag{6.3.1}$$

Equations of a wavefront at a fixed time t (obtained from (3.3.11)):

$$x_\xi = -g\sin\theta \ , \ y_\xi = g\cos\theta \tag{6.3.2}$$

Kinematical conservation laws:

$$(g\sin\theta)_t + (m\cos\theta)_\xi = 0, \ \ (g\cos\theta)_t - (m\sin\theta)_\xi = 0 \tag{6.3.3}$$

Energy transport relation along a nonlinear ray :

$$g = (m-1)^{-2}e^{-2(m-1)} \tag{6.3.4}$$

Given an initial position of a nonlinear wavefront:

$$x(\xi,0) = x_0(\xi), \;\; y(\xi,0) = y_0(\xi) \tag{6.3.5}$$

The initial values of the angle θ (obtained from (6.3.5)), the amplitude m and the metric g is obtained from (6.3.4):

$$\theta(\xi,0) = \theta_0(\xi), \;\; m(\xi,0) = m_0(\xi), \;\; g(\xi,0) = g_0(\xi) \tag{6.3.6}$$

There are two ways to set up the initial data $x_0(\xi)$ and $y_0(\xi)$ in terms of ξ.

Procedure 1 An initial position of the wavefront is generally given in terms of its arc length η measured from a fixed point: $x = \tilde{x}_0(\eta)$, $y = \tilde{y}_0(\eta)$ on which $\tilde{m}_0(\eta)$ is prescribed. $\tilde{\theta}_0(\eta)$ is calculated from $(\tilde{x}_0(\eta), \tilde{y}_0(\eta))$. Initially, we choose a variable $\xi' = \eta$ so that $g' = 1$ at $t = 0$ (see the description leading to (6.1.14)). Then we calculate $f(\xi')$ from (6.1.17). Finally, we introduce $\xi = \int_0^{\xi'} f(\xi')d\xi$ (see the discussion leading to (6.1.18). Now we convert the data given as functions of $\eta(=\xi')$ in terms of ξ.

Procedure 2 First prescribe $m_0(\xi)$ and $\theta_0(\xi)$. Then calculate $g_0(\xi)$ from (6.3.4) and finally $x_0(\xi)$ and $y_0(\xi)$ by integrating (6.3.2) with respect to ξ. Using this procedure, we have no control on the shape of the initial wavefront. Distribution of $\theta_0(\xi)$ only tells whether the initial front is concave or convex.

When prescription of the initial data is completed, we solve the system of conservation laws (6.3.3). To get a particular ray, say starting from a point ξ_0 on the initial front, we integrate (6.3.1) with respect to t with initial condition $x(\xi_0, 0) = x_0(\xi_0), y(\xi_0, 0) = y_0(\xi_0)$. To get the nonlinear wavefront at any time, we first trace a particular ray, say corresponding to ξ_0, up to the point $(\tilde{x}(\xi_0,t), \tilde{y}(\xi_0,t))$ and then integrate (6.3.2) with respect to ξ with condition $x(\xi_0,t) = \tilde{x}(\xi_0,t)$ and $y(\xi_0,t) = \tilde{y}(\xi_0,t)$. A shock in a solution of (6.3.3) will be mapped onto a kink. A ray (or the wavefront) will have a discontinuity in its direction at the kink.

In section 6.2.1, we discussed elementary wave solutions corresponding to a centered wave, which is a particular case of a simple wave solution of (6.3.3). Here, we continue discussion of a general simple wave solution, which we left incomplete in section 6.1.2. Let us consider an initial wavefront with a distribution of m_0 and θ_0

such that Q in (6.1.28) is constant on it. The solution of (6.3.3) with this initial data is a simple wave of the second family. Consider a characteristic of the c_2 family passing through $\xi = \xi_0$ at $t = 0$. Here

$$m = \text{ constant } = m_0(\xi_0) \text{ along } \frac{d\xi}{dt} = \sqrt{\frac{m-1}{2g^2}} \tag{6.3.7}$$

The equation of this characteristic in (ξ, t)-plane is

$$\xi - \frac{1}{\sqrt{2}}(m_0(\xi_0) - 1)^{5/2} t e^{2(m_0(\xi_0)-1)} = \xi_0 \tag{6.3.8}$$

If the members of this family of characteristics are non-intersecting, then for a given value of ξ and t, we can find the unique characteristic through (ξ, t) and the corresponding value of ξ_0 is unique. This implies that there exists a function $H(\xi, t)$, such that

$$\xi_0 = H(\xi, t) \tag{6.3.9}$$

The solution at (ξ, t) is given by

$$m(\xi, t) = m_0(\xi_0) = m_0(H(\xi, t)), \theta(\xi, t) = \theta_0(\xi_0) = \theta_0(H(\xi, t)) \tag{6.3.10}$$

The time t_c, when a shock appears in the above exact solution, is given by (6.1.39). After that, it is not possible to get an exact solution. However, in initial stages the shock is weak. In addition to this our theory WNLRT is valid only for $0 < m - 1 << 1$ and we can prescribe only such an initial data. In this case we can easily continue the exact solution valid up to t_c beyond this time as an approximate solution by fitting the shock into the above simple wave solution on both sides of the shock or get a numerical approximation of the exact solution by solving the conservation laws (6.3.3) using a suitable finite difference scheme.

Composite simple wave solution Of special interest are composite simple wave solutions in (ξ, t)-plane, *which we define to be two simple waves* (not necessarily centered) *separated by a constant state.* An example is the solution depicted in Fig. 6.2.5(a). Here, we choose each of the two simple waves to be of finite extent in ξ-direction. The initial value of $\theta(\xi, t)$ is prescribed in the following way

$$\theta_0(\xi) = \begin{cases} B\sin((\pi k\xi)/2) & , \quad |\xi| \le \frac{1}{k} \\ B\xi/|\xi| & , \quad |\xi| > \frac{1}{k} \end{cases} \tag{6.3.11}$$

where k is a non-zero constant. It is curved in its central part and extended on its two sides by the tangents at points corresponding to $\xi = \pm 1/k$. Now we prescribe the value of $m_0(\xi)$ such that Q is constant on the upper part ($\xi > 0$) of the initial wavefront, R is constant on the lower part ($\xi < 0$) and $m_0(\xi)$ is an even function of ξ and continuous at $\xi = 0$. Fixing of the value $m_0(0) = m^*$ uniquely determines $m_0(\xi)$. Now we use a procedure to construct the initial wavefront which is symmetric with the x-axis as the line of symmetry and having continuously turning tangent. Then we solve the problem (Sangeeta (1996)).

The values of θ_0 at $\xi = \pm\frac{1}{k}$ are $\pm B$. Let the value of m at $\xi = \pm k$ by $\bar{m}$. Then

$$\bar{m} = \frac{1}{8}\left\{1 + \left(B + \sqrt{8(m^* - 1)}\right)^2\right\} \tag{6.3.12}$$

We also introduce notations

$$\bar{c}_2 = \sqrt{\frac{\bar{m} - 1}{2\bar{g}^2}} = -\bar{c}_1, \;\; c_2^* = c_2(m^*) = \sqrt{\frac{m_* - 1}{2g^{*2}}} \tag{6.3.13}$$

$$\bar{g} = (\bar{m})\,, \;\; g^* = g(m^*) \tag{6.3.14}$$

The composite simple wave solution in (ξ, t)-plane, valid till a shock appears, consists of

(i) a constant state $(\bar{m}, B)$ to the left of the characteristic $\xi = \bar{c}_1 t$ (i.e., $\xi = -\bar{c}_2 t$), a second constant state $(\bar{m}, -B)$ to the right of the characteristic $\xi = \bar{c}_2 t$ and a third central constant state $(m^*, 0)$ in $-c_2^* t < \xi < c_2^* t$; and

(ii) two simple waves (with diverging characteristics for $B > 0$ and converging characteristics for $B < 0$), one belonging to the first characteristic family in $-\bar{c}_2 t - \frac{1}{k} < \xi < -c_2^* t$ and another belonging to the second characteristic family in $c_2^* t < \xi < \bar{c}_2 t + \frac{1}{k}$.

The central constant state between the two simple waves in (ξ, t)-plane is mapped onto a central region in (x, y) where the nonlinear wavefront is a straight line and rays are parallel to the x-axis. A figure representing such a solution in (x, y)-plane is Fig. 6.1 on page 114 of Prasad (1993).

Case 1: $B = 0.3$ and $k = 2.5$. This corresponds to an initially convex wavefront (Fig. 6.3.1). The nonlinear rays become straight

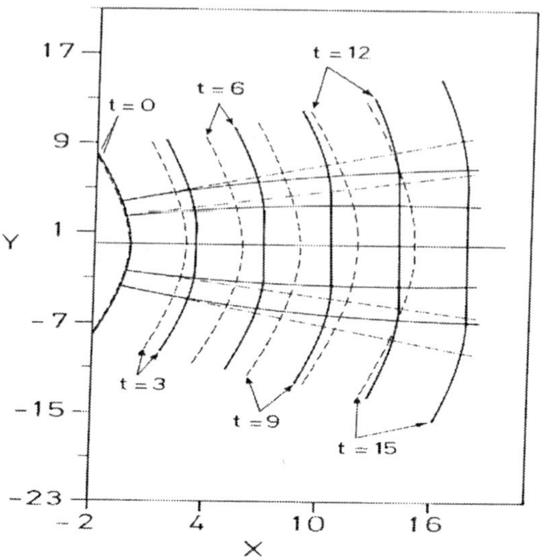

Fig. 6.3.1: $B = 0.3$, $k = 2.5$, $m^* = 1.2628$.
– – – Linear wavefront and linear rays
——— Nonlinear wavefront and nonlinear rays.

lines parallel to the x-axis as soon as they enter into the central constant state region and the nonlinear wavefront also becomes straight and perpendicular to the x-axis in the central region. The front remains smooth and without any kinks.

Case 2: $B = -0.3$ and $k = 2.5$. This corresponds to an initially concave wavefront (Fig. 6.3.2). The nonlinear rays are initially close to the corresponding linear rays; they then deviate significantly and ultimately become parallel to the x-axis. The linear caustic (see also Fig. 3.3.1) is completely resolved. The nonlinear wavefront emerges flattened without any fold but a pair of kinks appear. The asymptotic shape of the nonlinear wavefront is the same as that in the Fig. 6.2.7.

6.4 Conclusion and validity of WNLRT

The most important feature which the exact solution in the section 6.3 shows is that the amplitude of the wave (which tends to infinity in the linear theory at the caustic) on the central flat part remains

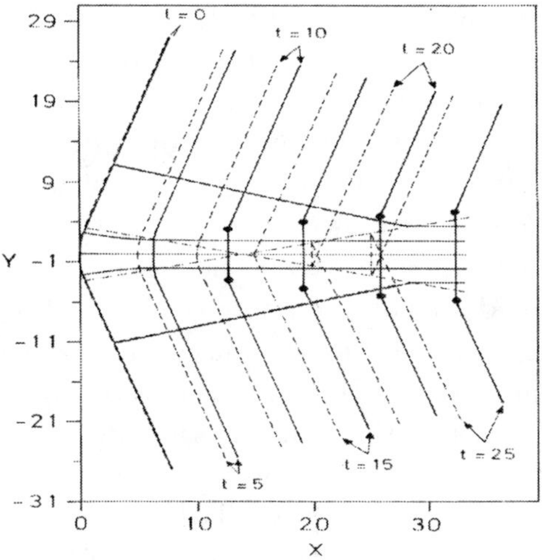

Fig. 6.3.2: $B = -0.3$, $k = 2.5$, $m^* = 1.2628$
$---$ linear wavefront and linear rays
———linear wavefront and nonlinear rays.

constant equal to m^* until it starts getting modified by the reflected waves (on the wavefront) from the kinks. Since the kinks appear on the nonlinear wavefront only a little before the front reaches the arête, the wave amplitude at the focus is m^*. Slowly, as it increases further, the value at the center of the front changes to m_i (given by (6.2.16)) which is another finite value of the order of m^* or $\bar{m}$. These results lead to a conclusion that the WNLRT gives an amplitude $m - 1$ of the wavefront which remains small and of the same order as that of the amplitude of the initial wavefront. This is in contrast to the linear theory where the amplitude tends to infinity in the first approximation and remains large in the next approximation (Buchal and Keller, 1960). This conclusion agrees with the results obtained in a large number of examples worked out by Prasad and Sangeeta (1999) using numerical integration of initial value problems of the conservation laws (6.2.1 - 2). Thus, the condition of small amplitude required in the derivation of WNLRT remains valid as the nonlinear wavefront propagates.

The reason for $m - 1$ to be of the order of the initial value of the amplitude can be seen from the following argument. Suppose $m - 1$ becomes large on a small part of the nonlinear wavefront, then we shall have a large gradient $\frac{\partial m}{\partial \xi}$ either on that part or near it. Then the equation (6.1.21) will imply a significant change in the direction of the rays which would immediately cause the amplitude $m-1$ to change from a large value to a small value. This has been clearly observed in the numerical computation (see chapter 10) when we prescribed minimum value at the center of a dent. This would have caused the nonlinear rays to converge more strongly. But it did not happen. The value of $m-1$ at the center of the dent increased rapidly and attained a local maximum so that the rays soon stopped converging. Since $m-1$ remains small, we also notice that g can not become zero on any part of the nonlinear wavefront.

Some of the cases worked out by Prasad and Sangeeta (1999) show that though the nonlinear rays ultimately tend to become parallel, there are intermediate stages when the converging rays are so much pushed away that they start diverging before converging again. This is more clearly seen when the initial wavefront has sinusoidal (or periodic) shape and has periodic amplitude distribution on it. We shall discuss such cases in chapter 10 in context of weak shock propagation and point out the property of corrugational stability of a nonlinear wavefront and a shock front.

Now an important question arises. In the geometric derivation of the WNLRT, we used the ratio of δ (the short wave length of the perturbation) to a length scale R of the order of the principal radii of curvature of the wavefront. We note that except at the kinks, the nonlinear wavefront tends to become plane so that ultimately R tends to infinity. We have also seen from the exact solution of section 6.3 that the initial radii of curvature play no role because the distance of the arête of a caustic from the initial wavefront plays no role in the solution. Thus, the WNLRT is valid over a very large distance of propagation. We observe this when the initial wavefront is periodic. Prasad and Sangeeta (1999) compute the solution for a very long time and the solution seems to remain valid. Thus, the validity of WNLRT is far beyond what was initially expected.

Chapter 7

Compatibility conditions on a shock: single conservation law

7.1 Derivation of the infinite system of compatibility conditions

Consider a solution $u(x,t)$ of the single conservation law (1.3.1)

$$u_t + (\frac{1}{2}u^2)_x = 0 \ , \ (x,t) \in \mathbf{R} \times \mathbf{R}_+ \tag{7.1.1}$$

with initial condition

$$u(x,0) = \varphi(x) \ , \ x \in \mathbf{R} \tag{7.1.2}$$

such that the solution is sufficiently smooth, except for a single shock curve

$$\Omega : x = X(t) \ , \ t \in \mathbf{R}_+ \tag{7.1.3}$$

The solution can be represented in the form *

$$u(x,t) = u_\ell(x,t) + H(s)(u_r(x,t) - u_\ell(x,t)), \ s = x - X(t) \tag{7.1.4}$$

*We make a little departure from the notations used in chapter 1, definition (1.3.7), where $u_\ell(t)$ and $u_r(t)$ represent the limiting values of u from the two sides of the shock. From now onwards, u_ℓ and u_r represent functions defined on $\mathbf{R}^2$; and the limiting values from the two sides are denoted by $u_\ell(X(t),t)$ (or $u_\ell|_\Omega$) and $u_r(X(t),t)$ (or $u_r|_\Omega$).

where H is the Heaviside function, and u_ℓ and u_r are sufficiently smooth functions defined on $\mathbf{R} \times \mathbf{R}_+$. The state u_r on the right of the shock $(s > 0)$, the state u_ℓ on the left of the shock $(s < 0)$ and the shock path $X(t)$ are uniquely determined by the initial condition (2.2). The jump condition (1.3.10) across the shock gives

$$\frac{dX(t)}{dt} = \frac{1}{2}\{u_\ell(X(t),t) + u_r(X(t),t)\} \equiv C, \text{ say} \tag{7.1.5}$$

The solution on the left of the shock $x < X(t)$ satisfies the partial differential equation (1.1.6) which we write in the form

$$\frac{\partial u_\ell}{\partial t} + \frac{1}{2}(u_\ell + u_r)\frac{\partial u_\ell}{\partial x} = \frac{1}{2}(u_r - u_\ell)\frac{\partial u_\ell}{\partial x}$$

Taking the limit of this equation as $x \to X(t) - 0$ we get

$$\frac{du_\ell(X(t),t)}{dt} = \frac{1}{2}\{u_r(X(t),t) - u_\ell(X(t),t)\}\frac{\partial u_\ell}{\partial x}|_\Omega \tag{7.1.6}$$

where

$$\frac{d}{dt} = \frac{\partial}{\partial t} + C\frac{\partial}{\partial x} \tag{7.1.7}$$

(2.1.6 - 7) is the first compatibility condition along the shock path. This equation giving the time-rate of change of u_ℓ along the shock path contains $\frac{\partial u_\ell}{\partial x}|_\Omega$. To get an equation for this new quantity $\frac{\partial u_\ell}{\partial x}|_\Omega$, we need to derive the second compatibility condition. We differentiate (1.1.6) with respect to x and write the result in the form

$$\frac{\partial}{\partial t}\left(\frac{\partial u_\ell}{\partial x}\right) + \frac{1}{2}(u_\ell + u_r)\frac{\partial}{\partial x}\left(\frac{\partial u_\ell}{\partial x}\right) = -\left(\frac{\partial u_\ell}{\partial x}\right)^2 + \frac{1}{2}(u_r - u_\ell)\frac{\partial^2 u_\ell}{\partial x^2}$$

Taking the limit as $x \to X(t) - 0$, we get the second compatibility condition

$$\frac{d}{dt}\left(\frac{\partial u_\ell}{\partial x}|_\Omega\right) = -\left(\frac{\partial u_\ell}{\partial x}|_\Omega\right)^2 + \frac{1}{2}\left\{(u_r - u_\ell)\frac{\partial^2 u_\ell}{\partial x^2}\right\}|_\Omega, \tag{7.1.8}$$

which also contains a new quantity, namely, $\frac{\partial^2 u_\ell}{\partial x^2}|_\Omega$

To derive the ith compatibility conditions, we differentiate (1.1.6) i times, arrange the terms appropriately and take the limit as $x \to X(t) - 0$ to get

$$\frac{d}{dt}\left(\frac{\partial^i u_\ell}{\partial x^i}|_\Omega\right) = \frac{1}{2}\left\{(u_r - u_\ell)\frac{\partial^{i+1} u_\ell}{\partial x^{i+1}}\right\}|_\Omega$$

$$- \left\{ \sum_{j=1}^{i} {}^iC_j \frac{\partial^j u_\ell}{\partial x^j} \frac{\partial^{i-j+1} u_\ell}{\partial x^{i-j+1}} \right\} |_\Omega \qquad (7.1.9)$$

where ${}^iC_j = i!/(j!(i-j)!)$. The number of compatibility conditions which we can derive on the shock path depends on the degree of smoothness of the solution in the left subdomain of the shock. For solutions which are infinitely differentiable, we get an infinite system of compatibility conditions. However, if the solution in the left subdomain has a Taylor's series, the infinite system of equations can be derived by substituting the Taylor's series

$$u_\ell(x,t) = \sum_{i=0}^{\infty} \frac{1}{i!} \left(\frac{\partial^i u_\ell}{\partial x^i} |_\Omega \right) (x - X(t))^i \qquad (7.1.10)$$

in the equation (1.1.6), equating the coefficients of the various powers of $(x - X(t))$ equal to zero and using (7.1.5).

Let us denote the value of the ith spatial derivative at the shock divided by $i!$ by $v_i(t)$ i.e.,

$$v_i(t) = \frac{1}{i!} \frac{\partial^i u}{\partial x^i} |_\Omega \, , \; i = 1, 2, 3, \ldots \qquad (7.1.11)$$

and let $u_0(t) = u_\ell(X(t), t)$ represent the state just behind the shock. The equation for the shock path and the compatibility conditions are

$$\frac{dX}{dt} = \frac{1}{2}(u_0 + u_r) \qquad (7.1.12)$$

$$\frac{du_0}{dt} = -\frac{1}{2}(u_0 - u_r)v_1 \qquad (7.1.13)$$

and

$$\frac{dv_i}{dt} = -\frac{(i+1)}{2}(u_0 - u_r)v_{i+1} - \frac{i+1}{2} \sum_{j=1}^{i} v_j v_{i-j+1}, \; i = 1, 2, 3, \ldots \qquad (7.1.14)$$

In the derivation of the second term on the right hand side of (7.1.14) we have used the identity

$$\sum_{j=1}^{i} (i - j + 1) v_j v_{i-j+1} = \frac{i+1}{2} \sum_{j=1}^{i} v_j v_{i-j+1} \qquad (7.1.15)$$

The compatibility conditions (7.1.13 - 14) were derived independently by Grinfel'd (1978) and Maslov (1980). The shock position $X(t)$, the shock strength u_0 and the spatial derivatives $i!v_i$, $i \geq 1$ can be obtained if we can solve the infinite system of ordinary differential equations (7.1.12 - 14).

The initial values of X, u_0 and v_i are given by the initial data (7.1.2)

$$\left.\begin{aligned} X(0) &= X_0 \quad ; \quad u_0(0) = \varphi(X_0 - 0) \equiv u_{00}, \text{ say} \\ v_i(0) &= \lim_{x \to X_0 - 0} \left\{ \frac{1}{i!}\frac{d^i\varphi}{dx^i} \right\} \equiv v_{i0}, \text{ say, } i = 1, 2, 3, \ldots \end{aligned}\right\} \qquad (7.1.16)$$

where X_0 is the value of x at which $\varphi(x)$ has the discontinuity.

Equations (7.1.13 - 14) form a coupled system. The ith equation for $i!\, v_i$ contains $(i+1)!v_{i+1}$ and hence is coupled to the next equation. The coupling coefficient in the equation, namely $-\frac{1}{2}(u_0 - u_r)$ is the same for all equations. The theory of an infinite system of ordinary differential equations (an infinite-dimensional problem) is not necessarily simpler than that of a partial differential equation. However, unlike in the original initial value problem (7.1.1 - 2), we have to deal only with smooth functions in the new initial value problem for (7.1.12 - 14). The coupling coefficient is small for a weak shock.

It is interesting to note that when u_r is a constant

$$u_0(t) = u_r + \frac{A}{\sqrt{t+\alpha}}, \quad v_1(t) = \frac{1}{t+\alpha}, \quad v_i(t) = 0, \ 1 \geq 2 \qquad (7.1.17)$$

is an exact solution of the infinite system of compatibility conditions for all values of A and $\alpha > 0$. It represents the solution of an initial value problem in which the initial data behind the shock is linear.

A formal solution of the system approaching (7.1.17) with $u_r = 0$ as $t \to \infty$ can be obtained in powers of $(t+\alpha)^{1/2}$ as :

$$\left.\begin{aligned} u_0(t) &= \frac{A_0}{\sqrt{t+\alpha}} + \frac{B_0}{t+\alpha} + \frac{C_0}{(t+\alpha)^{3/2}} + \ldots \\ v_1(t) &= \frac{1}{t+\alpha} + \frac{B_1}{(t+\alpha)^{3/2}} + \frac{C_1}{(t+\alpha)^2} + \ldots \\ v_i(t) &= \frac{A_i}{(t+\alpha)^{\frac{i+1}{2}}} + \frac{B_i}{(t+\alpha)^{\frac{i+3}{2}}} + \frac{C_i}{(t+\alpha)^{\frac{i+5}{2}}} + \ldots, i = 2, 3, \ldots \end{aligned}\right\} (7.1.18)$$

where $A_i, B_i, C_i, \ldots, i \geq 1$ can be uniquely determined successively in terms of $A_0, B_0, C_0, \ldots$. This formal series, giving higher order terms, shows that as $t \to \infty$ the higher order spatial derivatives of u at the shock tend to zero faster than the lower order derivatives.

7.2 Existence and uniqueness of the solution of the infinite system

Let us first mention a theorem on the existence and uniqueness of the weak solution of the initial value problem (7.1.1 - 2). This result, which we call "theorem on entropy solution" (Smoller, 1983, Chapter 16) requires the initial data $\varphi(x)$ to belong to a very general class of functions : $L_\infty(\mathbf{R})$. Then the theorem asserts that the initial value problem has a unique solution which is stable with respect to perturbations in the initial data, which exists for all $t > 0$ and for which all discontinuities are shocks. Further, at any fixed time $t > 0$, the solution is of locally bounded total variation with respect to x and $|u(x,t)| \leq ||\phi_0||$ for $(x,t) \in I\!R \times I\!R_+$.

Let us apply this theorem to two problems when the initial data is sufficiently smooth except for a single discontinuity at $x = 0$

$$P_1 : \varphi(x) = \begin{cases} 1 \text{ for } & x \leq 0 \\ 0 \text{ for } & x > 0 \end{cases} \tag{7.2.1}$$

and

$$P_2 : \varphi(x) = \begin{cases} 1 - e^{\frac{1}{x}} & \text{for} \quad x < 0 \\ 0 & \text{for} \quad x > 0 \end{cases} \tag{7.2.2}$$

The solution (1.4.7) of the problem P_1 valid for $t > 0$ has already been discussed. The application of the theorem on an entropy solution shows that the solution of the problem P_2 also exists for all $t \geq 0$ and since the state behind the shock satisfies $\varphi'(x) > 0$, no shock other than the one starting from $X_0 = 0$ appears in the solution. If we use $\xi = x - ut$ and t as independent variables instead of x and t, then the value $\xi(t)$ corresponding to the shock is given by $\xi_0(t) = X(t) - tu_0(t)$. We note $\xi_0(0) = 0$ and $\xi_0(t) < 0$ for $t > 0$ and

$$\frac{\partial}{\partial x} = (1 - tu_x)\frac{\partial}{\partial \xi} = \frac{1}{1 + t\varphi'(\xi)}\frac{\partial}{\partial \xi}$$

Hence, differentiating the relation $u = \varphi(x - ut)$ successively we deduce

$$\lim_{x \to X(t)-0} \frac{\partial^i u}{\partial x^i} \equiv i! v_i(t) = \lim_{\xi \to \xi_0(t)} \left\{ \left(\frac{1}{1 + t\varphi'(\xi)} \frac{\partial}{\partial \xi} \right)^{i-1} \frac{\varphi'(\xi)}{1 + t\varphi'(\xi)} \right\} \tag{7.2.3}$$

The expression (7.2.3) is a polynomial in t of degree $i - 2$ and the derivatives of φ at $\xi_0(t)$ divided by $(1 + t\varphi'(\xi))^{2i-1}$. Since $\varphi'(\xi) > 0$ for $\xi < 0$, all derivatives of ϕ at $\xi_0(t)$ are finite and hence all $v_i(t)$ remain finite for $t > 0$. This result shows that the solution of P_2, guaranteed by the theorem on the entropy solution, has the property that the functions $u_0(t)$ and $v_i(t), i = 1, 2, 3, \ldots$, derived from it exists for all $t > 0$. Since the initial data behind the shock in P_1 and P_2 are different, the functions $u_0(t)$ and $v_i(t), i = 1, 2, 3, \ldots$ obtained from the solution of the problems P_1 and P_2 are different for all $t > 0$. Thus, each of the two distinct sets of functions $u_0(t)$ and $v_i(t)$ obtained from these problems contain C^∞ functions (since φ is smooth except at $x = 0$) and satisfy the infinite system of equations (7.1.13 - 14), and the same initial condition

$$u_0(0) = 1, \quad v_i(0) = 0 \tag{7.2.4}$$

Now we have finished showing that the infinite system of compatibility conditions with initial data (7.2.4) does not have a unique solution in the class of C^∞ functions.

The non-uniqueness of a solution of the initial value problem in the class of C^∞ functions appears to severely restrict our capacity to use the infinite system of equations for many practical problems. It is now necessary to find a suitable function space in which uniqueness can be guaranteed. This class is the class of analytic functions.

Theorem 7.2.1 Let the initial data

$$\varphi(x) = \varphi_r(x) + H(X_0 - x)(\varphi_\ell(x) - \varphi_r(x)) \tag{7.2.5}$$

be such that $\varphi_\ell(x)$ and $\varphi_r(x)$ are analytic functions in a neighbourhood of the point X_0. Then the initial value problem (7.1.12 - 14 and 16) has a unique analytic solution in a neighbourhood of $t = 0$.

Proof Given the functions $\varphi_\ell(x)$ and $\varphi_r(x)$ to be analytic in a neighbourhood of the point X_0, the Cauchy–Kowalesky theorem gives unique analytic solutions $u_\ell(x,t)$ and $u_r(x,t)$ of the initial value problem (7.1.1 - 2) with $\varphi = \varphi_\ell$ and $\varphi = \varphi_r$, respectively. We can always choose a neighbourhood of $(X_0, 0)$ in the (x,t)-plane where both solutions are analytic.

Integrating the ordinary differential equation (7.1.5) subject to the condition $X(0) = X_0$, we get a unique analytic function $X(t)$ for small t. Then we find the analytic functions $u_0(t)$, $v_i(t)$ from the expressions

$$u_0(t) = u_\ell(X(t), t), \; v_i(t) = \frac{1}{i!}\frac{\partial^i u_\ell}{\partial x^i}|_{x=X(t)} \;, \; i = 1, 2, 3, \ldots \quad (7.2.6)$$

Thus, we have determined analytic functions which satisfy the infinite system (7.1.12 - 14) and the initial conditions (7.1.16) proving the existence of an analytic solution for small t. To prove the uniqueness of the analytic solution, we substitute the Taylor's series expansion at $t = 0$:

$$X(t) = \sum_{j=0}^{\infty} \frac{X_j}{j!} t^j, u_0(t) = \sum_{j=0}^{\infty} \frac{u_{0j}}{j!} t^j, v_i(t) = \sum_{j=0}^{\infty} \frac{v_{ij}}{j!} t^j, i = 1, 2, 3, \ldots \quad (7.2.7)$$

and

$$u_r(X, t) = \sum_{i=0}^{\infty}\sum_{j=0}^{\infty} a_{ij} X^i t^j \quad (7.2.8)$$

in the system (7.1.12 - 14) and equate the coefficients of powers of t on the two sides in each equation. Note that we assume a_{ij} to be known. This provides a system of equations from which the coefficients X_j, u_{0j} and $v_{ij}, j = 0, 1, 2, \ldots$ can be uniquely determined step-by-step (see also the proof of Theorem 7.3.1). The unique determination of the coefficients in (7.2.7) in terms of the initial values X_0, u_{00}, v_{i0} and the coefficients appearing in the expansion of the known function u_r proves the uniqueness of the analytic solution of the initial value problem.

Theorem 7.2.1, combined with the theorem on the entropy solution, provides a method for finding an approximate solution of an initial value problem for the single conservation law (7.1.1). Assume that the initial value (7.1.2) is such that the function $\varphi(x)$ can be approximated by an analytic function $\varphi_\ell(x)$ in an open interval containing $(X_0 - m, X_0)$ and by another analytic function $\varphi_r(x)$ on an open interval containing (X_0, X_0+m), where $m > 0$. Then, from the continuous dependence of the solution on the initial data guaranteed by the theorem on the entropy solution, it follows that by solving the initial value problem with the initial data (7.2.5) we get a local approximate solution in the L_∞ norm of the original solution. This approximate solution, which we call the first approximation to the original problem, can be obtained by summing the Taylor's series (7.2.7) after the analytic solution $X(t), u_0(t), v_i(t)$ of the infinite dimensional problem (7.1.12 - 14 and 16) has been obtained:

$$u(x,t) = u_0(t) + \sum_{i=1}^{\infty} v_i(t)(x - X(t))^i \tag{7.2.9}$$

The solution of an infinite system of ordinary differential equations, even numerically, is not simple. Our problem is now reduced to finding a function $\bar{u}(x,t)$ which is an approximation to the function $u(x,t)$ in (7.2.9). We expect that in the construction of $\bar{u}(x,t)$, which we call the second approximation to the original problem, the compatibility conditions (7.1.12 - 14) would play an important role. This leads to our new theory of shock dynamics (Ravindran and Prasad (1990), and Prasad and Ravindran (1990)).

7.3 A new theory of shock dynamics: analytic considerations

If we set $v_{n+1} = 0$ in the nth equation in (7.1.14), then the first $(n+1)$ equations in (7.1.13-14) form a closed system. Let $\bar{X}(t), \bar{u}_0(t), \bar{v}_i(t)$, $i = 1, 2, 3, \ldots, n$ is the solution of the truncated system of $n+2$ equations.

$$\frac{d\bar{X}}{dt} = \frac{1}{2}(\bar{u}_0 + u_r), \;\; u_r = u_r(\bar{X}, t) \tag{7.3.1}$$

$$\frac{d\bar{u}_0}{dt} = -\frac{1}{2}(\bar{u}_0 - u_r)\bar{v}_1 \tag{7.3.2}$$

$$\frac{d\bar{v}_i}{dt} = -\frac{i+1}{2}(\bar{u}_0 - u_r)\bar{v}_{i+1} - \frac{(i+1)}{2}\sum_{j=1}^{i}\bar{v}_j\bar{v}_{i-j+1}, \quad i = 1, 2, \ldots, n-1 \tag{7.3.3}$$

and

$$\frac{d\bar{v}_n}{dt} = -\frac{(n+1)}{2}\sum_{j=1}^{n}\bar{v}_j\bar{v}_{n-j+1} \tag{7.3.4}$$

with initial conditions for these $n+2$ quantities as in (7.1.16), i.e.,

$$\bar{X}(0) = X_0,\ \bar{u}_0(0) = u_{00},\ \bar{v}_i(0) = v_{i0},\ i = 1, 2, 3, \ldots, n \tag{7.3.5}$$

We assume that the function u_r, giving the state ahead of the shock to be a known function. The right hand sides of (7.3.1 - 4) are entire analytic functions of $u_r, \bar{u}_0, \bar{v}_i (i = 1, 2, 3, \ldots, n)$. Hence, the existence and uniqueness of a local solution of the initial value problem (7.3.1 - 5) (now a system of a finite number of ordinary differential equations) requires a very mild condition on $u_r(\bar{X}, t)$, a common requirement being the continuity of $u_r(\bar{X}, t)$ in a neighbourhood of $(X_0, 0)$ and its Lipschitz continuity with respect to $\bar{X}$. We have seen in section 1.7 that a shock, once formed, persists for all time. Hence, it would be interesting to study the conditions under which a local solution of (7.3.1 - 5) can be continued for all time. For $u_r = 0$, a sufficient condition for this is that $\varphi'_\ell(x) > 0$ which implies $v_1 > 0$. This has been shown by Ravindran, Sunder and Prasad (1993). In the case where $\varphi'_\ell(x) < 0$ on a sub-interval of $x < X_0$, the situation is complicated and the new theory of shock dynamics has to be carefully used even if the solution can be continued for all time (Sunder, Prasad and Ravindran (1992)). In the following discussion, we require analyticity of $u_r(\bar{X}, t)$ in a neighbourhood of $(X_0, 0)$.

When $u_r(\bar{X}, t)$ is analytic in a neighbourhood of $(X_0, 0)$, we can use Cauchy's existence theorem (Goursat (1917)) to prove the existence of a unique analytic solution of the initial value problem (7.3.1 - 5) valid in a neighbourhood of $t = 0$. This solution can be expressed in the form

$$\bar{X}(t) = \sum_{j=0}^{\infty}\frac{\bar{X}_j}{j!}t^j,\ \bar{u}_0(t) = \sum_{j=0}^{\infty}\frac{\bar{u}_{0j}}{j!}t^j,\ \bar{v}_i(t) = \sum_{j=0}^{\infty}\frac{\bar{v}_{ij}}{j!}t^j,\ i = 1, 2, 3, \ldots, n \tag{7.3.6}$$

Having obtained the solution of the problem (7.3.1 - 5), we construct a function $\bar{u}(x, t)$ by

$$\bar{u}(x,t) \begin{array}{ll} = \bar{u}_0(t) + \sum_{i=1}^{n}\bar{v}_i(t)(x - X(t))^i, & x < \bar{X}(t) \\ = u_r(x,t), & x > \bar{X}(t) \end{array} \tag{7.3.7}$$

We claim that this function is an approximate solution of the partial differential equation (7.1.1) in the left subdomain $x < X(t)$. Substituting (7.3.7) in (7.1.1) and using the equations (7.3.1 - 5) we get

$$\bar{u}_t + \bar{u}\bar{u}_x = (x - \bar{X}(t))^{n+1} h(x,t) \; , \; x < \bar{X}(t) \tag{7.3.8}$$

where

$$h(x,t) = \sum_{i=1}^{n-2} \frac{n+2+i}{2} (x - \bar{X}(t))^i \left\{ \sum_{j=i+2}^{n} \bar{v}_{n-j+i+2} \bar{v}_j \right\}$$

This shows, at least in a formal way, that the equation (7.1.1) approximately satisfies near the shock $x = \bar{X}$ and the accuracy with which it is satisfied increases as n increases. If we set

$$u(x,t) = \bar{u}(x,t) + \mathcal{R}(x,t) \tag{7.3.9}$$

then the error function $\mathcal{R}$ satisfies the non-homogeneous partial differential equation

$$\mathcal{R}_t + \mathcal{R}\mathcal{R}_x + \bar{u}(x,t)\mathcal{R}_x + \bar{u}_x(x,t)\mathcal{R} + F(x,t) = 0 \tag{7.3.10}$$

where

$$F(x,t) = (x - \bar{X}(t))^{n+1} h(x,t) \tag{7.3.11}$$

All these results are still local and are valid as long as no other shock appears in the solution.

For small t, we can make a more precise statement regarding the manner in which the function $\bar{u}(x,t)$ given by (7.3.7) tends to $u(x,t)$ given by (7.2.9). We express this relation not so much between the functions $\bar{u}(x,t)$ and $u(x,t)$ but between the coefficients $u_i(t)$ and $\bar{u}_i(t)$ in the expressions for these functions.

Theorem 7.3.1 For small t, the analytic solution of the initial value problem (7.3.1 - 5) tends to the analytic solution of the initial value problem (7.1.12 - 14 and 16) as n tends to infinity.

Proof We have already seen that when $\varphi_\ell(x)$ and $\varphi_r(x)$ in the initial data (7.2.5) are analytic in a neighbourhood of X_0, the infinite system has a unique analytic solution (7.2.7) for small t. The finite system (7.3.1 - 5) also has a unique analytic solution (7.3.6) for small t. Substituting (7.2.7) in the infinite system and (7.3.6) in the finite system and equating coefficients of like powers of t^j, we get a recursive system of algebraic equations for the coefficients. A step-by-step evaluation of the coefficients shows that the values of X_1 and $\bar{X}_1$ depend only on the initial value u_{00}, and the values of $X_2, \bar{X}_2$ and $u_{01}, \bar{u}_{01}$ depend only on the initial values u_{00}, v_{10}. In general, the values of X_{j+1}, $\bar{X}_{j+1}$, u_{0j}, $\bar{u}_{0j}$, $v_{\alpha\beta}$, $\bar{v}_{\alpha\beta}$ $(\alpha+\beta=j)$ depend only on the initial values u_{00}, v_{10}, v_{20}, $\ldots$, v_{j0} for all j.

Further, for $j=1,2,\ldots,n$, the expressions for X_{j+1} and $\bar{X}_{j+1}$; u_{0j} and $\bar{u}_{0j}$; and $v_{\alpha\beta}$ and $\bar{v}_{\alpha\beta}(\alpha+\beta=j)$ are exactly the same functions of $u_{00}, v_{10}, v_{20}, \ldots, v_{j0}$. It follows that

$$\left.\begin{array}{l} \bar{X}_{j+1} = X_{j+1}\ ,\ \bar{u}_{0j} = u_{0j}\ ,\ j=0,1,2,\ldots,n \\ \text{and} \\ \bar{v}_{ij} = v_{ij}\ ,\ (i=1,2,\ldots,n; i+j\leq n) \end{array}\right\} \tag{7.3.12}$$

Thus, if the $n+2$ equations in the finite system are considered, then

$$\left.\begin{array}{l} X-\bar{X} = 0(t^{n+2}), u_0-\bar{u}_0 = 0(t^{n+1}) \\ \text{and} \\ v_i - \bar{v}_i = 0(t^{n-i+1}), i=1,2,\ldots,n \end{array}\right\} \tag{7.3.13}$$

for small t. This proves the statement in the theorem.

We call the procedure of constructing the function $\bar{u}(x,t)$ by (7.3.7) after solving the initial value problem (7.3.1 - 5), the *"new theory of shock dynamics* (NTSD)". The theorem (7.3.1) shows that the new theory of shock dynamics will certainly give a good approximate solution for small t and the result (7.3.8) shows this approximate solution is quite likely to remain a good approximation near the shock even for large values of t. A convincing proof of this statement will require an analysis of the partial differential equation (7.3.10) with an appropriate initial data.

7.4 A new theory of shock dynamics: comparison of numerical results with the exact solution

The infinite system of equations reduces the problem (7.1.1 - 2) to a new problem in which the quantities involved are the values of u and its spatial derivatives on the shock path. In the case of a *single conservation law* which we shall consider, something more can be achieved. It is possible to reduce the problem to a finite-dimensional problem of integrating two first order ordinary differential equations on the shock path or an integral equation, which in an important particular case reduces simply to an algebraic or transcendental equation. This also leads to exact solutions.

Exact solution The solution of the initial value problem in the left subdomain is given by $u_\ell(x,t) = \varphi_\ell(x - u_\ell t)$. This gives

$$v_1(t) = \frac{\varphi_\ell'(\eta_\ell)}{1 + t\varphi_\ell'(\eta_\ell)} \tag{7.4.1}$$

where

$$\eta_\ell(t) = \lim_{x \to X-0} (x - u_\ell t) = X(t) - tu_0(t) \tag{7.4.2}$$

Also

$$u_0(t) \equiv u_\ell(X(t), t) = \varphi_\ell(\eta_\ell) \tag{7.4.3}$$

Using (7.1.12 - 13) in

$$\frac{d\eta_\ell}{dt} = \frac{dX}{dt} - u_0 - t\frac{du_0}{dt}$$

we get

$$\frac{d\eta_\ell}{dt} = \frac{u_r(X(t), t) - \varphi_\ell(\eta_\ell)}{2(1 + t\varphi_\ell'(\eta_\ell))} \tag{7.4.4}$$

We write (7.1.12) in the form

$$\frac{dX}{dt} = \frac{1}{2}(\varphi_\ell(\eta_\ell) + u_r(X(t), t)) \tag{7.4.5}$$

The function $\varphi_\ell(\eta_\ell)$ is given and the solution $u_r(x,t)$ ahead of the shock is assumed to be known solution by the method characteristics. Thus, the right hand sides of the above two equations are known functions of X, t and η_ℓ. Therefore, these two equations with the initial conditions

$$X(0) = X_0 \quad \text{and} \quad \eta(0) = X_0 \tag{7.4.6}$$

can be solved to give $\eta_\ell(t)$ and the shock position $X(t)$. The shock amplitude $u_0(t)$ is then given by (7.4.3).

Multiplying (7.4.4) by $2\varphi_\ell(1+t\varphi'_\ell)$ and integrating we get

$$2\int_{X_0}^{\eta_\ell(t)} \varphi_\ell(s)ds + t\varphi_\ell^2(\eta_\ell(t)) = \int_0^t u_r(x(\tau),\tau)\varphi_\ell(\eta_\ell(\tau))d\tau \qquad (7.4.7)$$

From (7.4.5) we get

$$X(t) = X_0 + \frac{1}{2}\int_0^t \varphi_\ell(\eta_\ell(\tau))d\tau + \frac{1}{2}\int_0^t u_r(X(\tau),\tau)d\tau \qquad (7.4.8)$$

Equations (7.4.7 - 8) form a pair of integral equations for $\eta_\ell(t)$ and $X(t)$. In the particular case, when the state ahead of the shock is $u_r = 0$, the equation (7.4.7) becomes

$$2\int_{X_0}^{\eta_\ell(t)} \varphi_\ell(s)ds + t\varphi_\ell^2(\eta_\ell(t)) = 0 \qquad (7.4.9)$$

which decouples from the second equation. Equation (7.4.9) is no longer an integral equation but an algebraic or a transcendental equation for η_ℓ for a given value of t depending on whether φ is an algebraic or a transcendental function.

Let us consider two examples with $u_r = 0$:

$$E_1 : \varphi(x) = \begin{cases} \left(\frac{x+\delta}{1+\delta}\right)^2, & x \in [-\delta, 1],\ -1 < \delta < \infty \\ 0 & \text{elsewhere} \end{cases} \qquad (7.4.10)$$

The initial data is non-zero only on a closed bounded interval. There is a shock at $x = 1$ and

$$X_0 = 1,\ u_{00} = 1,\ v_{10} = \frac{2}{1+\delta},\ v_{20} = \frac{1}{(1+\delta)^2},\ v_{i0} = 0,\ i \geq 3 \qquad (7.4.11)$$

The equation (7.4.9) reduces to

$$t\left(\frac{\eta_\ell+\delta}{\delta+1}\right)^4 + \frac{2}{3}\,\frac{(\eta_\ell+\delta)^3 - (1+\delta)^3}{(1+\delta)^2} = 0 \qquad (7.4.12)$$

$$E_2 : \varphi(x) = \begin{cases} \alpha e^{\beta x} & , \quad x \le 0, \ \alpha, \beta > 0 \\ \\ 0 & , \qquad x < 0 \end{cases} \tag{7.4.13}$$

Without loss of generality, we can set $\alpha = 1, \beta = 1$ because a change of variable $x \to \beta x, t \to \beta t/\alpha, u \to \alpha u$ leads to this result. Hence, we take

$$X_0 = 0, \ u_{00} = 1, \ v_{i0} = \frac{1}{i!}, i \ge 1 \tag{7.4.14}$$

The equation (7.4.9) becomes

$$te^{2\eta_\ell} + 2(e^{\eta_\ell} - 1) = 0 \tag{7.4.15}$$

It is simple to solve numerically the equations (7.4.12) or (7.4.15) for $\eta_\ell(t)$. Then the exact value of the shock strength is found easily from (7.4.3) where φ_ℓ is given by (7.4.10) or (7.4.13).

It is interesting to note a few properties of the partial derivatives $\frac{\partial^n u}{\partial x^n}|_\Omega$ and $v_n(t)$ for the solution with initial data (7.4.10). Since $\frac{\partial^i \varphi}{\partial x^i} = 0$ for $i \ge 3$, the expression (7.2.3) reduces to

$$\frac{\partial^i u}{\partial x^i}|_\Omega = (-1)^i \frac{(2i-2)! t^{i-2} (\varphi''(\eta_\ell))^{i-1}}{2^{i-1}(i-1)!(1 + t\varphi'(\eta_\ell))^{2i-1}}, \quad i \ge 2 \tag{7.4.16}$$

Using Stirling's formula for the factorial of a large integer i : $i! \simeq i^{i+1} e^{-i} \left(\frac{2\pi}{i}\right)^{\frac{1}{2}}$, we get

$$\frac{\partial^i u}{\partial x^i}|_\Omega \simeq \frac{(-1)^i \sqrt{2}}{t(1 + t\varphi')} \left[\frac{2t\varphi''}{(1 + t\varphi')^2 e}(i-1)\right]^{i-1}, \quad i \text{ large} \tag{7.4.17}$$

The formula (7.4.16) verifies that for a fixed $i \ge 3$, $\frac{\partial^i u}{\partial x^i}|_\Omega \to 0$ as $t \to 0$. Since $\varphi'(\eta_\ell) > 0$, the formula also shows that $\frac{\partial^i u}{\partial x^i}|_\Omega \to 0$ as $t \to \infty$. However, the approximate result (7.4.17) for large i shows that

$$\left|\frac{\partial^i u}{\partial x^i}|_\Omega\right| \to \infty \quad \text{as} \quad i \to \infty \quad \text{for a fixed } t \text{ and } \varphi''(\eta) \ne 0 \tag{7.4.18}$$

since for any fixed t (including arbitrarily small values of t) we can always choose n so large that $\left|\frac{2t\varphi''}{(1+t\varphi')^2 e}(n-1)\right| \ge$ constant > 1. This result is indeed curious and interesting for the initial data (7.4.10), especially when we know that for a linear $\varphi_\ell = \varphi_0 + \varphi_1(x-1)$ behind the shock,

$$u_\ell(x,t) = \frac{\varphi_0}{1+t\varphi_1} + \frac{\varphi_1}{1+t\varphi_1}(x-1) \text{ i.e., } \frac{\partial u_\ell^i}{\partial x^i} = 0 \quad \text{for } i \geq 2$$

This was observed in numerical computation by the new theory of shock dynamics. However, it is simple to see that for a fixed small t, $v_i = \frac{1}{i!}\frac{\partial^i u}{\partial x^i}|_\Omega$ tends to zero as $i \to \infty$.

We give now an alternative derivation of the result (1.7.5). In deducing (7.4.7), we assumed the state $u_r(x,t)$ to be known and concentrated on the variable $\eta_\ell(t) = X(t) - u_\ell(X(t),t)t$. However, the functions $u_\ell(x,t)$ and $u_r(x,t)$ appear symmetrically in the problem (see (7.1.2) and (7.1.4)) and it is interesting to use the variable

$$\eta_r(t) = X(t) - u_r(X(t),t)t \;=\; X(t) - \varphi_r(\eta_r)t \tag{7.4.19}$$

(where we have used the fact $u_r(X(t),t) = \varphi(\eta_r)$) and deduce some further results. We can write (7.4.7) in the form

$$2\int_{X_0}^{\eta_\ell(t)} \varphi_\ell(s)ds + t\varphi_\ell^2(\eta_\ell(t)) \;=\; \int_0^t \varphi_r(\eta_r(\tau))\varphi_\ell(\eta_\ell(\tau))d\tau \tag{7.4.20}$$

Similarly, we can deduce

$$2\int_{X_0}^{\eta_r(t)} \varphi_r(s)ds + t\varphi_r^2(\eta_r(t)) \;=\; \int_0^t \varphi_r(\eta_r(\tau))\varphi_\ell(\eta_\ell(\tau))d\tau \tag{7.4.21}$$

Subtracting (7.4.20) from (7.4.21), we get the relation (1.7.5)

$$t = \frac{2}{\varphi_\ell^2(\eta_\ell(t)) - \varphi_r^2(\eta_r(t))}\left[\int_{\eta_\ell(t)}^{X_0} \varphi_\ell(s)ds + \int_{X_0}^{\eta_r(t)} \varphi_r(s)ds\right] \tag{7.4.22}$$

where we note that $\eta_\ell(t) < X_0$ and $\eta_r(t) > X_0$ for $t > 0$. Any two of the three relations (7.4.20 - 22) can be used to solve η_ℓ and η_r as functions of t. The shock path $\Omega : x = X(t)$ can be obtained from (7.4.8), i.e.,

$$X(t) = X_0 + \frac{1}{2}\int_0^t \varphi_\ell(\eta_\ell(\tau))d\tau + \frac{1}{2}\int_0^t \varphi_r(\eta_r(\tau)d\tau \tag{7.4.23}$$

The shock strength and all other quantities on the shock path can now be easily obtained.

Numerical solution In the case of the initial condition (7.4.10), the state behind the shock approaches a constant state $u = 1$ as $\delta \to \infty$. For δ close to -1, φ increases rapidly from 0 to 1 over a very short distance; thus the spacial derivatives of u play an important role in this case. In Table 7.4.1 we present numerical computation only for $\delta = -0.5$. For $\delta = 1$ and $\delta = 50$, the result obtained from the new theory of shock dynamics were very close to the exact solution even for $n = 2$.

Tables 7.4.1 and 7.4.2 give the values of u for initial values corresponding to (7.4.11) and (7.4.14), respectively. The results are given at $t = 1, t = 5$ and $t = 10$. The result for $n = k$ assumes that v_j for $j \geq k + 1$ are set equal to zero and $k + 2$ equations (7.3.1 - 4) with (7.3.5) are considered. For $n = 1$, the percentage of error $(100^{-1}) \times (\bar{u}_0 - u_0)/ \mid u_0 \mid$ is sizeable, but even for $n = 2$, the error drops rapidly (to less than 1% in case E_2); while for $n = 3$, it is uniformly small, as for $n = 5,\ 8,\ 25$ as well. Computation was done for E_2 with $\alpha = 1,\ \beta = 1$.

Table 7.4.1: Solution with initial value $E_1, \delta = -0.5$

	t = 1.0		t = 5.0		t = 10.0	
	u_0	% error	u_0	% error	u_0	% error
Exact	.47390445	-	.24231081	-	.17572092	-
$n=1$	.44721360	-5.6	.21821789	-9.9	.15617376	-11.0
$n=2$	.47171239	-.46	.23787367	-1.8	.17140803	-2.5
$n=3$	.47366942	$-.50\times10^{-1}$	.24120493	-.46	.17440411	-.75
$n=5$	.47390183	$-.56\times10^{-3}$	.24221465	$-.40\times10^{-1}$	.17554484	-.10
$n=8$	.47390560	$.24\times10^{-3}$	.24230783	$-.12\times10^{-2}$	.17570887	$-.68\times10^{-2}$
$n=25$	.47390561	$.24\times10^{-3}$	.24231136	$.24\times10^{-3}$	.17572129	$.22\times10^{-3}$

Table 7.4.2: Solution with initial value $E_2, \alpha = 1, \beta = 1$

	t = 1.0		t = 5.0		t = 10.0	
	u_0	% error	u_0	% error	u_0	% error
Exact	.73205081	-	.46332497	-	.35825757	-
$n=1$	.70710678	-3.4	.40824829	-12	.30151134	-16
$n=2$	.73372900	-.23	.46777169	9.6	.36157950	9.3
$n=3$	.73200502	$-.63\times10^{-2}$	.46355666	$-.50\times10^{-1}$	.35872978	.13
$n=5$	.73205096	$.26\times10^{-4}$	.46331988	$-.11\times10^{-2}$	.35825020	$-.21\times10^{-2}$
$n=8$	.73205081	0	.46332497	0	.35825765	$.22\times10^{-4}$
$n=25$	.73205081	0	.46332496	0	.35825757	0

7.5 Conclusion

Theoretical justification in the previous section and comparison of numerical results with an exact solution show that the new theory of shock dynamics is a good theory for computation of the shock position and the shock strength. This is especially so if the initial data is such that $\phi'(\xi) > 0$ on the x-axis behind the shock and as long as the shock under consideration does not interact with another shock. Unlike a finite difference method, the shock position is determined exactly, as the shock transition is not spread over a few mesh points. In addition to this, the new theory of shock dynamics also gives a number of spatial derivatives. Since the infinite-dimensional problem is replaced by a finite-dimensional problem of integrating only 4 or 5 equations for reasonably good accuracy, the new theory of shock dynamics achieves a lot of computational efficiency.

It may be tempting to compare the new theory of shock dynamics with the original shock dynamics of Whitham developed in 1957 and 1959 (see Whitham (1974)). For a single conservation law, Whitham's shock dynamics is equivalent to the new theory of shock dynamics for $n = 0$, which is exact only when the state behind the shock is uniform. Whitham's shock dynamics is based on intuitive arguments and cannot be justified theoretically (Srinivasan and Prasad (1985), Prasad (1990)). It has also been shown that Whitham's shock dynamics can give very large errors if the state behind the shock is not uniform (Prasad, Ravindran and Sau (1991). Comparison of results obtained for a converging shock by Whitham's shock dynamics and NTSD is given in Chapter 10.

Chapter 8

One-dimensional piston problem : an application of NTSD

In this chapter we shall present an application of the new theory of shock dynamics (NTSD) to a physically realistic problem, namely the one-dimensional piston problem in a polytropic gas. With this application we wish to study two aspects of the NTSD: (i) is the theory applicable to a wide range of velocities and accelerations of the piston? and (ii) what is the computational efficiency of the theory? In spite of the fact that the NTSD appears to be of doubtful accuracy for a single conservation law (Sundar, Prasad and Ravindran (1992)) when the initial data behind the shock has negative gradients the results in this chapter show that the theory gives a very good result when the piston is accelerating. This is achieved by a special formulation of the equations of the NTSD (Lazarev, Prasad and Singh, 1995). Though the equations of the NTSD with just three compatibility conditions have quite lengthy expressions, the theory has very high computational efficiency – it takes less than 0.5% of the computational time (for large times) of a typical finite difference method applied to full Euler's equations of the gas dynamics. In section 3.1.1, we discussed the piston problem when the piston velocity gradually increased or decreased from zero initial velocity. The resulting motion of the fluid was given by a simple wave solution (valid only up to a critical time in the case of an accelerating piston). We can also get an exact solution to this problem when the piston starts suddenly with a non-zero velocity and then moves with a constant velocity. However, when the piston moves into the gas on the right and its initial velocity is positive and non-zero and it has a non-zero acceleration or deceleration, there is no way to solve this problem other than a finite difference numerical method applied to Euler's equations or the NTSD developed here. Therefore, a comparison of these two methods following Lazarev, Prasad and Singh (1995) is important.

8.1 Formulation of the problem

Consider the one-dimensional motion of a gas in a shock tube produced by an accelerating or decelerating piston starting with a non-zero initial velocity. The medium is assumed to be an ideal gas with constant specific heats. It is well known that the disturbed flow in front of the piston moving *into* the gas with non-zero initial velocity is separated from the undisturbed flow by a shock front. Unless the piston velocity is very small, the shock is quite strong and the weak shock theory of Chandrasekhar and Friedrichs (see Courant and Friedrichs (1948)) is not applicable.

Since the shock produced in the flow may be strong, the flow can not be assumed to be isentropic as in section 3.1.1. Therefore, we use the equations of motion in the form (2.1.12 - 14). Let the piston position at time t be $X_p(t)$. We take without loss of any generality $X_p(0) = 0$. Thus, we are looking for a solution of (2.1.8 - 20) in the domain $(x > X_p(t), t > 0)$ of the (x, t)-plane satisfying the following initial and boundary conditions

$$q(x,0) = 0, \;\; p(x,0) = p_0, \;\; \rho(x,0) = \rho_0 \;\text{ for }\; x > 0 \qquad (8.1.1)$$

and

$$q(X_p(t), t) = X_p'(t)(= \text{ piston velocity}), \text{ for } t > 0. \qquad (8.1.2)$$

When the initial velocity of the piston is non-zero and positive: $X'(0) > 0$, the solution contains a non-constant state in a domain $G : X_p(t) < x < X(t)$, where the shock path $x = X(t)$ separates this domain from the undisturbed region $x > X(t)$. We assume that the piston velocity is such that no other shock is produced in the domain G. The functions u, p and ρ and their partial derivatives as we approach the shock from the domain G approach finite limits.

We note the kinematical compatibility conditions in section 3.3.4 and start now with the derivation of the dynamical compatibility conditions on the shock path.

8.2 Dynamical compatibility conditions

We introduce the following notations

$$D_0 = [\rho], \; H_0 = [q], \; S_0 = [p] \tag{8.2.1}$$

where [ ] on a quantity h denotes the jump across the shock front $\Omega : x = X(t)$, i.e., $[h] = h_r(t) - h_\ell(t)$. For the shock moving into the uniform medium at rest $q_r = 0$, $p_r = p_0$ and $\rho_r = \rho_0$, the Rankine–Hugoniot conditions give the following expressions for H_0, S_0 and the shock velocity C in terms of a single parameter D_0:

$$\left.\begin{aligned} H_0 = D_0 C(\rho_0 - D_0)^{-1}, \; S_0 = \rho_0 D_0 C^2 (\rho_0 - D_0)^{-1} \\ C = a_0 \left\{ \frac{2(\rho_0 - D_0)}{2\rho_0 + D_0(\gamma - 1)} \right\}^{\frac{1}{2}} \end{aligned}\right\} \tag{8.2.2}$$

where a_0 is the local sound velocity in the state ahead of the shock.

The derivatives of the quantities on the shock satisfy the following relations

$$\begin{aligned} \frac{dh_{\ell,r}}{dt} &= \left.\frac{\partial h}{\partial t}\right|_{\ell,r} + C \left.\frac{\partial h}{\partial x}\right|_{\ell,r} \\ \frac{d}{dt}\left(\frac{\partial^N h}{\partial x^N}\Big|_{\ell,r} \right) &= \left.\frac{\partial^{N+1} h}{\partial t \partial x^N}\right|_{\ell,r} + C \frac{\partial^{N+1} \rho}{\partial x^{N+1}}\Big|_{\ell,r} \end{aligned} \tag{8.2.3}$$

for $N = 1, 2, 3, \ldots$, where h is any one of q, p and ρ and the relation is valid either with the subscript ℓ or r. From these we derive

$$\left[\frac{\partial h}{\partial t}\right] = \frac{d}{dt}[h] - C\left[\frac{\partial h}{\partial x}\right]$$

$$\left[\frac{\partial^{N+1}h}{\partial t\partial x^N}\right] = \frac{d}{dt}\left[\frac{\partial^N h}{\partial x^N}\right] - C\left[\frac{\partial^{N+1}h}{\partial x^{N+1}}\right] \tag{8.2.4}$$

We can express the relation (8.2.4) in terms of

$$D_N = \left[\frac{\partial^N \rho}{\partial x^N}\right], \quad H_N = \left[\frac{\partial^N q}{\partial x^N}\right], \quad S_N = \left[\frac{\partial^N p}{\partial x^N}\right] \tag{8.2.5}$$

by choosing $h = \rho, q$ or p.

We consider the jump of the expressions on the left hand side of (2.1.12 - 14) on the shock front Ω_t and equate them to zero. This leads to the vector form of the first set of compatibility conditions

$$\frac{d}{dt}U_0 \; + \; PU_1 = 0 \tag{8.2.6}$$

where we have used the matrix notation

$$U_N = \begin{pmatrix} D_N \\ H_N \\ S_N \end{pmatrix}, \quad P = \begin{pmatrix} -(C+H_0) & \rho_0 - D_0 & 0 \\ 0 & -(C+H_0) & (\rho_0 - D_0)^{-1} \\ 0 & \gamma(p_0 - S_0) & -(C+H_0) \end{pmatrix}$$

For the derivation of the second set of compatibility conditions, we differentiate the equations (2.1.12 - 14) with respect to x and take its jump across Ω to get

$$\frac{d}{dt}U_1 + PU_2 = f_1 \tag{8.2.7}$$

where

$$f_1 = \begin{pmatrix} 2D_1H_1 \\ \\ H_1^2 - D_1S_1(\rho_0 - D_0)^{-2} \\ \\ (1+\gamma)H_1S_1 \end{pmatrix} \tag{8.2.8}$$

For the derivation of the third set of compatibility conditions, we differentiate (2.1.12 - 14) twice with respect to x and take its jump across Ω. This gives

$$\frac{d}{dt}U_2 + PU_3 \;=\; f_2 \tag{8.2.9}$$

where

$$f_2 = \begin{pmatrix} 3D_2H_1 + 3D_1H_2 \\ 3H_1H_2 - (2D_1S_2 + D_2S_1)(\rho_0 - D_0)^{-2} - 2D_1^2 S_1(\rho_0 - D)^{-3} \\ S_1H_2 + 2S_2H_1 + \gamma(S_2H_1 + 2S_1H_2) \end{pmatrix} \tag{8.2.10}$$

Proceeding in this way, we can derive an infinite system of the vector form of dynamical compatibility conditions.

The infinite system of the compatibility conditions in vector form can be reduced to an infinite system of compatibility conditions in scalar form. We first note that the eigenvalues of the matrix P are

$$\lambda_1 = -(C + H_0) \ , \ \lambda_{2,3} = -(C + H_0) \pm a_\ell \tag{8.2.11}$$

where a_ℓ is the sound velocity behind the shock. As the shock strength H_0 tends to zero, the shock velocity C and the local sound velocity tend to a common value a_0 (note $q_0 = 0$), hence λ_2 tends to zero. Also, in this limit, the first set of compatibility conditions (8.2.6) must lead to the scalar form of the characteristic compatibility condition in which the first order derivative terms arising out of PU_1 must be zero. Hence, we choose the left eigenvector of P corresponding to the eigenvalue λ_2. This eigenvector is

$$L = \left(0, \ \frac{\rho_0 - D_0}{2} \ , \ \frac{1}{2a_\ell}\right) \tag{8.2.12}$$

We now introduce scalar variables $\pi_0, \pi_1, \pi_2, \ldots$ on the shock

$$\pi_0(t) = D_0 \ , \ \pi_i(t) = LU_i = \frac{\rho_0 - D_0}{2} H_i + \frac{1}{2a_\ell} S_i, \ i = 1, 2, 3, \ldots \tag{8.2.13}$$

which are functions of time t only.

Using the Rankine–Hugoniot conditions in (8.2.6), we get

$$\begin{pmatrix} 1 \\ A \\ B \end{pmatrix} \frac{d\pi_0}{dt} = -PU_1 \tag{8.2.14}$$

where

$$A \equiv \frac{\partial H_0}{\partial \pi_0} = \frac{C\rho_0(4\rho_0 + \pi_0(\gamma - 3))}{2(\rho_0 - \pi_0)^2(2\rho_0 + \pi_0(\gamma - 1))} \tag{8.2.15}$$

$$B \equiv \frac{\partial S_0}{\partial \pi_0} = \frac{4a_0^2\rho_0^2}{(2\rho_0 + \pi_0(\gamma - 1))^2} \tag{8.2.16}$$

It is possible to eliminate $\frac{d\pi_0}{dt}$ from (8.2.14) and express any two of D_1, H_1 and S_1 in terms of the third or we can express all of them in terms of π_1. We shall do this a little later. We first note that pre-multiplication of (8.2.14) by L gives the scalar form of the first compatibility condition

$$\frac{d\pi_0}{dt} = -G\lambda_2\pi_1 \tag{8.2.17}$$

where

$$G = 2\{A(\rho_0 - \pi_0) + Ba_\ell^{-1}\}^{-1} \tag{8.2.18}$$

Pre-multiplying (8.2.14) by P^{-1} we get

$$U_1 = -\frac{d\pi_0}{dt}P^{-1}\begin{pmatrix} 1 \\ A \\ B \end{pmatrix}$$

in which we substitute $\frac{d\pi_0}{dt}$ from (8.2.17) to deduce

$$D_1 = d_1\pi_1, \;\; H_1 = h_1\pi_1, \;\; S_1 = s_1\pi_1 \tag{8.2.19}$$

where d_1, h_1 and s_1 are functions of π_0 only :

$$d_1 = G(\lambda_2\lambda_3 + \rho_0 CA + B)/(\lambda_1\lambda_3) \tag{8.2.20}$$

$$h_1 = -G(\rho_0 CA + B)/\{\lambda_3(\rho_0 - \pi_0)\} \tag{8.2.21}$$

and

$$s_1 = -G\{a_\ell^2 a_0(\rho_0 - \pi_0) - \lambda_1 B\}/\lambda_3 \tag{8.2.22}$$

Following the procedure of the derivation of (8.2.17), the equation (8.2.7) leads to

$$\frac{d\pi_1}{dt} = K^{(1)}\pi_1^2 - \lambda_2\pi_2 \tag{8.2.23}$$

where

$$K^{(1)} = \frac{1}{2}\{(\rho_0 - \pi_0)h_1^2 - (\rho_0 - \pi_0)^{-1}d_1 s_1 + (\gamma + 1)a_0^{-1}s_1 h_1 + G\lambda_2(h_1 - es_1)\} \tag{8.2.24}$$

with

$$e = -a_\ell^{-2}\frac{\partial a_\ell}{\partial \pi_0} = \frac{1}{2a_\ell(\rho_0 - \pi_0)}\left\{\frac{\gamma C^4 \rho_0^2}{a_0^2 a_\ell^2(\rho_0 - \pi_0)^2} - 1\right\} \tag{8.2.25}$$

The expressions for D_2, H_2 and S_2 in terms of π_2 are

$$D_2 = \delta_1\pi_1^2 + \delta_2\pi_2, \;\; H_2 = \omega_1\pi_1^2 + \omega_2\pi_2, \;\; S_2 = \sigma_1\pi_1^2 + \sigma_2\pi_2 \tag{8.2.26}$$

where

$$\begin{pmatrix}\delta_1\\ \omega_1\\ \sigma_1\end{pmatrix} = P^{-1}\begin{pmatrix} 2d_1h_1 + d_1'G\lambda_2 - K^{(1)}d_1\\ h_1^2 - (\rho_0 - \pi_0)^{-2}d_1 s_1 + h_1'G\lambda_2 - K^{(1)}h_1\\ (\gamma+1)s_1h_1 + s_1'G\lambda_2 - K^{(1)}s_1\end{pmatrix} \tag{8.2.27}$$

$$\begin{pmatrix}\delta_2\\ \omega_2\\ \sigma_2\end{pmatrix} = \lambda_2 P^{-1}\begin{pmatrix}d_1\\ h_1\\ s_1\end{pmatrix} \tag{8.2.28}$$

$$d_1' = \frac{dd_1}{d\pi_0}, \; h_1' = \frac{dh_1}{d\pi_0}, \; s_1' = \frac{ds_1}{d\pi_0} \tag{8.2.29}$$

In the above expressions, $K^{(1)}, \delta_1, \delta_2, \omega_1, \omega_2$ and σ_1, σ_2 are functions of π_0 only.

Proceeding in the same way, we get the following equation for π_2

$$\frac{d\pi_2}{dt} = K_1^{(2)}\pi_1^3 + K_2^{(2)}\pi_1\pi_2 - \lambda_2\pi_3, \tag{8.2.30}$$

where

$$\begin{aligned} K_1^{(2)} &= \frac{1}{2}[G\lambda_2(\omega_1 - e\sigma_1) + (\rho_0 - \pi_0)\\ &\quad \left\{3h_1\omega_1 - 2(\rho_0-\pi_0)^{-3}d_1^2 s_1 - (\rho_0 - \pi_0)^{-2}(2d_1\sigma_1 + s_1\delta_1)\right\}\\ &+ a_0^{-1}\{(\gamma+2)h_1\sigma_1 + (1+2\gamma)s_1 w_1\}\,] \end{aligned} \tag{8.2.31}$$

and

$$\begin{aligned} K_2^{(2)} &= \frac{1}{2}[G\lambda_2(\omega_2 - e\sigma_2) + (\rho_0 - \pi_0)\\ &\quad \left\{3h_1\omega_2 - (\rho_0 - \pi_0)^{-2}(2d_1\sigma_2 + s_1\delta_2)\right\}\\ &+ a_\ell^{-1}\{(\gamma+2)h_1\sigma_2 + (2\gamma+1)s_1\omega_2\}\,] \end{aligned} \tag{8.2.32}$$

The coefficients $K_1^{(2)}$ and $K_2^{(2)}$ are functions of π_0 only.

The new theory of shock dynamics, using compatibility conditions up to the third, is obtained by setting $\pi_3 = 0$ in (8.2.30). The equations (8.2.17 and 23) and (8.2.30) with $\pi_3 = 0$, i.e.,

$$\frac{d\pi_2}{dt} = K_1^{(2)}\pi_1^3 + K_2^{(2)}\pi_1\pi_2 \tag{8.2.33}$$

form a closed system of equations for the shock strength π_0, a quantity π_1 (which is a linear combination of the jumps in the first derivatives) and a quantity π_2 (a linear combination of the jumps in the second derivatives). The coefficients $G, K^{(1)}, K_1^{(2)}$ and $K_2^{(2)}$ are known functions of π_0. The shock position $X(t)$ is determined by numerically integrating the expression

$$X(t) = X_0 + \int_0^t C dt \tag{8.2.34}$$

where C is a known function of π_0 from (8.2.2) with $D_0 = \pi_0$. We can also determine the values of jumps in the first and second derivatives of the density, velocity and pressure : $D_1, D_2; H_1, H_2; S_1, S_2$ from (8.2.19) and (8.2.26). These can be used to find the values of the density, velocity and pressure near the shock by using the first three terms of the Taylor's series.

In order to find a solution of the equations (8.2.17, 23 and 33), we need initial values of π_0, π_1 and π_2. We obtain these values for the piston problem in the next section. However, before we derive these values we shall make a few comments on the existence and uniqueness of the solution.

As in the case of a single conservation law, it is simple to verify that the solution of an initial value problem for the infinite system of compatibility conditions is non-unique in the class of C^∞ functions. The proof given in section 7.2 of the existence of an analytic solution of the infinite system for a single conservation law is no longer valid. In the case of a system of conservation laws, the shock influences the solution behind the shock and hence cannot be determined only from the initial data behind the shock. However, assuming that an analytic solution of the infinite system exists for small t, it is possible to show that it is unique by determining the coefficients in the expansion successively. As in section 7.3, it is also possible to show that the analytic solution of the truncated finite system of equations of the new theory of shock dynamics for small t tends to the analytic solution of the infinite system as the order of the finite system tends to infinity. It is also interesting to note from results in the next section that the new theory of shock dynamics gives good results for a shock of arbitrary strength.

8.3 Initial conditions for the piston problem

In this section, we shall obtain initial values of π_0, π_1 and π_2 for the piston problem for which the initial and boundary conditions are given by (8.1.1 - 2). We first introduce non-dimensional variables (denoted by a bar above the variables) by

$$x = L\bar{x}, \; t = \frac{L}{a_0}\bar{t}, \; q = a_0\bar{q}, \; \rho = \rho_0\bar{\rho}, \; p = \gamma p_0\bar{p}, \; C = a_0\bar{C} \quad (8.3.1)$$

where the characteristic length L is the distance travelled by sound in unit-time in the undisturbed medium ahead of the shock.

The piston motion starting from $x = 0$, is prescribed such that it is representable as a power series in t

$$\bar{X}_p(\bar{t}) = X_{p1}\bar{t} + x_{p2}\bar{t}^2 + X_{p3}\bar{t}^3 + \dots \quad (8.3.2)$$

The flow field in the region $\bar{X}_p(t) < \bar{x} < \bar{X}(t)$ is infinitely differentiable and it is assumed that the solution in this region is also representable by the power series

$$\bar{h}(\bar{x}, \bar{t}) = \bar{h}_{\ell 0} + h_{11}\bar{x} + h_{12}\bar{t} + h_{21}\bar{x}^2 + h_{22}\bar{x}\bar{t} + h_{23}\bar{t}^2 + \dots \quad (8.3.3)$$

where h stands for any one of q, p, ρ. We note $\bar{q}_0 = 0$. From $X'(t) = C(X, t)$ and (8.3.3), it follows that the shock path has a power series expansion

$$\bar{X}(\bar{t}) = X_1\bar{t} + X_2\bar{t}^2 + X_3\bar{t}^3 + \dots \quad (8.3.4)$$

Given the coefficients $X_{p1}, X_{p2}, X_{p3}, \dots$ in (8.3.2), we need to find the coefficients in (8.3.3) for $h = \rho, q$ and p and those in the shock position (8.3.4). This is quite simple and straightforward, though it involves lengthy algebraic calculations. We first substitute the expansion (8.3.3) in the differential equations (2.1.12 - 14), collect the coefficients of various powers and products of x and t and equate them to zero. This leads to an undetermined system of equations for the coefficients in (8.3.3). The expansion (8.3.3), with $\bar{x}$ replaced by the expansion for $\bar{X}(\bar{t})$, is then substituted in the non-dimensional form of Rankine–Hugoniot conditions

$$\left.\begin{array}{c}\bar{\rho}_\ell(\bar{q}_\ell - \bar{C}) = -\bar{C},\ \bar{p}_\ell + \bar{\rho}_\ell(\bar{q}_\ell - \bar{C})^2 = \frac{1}{\gamma} + \bar{C}^2 \\ \bar{a}_\ell^2/(\gamma-1)\} + (\bar{q}_\ell - \bar{C})^2/2 = 1/(\gamma-1) + \bar{C}^2/2\end{array}\right\} \quad (8.3.5)$$

Equating the coefficients of various powers of $\bar{t}$, we get further equations. However, the system of equations still remains undetermined. Finally, using the expansion (8.3.3) with $\bar{h} = \bar{q}$ and the expansion (8.3.2) for $\bar{X}_p(\bar{t})$ in the boundary condition (8.1.2) and equating various powers of t, we get additional equations with the help of which all unknowns are uniquely determined in terms of $\rho_0, p_0; X_{p1}, X_{p2}, \ldots$. For our purpose, to get initial values of $\bar{\pi}_0, \bar{\pi}_1$ and $\bar{\pi}_2$, we need only the values of ρ_{11} and ρ_{21}. The expressions giving these quantities are very long and are not given here.

The initial values of $\bar{\pi}_0, \bar{\pi}_1$ and $\bar{\pi}_2$ are finally given by

$$\bar{\pi}_0(0) = -\frac{(k^2-1)(1-\gamma\bar{p}_{\ell 0})}{1+k^2 0\bar{p}_{\ell 0}},\ \bar{\pi}_1(0) = -\frac{\rho_{11}}{\bar{d}_1},\ \bar{\pi}_2(0) = \frac{2\rho_{21} + \bar{\delta}_1\bar{\pi}_1^2(0)}{\bar{\delta}_2} \quad (8.3.6)$$

where $k^2 = \frac{\gamma-1}{\gamma+1}$ and $\bar{d}_1, \bar{\delta}_1$ and $\bar{\delta}_2$ are non-dimensional forms of the quantities d_1, δ_1 and δ_2 defined by (8.2.20, 27 and 28) and evaluated at $\bar{\pi}_0(0)$.

This completes, for the piston problem, the formulation of the new theory of shock dynamics using the first three compatibility conditions. We now have to solve numerically the non-dimensional forms of the three ordinary differential equations (8.2.17, 23 and 33) using the initial values (8.3.6). In principle, it is possible to take account of an arbitrary analytic path of the piston. However, we consider piston motion with only constant acceleration ($\bar{X}_{p2} > 0$) or deceleration ($\bar{X}_{p2} < 0$). Then $\bar{X}_{p2} \neq 0$ and all $\bar{X}_{pi} = 0, i \geq 3$.

This is the first application of the new theory of shock dynamics to a realistic practical problem. Therefore, we shall present a detailed discussion of the numerical results. Further, it would be interesting to compare the results with the solution obtained by another method. However, there is no exact solution with which the results can be compared. Hence, the problem has been solved also by a finite difference method using the Lagrangian form of the equations of the motion. The numerical method we use is Harten's high resolution TVD scheme (Harten (1983); Yee, Warming and Harten (1985)). The finite difference method has its own difficulty, since it becomes unstable for a large number of cases when the acceleration becomes even moderately large.

8.4 Results and discussions

We take $\gamma = 1.4$. The ordinary differential equations formulated in section 8.3 were numerically integrated using the Runge-Kutta-Gill method with the initial conditions (8.3.6).

Some of the results for the accelerating piston are shown in Figs. 8.4.1 and 8.4.2. The shock grows from the initial value of the shock strength $\bar{\pi}_0$ (corresponding to the case of uniform piston velocity) to the strong shock limit $\bar{\pi}_0 = 5.0$. The time taken to reach this limiting value depends on the values of X_{p_1} and X_{p_2}.

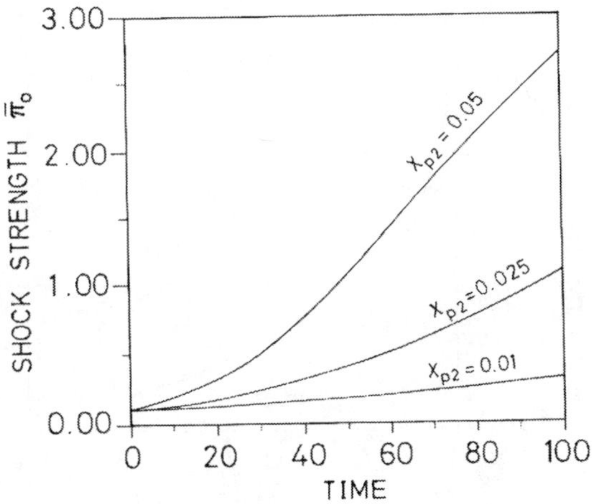

Fig. 8.4.1: Growth of the shock when the initial piston velocity is small $X_{p_1} = 0.1$

The results for a decelerating piston motion ($X_{p_2} < 0$), though worked out for many cases, has been shown in only two cases in Figs. 8.4.3 and 4. The shocks beginning with different initial values of $\bar{\pi}_0$

decay to the minimum value $\bar{\pi}_0 = 0$; the rate of decay depends on the values of X_{p_1} and X_{p_2}.

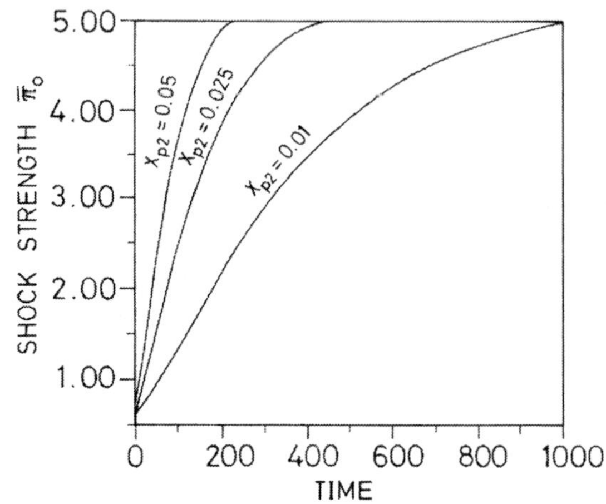

Fig. 8.4.2: Growth of a shock calculated until the shock becomes very strong, $X_{p_1} = 0.5$.

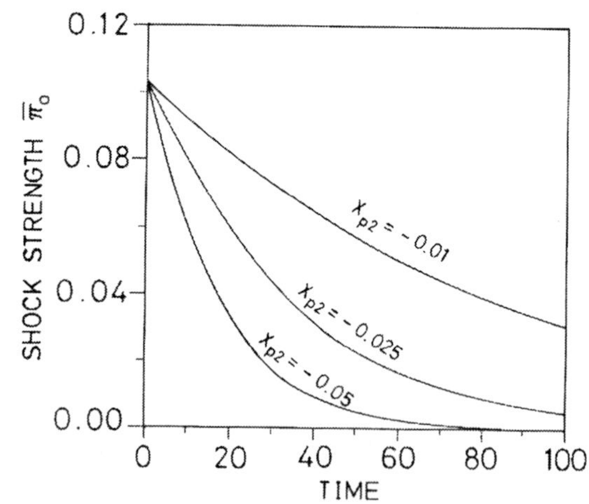

Fig. 8.4.3: Decay of a weak shock, $X_{p_1} = 0.1$.

The results obtained by Harten's finite difference scheme (FDM) for accelerating and decelerating pistons are shown in Figs. 8.4.5 and 6. The shock strength and the flow behind the shock as obtained by NTSD have also been plotted for a few points. The dots indicate the values of $\bar{\pi}_0 = \frac{\rho - \rho_0}{\rho_0}$ obtained by NTSD at and behind the shock front: '0' indicate the shock front and '1, 2, 3, 4 and 5' stand for the points behind the shock front at the distances given by $(x - X(t))/X(t) = 0.05$, 0.10, 0.15, 0.20 and 0.25, respectively. It is observed that at the shock front, there is very good agreement between the results obtained by NTSD and the finite difference scheme, but there is some deviation for the flow behind the shock which increases with time.

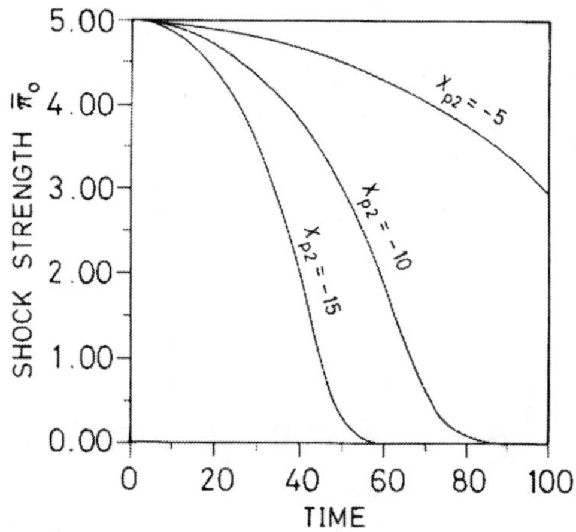

Fig. 8.4.4: Decay of a strong shock, $X_{p_1} = 50.0$.

This is, of course, expected because according to the equation for the error (7.3.8), the NTSD need not give good results away from the shock.

The results obtained by NTSD and FDM for the shock strength $\bar{\pi}_0$ are given in the Tables 8.4.1, 2 and 3. Three values for the piston velocity X_{p_1}: 0.1, 0.25 and 0.5 have been considered and for each value of X_{p_1} three values of acceleration and deceleration X_{p_2} $\pm 0.01, \pm 0.025$ and ± 0.05 have been chosen.

We define an error as

$$\varepsilon = ((\bar{\pi}_0)_{FDM} - (\bar{\pi}_0)_{NTSD})/(\bar{\pi}_0)_{FDM}$$

In some of the results by FDM, considerable error is observed initially i.e., up to $t \leq 1$ but disappears after some time. The FDM scheme is assumed to have failed if the algorithm fails due to some numerical instability when λ (where $\lambda = \Delta t/\Delta x$) is continuously decreased to the value 0.001.

The following trends are observed comparing the results by NTSD and those by FDM:

(a) **Accelerating piston**

(i) For $X_{p_1} = 0.1$, NTSD and FDM show very good agreement. For $X_{p2} = 0.01$, the error $\varepsilon < 0.04\%$ for all $t \leq 10.0$, whereas for $X_{p_2} = 0.025, \varepsilon < 0.08\%$, for $t \leq 5.0, \varepsilon < 1.2\%$ upto $t = 7.0$, beyond

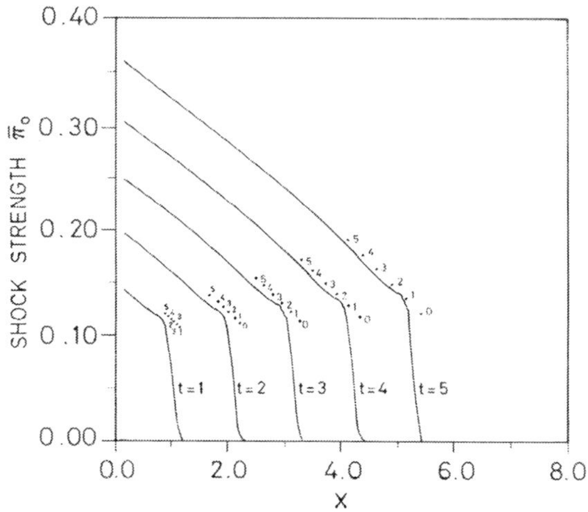

Fig. 8.4.5: Growth of a shock using Harten's high resolution TVD scheme for $X_{p_1} = 0.10$, $X_{p_2} = 0.025$ at the indicated times.

which FDM fails. For $X_{p_2} = 0.05$, FDM fails after $t = 4.0$ and except for the initial errors, $\varepsilon < 0.1\%$ which grows to 1.5% at the point of failure.

(ii) For $X_{p_1} = 0.25$ and $X_{p_2} = 0.01, \varepsilon < 0.05\%$ for $t \leq 10.0$. For $X_{p_2} = 0.025$, FDM fails for $t > 5.0$ and $\varepsilon < 0.02\%$ up to failure point. For $X_{p_2} = 0.05$, failure of FDM occurs after $t = 3.0$ and $\varepsilon < 0.06\%$ up to this point.

(iii) For $X_{p_1} = 0.5$ and $X_{p_2} = 0.01, \varepsilon < 0.4\%$ upto $t \leq 6.0$, beyond which FDM fails. For $X_{p2} = 0.025$, FDM fails after $t = 4.0$ and $\varepsilon < 0.8\%$ up to this point; whereas for $X_{p_2} = 0.05$, the failure of FDM occurs after $t = 2.0$ and $\varepsilon < 0.75\%$ up to the point of failure.

(b) Decelerating piston

(i) For $X_{p_1} = 0.1$ and $X_{p_2} = -0.01$, ε increases continuously from 1.7% to 5.9% in the time interval $1 \leq t \leq 10$ and for $X_{p_2} = -0.025$, the error grows from 4.4% to 13%; whereas for $X_{p2} = -0.05$, ε grows from 9.2% to 20.8% during the same time interval.

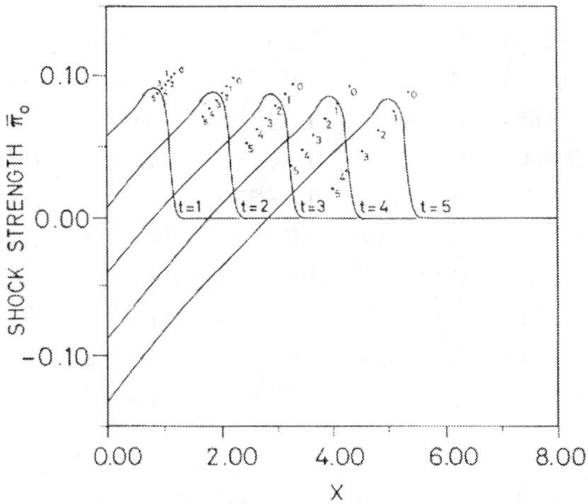

Fig. 8.4.6: Decay of a shock using Harten's high resolution TVD scheme for $X_{p_1} = 0.10, X_{p_2} = -0.025$ at the indicated times.

Table 8.4.1

Growth and decay of the shock strength $\bar{\pi}_0(t)$ for $X_{p_1} = 0.1$ and $X_{p_2} = \pm 0.01, \pm 0.025, \pm 0.5$

(a) Acceleration case $X_{p_1} = 0.10$ $\bar{\pi}_0(0) = 0.10397$

X_{p_2}	0.01		0.025		0.05	
T	NTSD	FDM	NTSD	FDM	NTSD	FDM
1.0	0.10521	0.10509	0.10710	0.10705	0.11031	0.11025
2.0	0.10646	0.10643	0.11031	0.11040	0.11700	0.11718
3.0	0.10773	0.10774	0.11361	0.11358	0.12407	0.12419
4.0	0.10901	0.10902	0.11700	0.11696	0.13154	0.13351
5.0	0.11031	0.11032	0.12048	0.12044		
6.0	0.11162	0.11165	0.12407	0.12422		
7.0	0.11294	0.11297	0.12775	0.12923		
8.0	0.11428	0.11430				
9.0	0.11563	0.11564				
10.0	0.11700	0.11705				

(b) Deceleration case

X_{p2}	-0.01		-0.025		-0.05	
T	NTSD	FDM	NTSD	FDM	NTSD	FDM
1.0	0.10275	0.10451	0.10093	0.10563	0.09798	0.10793
2.0	0.10153	0.10350	0.09798	0.10323	0.09231	0.10322
3.0	0.10033	0.10024	0.09510	0.10063	0.08695	0.09840
4.0	0.09915	0.10128	0.09231	0.09814	0.08188	0.09398
5.0	0.09798	0.10020	0.08959	0.09571	0.07709	0.08987
6.0	0.09682	0.09847	0.08695	0.09479	0.07258	0.08439
7.0	0.09567	0.09841	0.08438	0.09251	0.06832	0.08081
8.0	0.09545	0.09837	0.08188	0.09144	0.06429	0.07813
9.0	0.09342	0.09815	0.07945	0.09005	0.06049	0.07491
10.0	0.09231	0.09805	0.07709	0.08867	0.05690	0.07187

(ii) For $X_{p_1} = 0.025$ and $X_{p_2} = -0.01$, ε grows from 0.7% to 2.8% for $1.0 < t < 10.0$ and for $X_{p_2} = -0.025$, ε grows from 2% to 9% whereas for $X_{p_2} = -0.05$, ε grows from 3.8% to 21.6% in the same time interval.

(iii) For $X_{p_1} = 0.5$ and $X_{p_2} = -0.01$, ε grows from 0.1% to 2.8% for $1.0 < t < 10.0$ whereas for $X_{p_2} = -0.025$, it grows from 0.43% to 8.2% and for $X_{p_2} = -0.05$ it grows from 1.2% to 20.6% in the same time interval.

When $X_{p_2} \neq 0$, we do not have an exact solution of the problem to compare results obtained from NTSD and FDM. However, there is every reason to believe that the NTSD would give more accurate results for $X_{p_2} < 0$, since this corresponds to the case where the density versus x-coordinate has a positive slope. Therefore, strong deviation in the results by two methods in Tables 8.4.1(b), 8.4.2(b) and 8.4.3(b) is probably due to some deficiency in FDM.

A careful observation of large errors (as in the decelerating piston cases) and failure of the FDM algorithm (as in the cases with large acceleration) indicate that FDM gives good and stable results only when the perturbation from the uniform flow is small (characterized by small values of $\mid X_{p2}/X_{p_1} \mid$). The algorithm requires much smaller values of λ than that prescribed by the Courant-Friedrichs-Levi (CFL) condition (especially in the accelerating piston case), thus it becomes quite difficult to obtain the solution for larger time without an unreasonable grid refinement. Also, the performance of the FDM in the decelerating piston case is far worse than that in the corresponding acceleration cases. This is most striking for the case with small piston velocity and large deceleration (viz. $X_{p_1} = 0.1$ and $X_{p_2} = -0.025$ and -0.05), where the first few entries in the table show higher values of $\bar{\pi}_0$ than those for smaller deceleration (viz. $X_{p_2} = -0.01$), which is certainly not correct.

Table 8.4.2

Growth and decay of the shock strength $\bar{\pi}_0(t)$ for $X_{p_1} = 0.05$ and $X_{p_2} = \pm 0.01, \pm 0.025, \pm 0.05$

(a) Acceleration case $X_{p_1} = 0.25$ $\bar{\pi}_0(0) = 0.27437$

X_{p2}	0.01		0.025		0.05	
T	NTSD	FDM	NTSD	FDM	NTSD	FDM
1.0	0.27755	0.27745	0.28238	0.28257	0.29058	0.29051
2.0	0.28076	0.28071	0.29058	0.29050	0.30756	0.30744
3.0	0.28400	0.28387	0.29899	0.29899	0.32535	0.32516
4.0	0.28727	0.28728	0.30756	0.30761		
5.0	0.29057	0.29053	0.31634	0.31638		
6.0	0.29391	0.29391				
7.0	0.29728	0.29730				
8.0	0.30067	0.30070				
9.0	0.30410	0.30415				
10.0	0.30756	0.30763				

(b) Deceleration Case

X_{p_2}	-0.01		-0.025		-0.05	
T	NTSD	FDM	NTSD	FDM	NTSD	FDM
1.0	0.27122	0.27314	0.25898	0.25426	0.24444	0.24422
2.0	0.26810	0.27025	0.25891	0.26463	0.24419	0.25608
3.0	0.26500	0.26748	0.25146	0.25807	0.23018	0.24461
4.0	0.26194	0.26489	0.24419	0.25224	0.21685	0.23434
5.0	0.25891	0.26225	0.23709	0.24655	0.20420	0.22504
6.0	0.25591	0.26086	0.23018	0.24455	0.19221	0.22050
7.0	0.25293	0.25870	0.22343	0.24065	0.18082	0.21131
8.0	0.25000	0.25632	0.21685	0.23581	0.17003	0.20582
9.0	0.24707	0.25386	0.21044	0.22911	0.15981	0.19780
10.0	0.24418	0.25131	0.20419	0.22442	0.15014	0.19151

On the other hand, we never encounter such odd situations in NTSD, where even in the cases of large accelerations there is a smooth profile for the growth of the shock strength leading to $\bar{\pi}_0 = 5.0$; whereas the large deceleration cases give equally smooth profiles for decay of the shock strength leading to $\bar{\pi}_0 = 0.0$. From the computational point of view, NTSD proves to be far more economical. The NTSD takes less than 0.5% of the computational time (for a large time range) of the finite difference method.

Table 8.4.3

Growth and decay of shock strength $\bar{\pi}_0(t)$ for $X_{p_1} = 0.50, X_{p_2} = \pm0.01, \pm0.025, \pm0.05$

(a) Acceleration case $X_{p_1} = 0.50$, $\bar{\pi}_0(0) = 0.59240$

X_{p_2}	0.01		0.025		0.05	
T	NTSD	FDM	NTSD	FDM	NTSD	FDM
1.0	0.59879	0.59780	0.60847	0.60520	0.62481	0.62018
2.0	0.60523	0.60507	0.62481	0.61987	0.65827	0.65791
3.0	0.61171	0.61002	0.64141	0.63964		
4.0	0.61823	0.61694	0.65827	0.65764		
5.0	0.62480	0.62327				
6.0	0.63144	0.62888				

(b) Deceleration case

X_{p_2}	-0.01		-0.025		-0.05	
T	NTSD	FDM	NTSD	FDM	NTSD	FDM
1.0	0.58604	0.58535	0.57659	0.57912	0.56106	0.56805
2.0	0.57973	0.58235	0.56106	0.56434	0.53081	0.54772
3.0	0.57346	0.57792	0.54579	0.55625	0.50164	0.52831
4.0	0.56723	0.57273	0.53080	0.54358	0.47358	0.50968
5.0	0.56105	0.56663	0.51608	0.53251	0.44663	0.48681
6.0	0.55491	0.56395	0.50165	0.52857	0.42086	0.47563
7.0	0.54881	0.55795	0.48748	0.51415	0.39612	0.46003
8.0	0.54276	0.55377	0.47358	0.50607	0.37248	0.44341
9.0	0.53675	0.55016	0.45997	0.49541	0.34993	0.42885
10.0	0.53078	0.54626	0.44663	0.48664	0.32844	0.41393

We conclude that the NTSD is an extremely efficient and quite accurate method for solving difficult problems, especially up to the intermediate time range (i.e., not too large t). Since it takes very small computational time, it may be combined with a FDM scheme for shock fitting.

Chapter 9

Compatibility conditions on a shock manifold in multi-dimensions

For simplicity, we shall consider only the case of a two-dimensional shock. A discussion of the results of a shock front in 3-dimensions makes very heavy use of tensor analysis and can be found in a paper by Lazarev, Ravindran and Prasad (1998).

9.1 Shock rays

Since a two-dimensional shock front is also a propagating curve, all relevant results of the section 3.3 are applicable here. We ask here a more basic question. What are shock rays? The answer to this question is unambiguous for a shock front in a gas, where we can take the ray velocity to be

$$\chi = \mathbf{q}_r + \mathbf{N}A \tag{9.1.1}$$

where q_r is the fluid velocity ahead of the shock, $\mathbf{N}$ the unit normal to the shock front and A is normal speed of the shock relative to the gas ahead of it. If we represent the shock surface Ω in space-time by $s(\mathbf{x}, t) = 0$, then (9.1.1) can be derived from the Charpit (i.e., Hamilton canonical) ordinary differential equations of a shock manifold partial differential equation (SME)

$$s_t + \langle \mathbf{q}_r, \nabla \rangle s + A|\nabla s|^2 = 0 \tag{9.1.2}$$

When we try to derive (9.1.2) from the jump relations or Rankine – Hugoniot (RH) conditions, we run into difficulty. There are many other jump relations which can be derived from (3.5.22) and each one them would lead to an SME. For example, the well-known Prandtl relation for a curved shock, when expressed in terms of s, is

$$\{\langle \mathbf{q}_l, \nabla\rangle s\} \ \{\langle \mathbf{q}_r, \nabla\rangle s\} - a_*^2 |\nabla s|^2 = 0 \qquad (9.1.3)$$

where a_* is the common *critical speed* on the two sides of the shock

$$a_*^2 = (p_r - p_l)/(\rho_r - \rho_l) \qquad (9.1.4)$$

The Charpit equations of (9.1.3) give a different expression for the shock ray velocity $\boldsymbol{\chi}$. The question arises: "are the shock ray velocities (or more precisely the complete set of shock ray equations) obtained from different SMEs the same?"

The concept of SME and their equivalence in the above sense was first discussed by Prasad (1982). Making further use of the RH conditions, it was shown that the shock ray equations given by the two SMEs (9.1.1 and 2) are equivalent. This result was generalized by Roy and Ravindran (1988) for almost all SMEs.

9.2 Shock manifold equation for a weak shock

It is possible to write (Ramanthan, Prasad and Ravindran (1984)) in a very elegant form the approximate SME for a weak shock in a solution of a general system of hyperbolic conservation laws

$$\frac{\partial \mathbf{H}}{\partial t} + \frac{\partial \mathbf{F}^{(\alpha)}}{\partial x_\alpha} = 0, \quad \alpha = 1, 2, \ldots, m \qquad (9.2.1)$$

where $\mathbf{H}$ and $\mathbf{F}^{(\alpha)}$ are functions of $\mathbf{u} = (u_1, u_2, \ldots, u_n)^T$. For a genuine solution, this system is equivalent to

$$A\frac{\partial \mathbf{u}}{\partial t} + B^{(\alpha)}\frac{\partial \mathbf{u}}{\partial x_\alpha} = 0 \qquad (9.2.2)$$

where

$$A = \nabla_U \mathbf{H} \quad \text{and} \quad B^{(\alpha)} = \nabla_{\mathbf{u}} \mathbf{F}^{(\alpha)} \qquad (9.2.3)$$

The characteristic surfaces $\varphi(x_\alpha, t) =$ constant satisfy

$$Q_{ch}(\mathbf{u}; \varphi_{x_\alpha}, \phi_t) \equiv | A\varphi_t + B^{(\alpha)}\varphi_{x_\alpha} | = 0 \qquad (9.2.4)$$

Now, consider a weak solution of (9.2.1) containing a shock surface $\Omega : s(\mathbf{x}, t) = 0$, across which the jump conditions are

$$[\mathbf{H}]s_t + [\mathbf{F}^\alpha]s_{x_\alpha} = 0 \text{ on } s = 0 \tag{9.2.5}$$

For a weak shock, the Euclidean norm $\| \mathbf{u}_r - \mathbf{u}_\ell \| = \epsilon$ is small and hence

$$[\mathbf{H}] = \{A_{(\ell+r)/2}\}(\mathbf{u}_r - \mathbf{u}_\ell) + O(\epsilon^3) \tag{9.2.6}$$

$$[\mathbf{F}^\alpha] = \{B^{(\alpha)}_{(\ell+r)/2}\}(\mathbf{u}_r - \mathbf{u}_\ell) + O(\epsilon^3) \tag{9.2.7}$$

where

$$A_{(\ell+r)/2} = A\left(\frac{\mathbf{u}_\ell + \mathbf{u}_r}{2}\right), \quad \text{etc.} \tag{9.2.8}$$

Neglecting the terms of order $O(\epsilon^3)$, (9.2.5 - 8) combine to give a system of linear homogeneous relations in the components of the vector $\mathbf{u}_r - \mathbf{u}_\ell$. The condition, that a non-zero vector $\mathbf{u}_r - \mathbf{u}_\ell$ satisfies these relations, leads to the desired approximate condition on the manifold $s = 0$

$$Q_{sh} \equiv \left| A_{(\ell+r)/2} s_t + B^{(\alpha)}_{(\ell+r)/2} s_{x_\alpha} \right| = 0 \ , \text{ on } \Omega \tag{9.2.9}$$

Using the embedding theorem (see theorem 4.1, Prasad 1993) we can treat (9.2.9) as a partial differential equation for s. On comparison of the expressions in (9.2.4 and 9), we see that SME for a weak shock can be written down easily from the form of the characteristic partial differential equation:

$$Q_{sh} \equiv Q_{ch}\left(\frac{\mathbf{u}_\ell + \mathbf{u}_r}{2}; s_t, s_x\right) = 0 \tag{9.2.10}$$

Following the procedure of derivation of (2.4.6 - 7), we can use the Charpit equations of the first order partial differential equation (9.2.10) to give the equations for the shock position $\mathbf{X}$ and unit normal $\mathbf{N}$ of the shock giving shock rays. Neglecting terms of the third power in the shock strength ϵ, we can deduce the following results from the shock ray equations.

Theorem 9.2.1 For a weak shock, the shock ray velocity components are equal to the mean of the bicharacteristic velocity components just ahead and just behind the shock, provided we take the wavefronts generating the characteristic surface ahead and behind to be instaneously coincident with the shock surface. Similarly, the rate of turning of the shock front is also equal to the mean of the rates of turning of such wavefronts just ahead and just behind the shock.

If we choose the characteristic partial differential equation of the gas dynamic equations in the form

$$Q_{ch} \equiv \varphi_t + u_\alpha \varphi_{x_\alpha} + a \mid \nabla \varphi \mid = 0 \tag{9.2.11}$$

then (9.2.10) gives the following SME for shock fronts in two-space dimensions in the form

$$\begin{aligned} s_t + \frac{1}{2}(u_r + u_\ell)s_x + \frac{1}{2}(v_r + v_\ell)s_y \\ + \frac{1}{2}(a_r + a_\ell)(s_x^2 + s_y^2)^{\frac{1}{2}} = 0 \end{aligned} \tag{9.2.12}$$

Assuming the shock strength to be small, we can derive (9.2.12) also from the SME (9.1.3). The equation (9.2.12) (not interpreted as SME) was first derived by Kluwick (1971).

When a shock is weak, the effect of the waves reflected from it into the region behind it can be neglected. In this case, the solution behind the shock is approximately independent of the shock and can be calculated from the initial data behind the shock. Then the SME (9.2.12) can be used to fit the shock into the known solution ahead of and behind the shock. Ramanathan, Prasad and Ravindran (1984) have solved a problem using the shock rays of the SME (9.2.12).

9.3 Geometrical and kinematical compatibility conditions

In this section we continue the discussion of results in section 3.3. Some of the results are a rederivation of those in the earlier section but we now need them in forms written here.

9.3.1 Preliminary geometrical ideas for a moving curve in two-space-dimensions

Consider a smooth curve Ω_t in the (x, y)-plane depending on the time t as a parameter. If, at any time, t the position $\mathbf{x}$ of a point on Ω_t can be expressed in terms of the arc length s measured from a suitable point on Ω_t, then, in the neighbourhood of a point s_0, we can write

$$\begin{aligned}\mathbf{x}(s,t) &= \mathbf{x}(s_0,t) + (s-s_0)\boldsymbol{\tau}(s_0,t) \\ &+ \frac{1}{2}(s-s_0)^2\kappa(s_0,t)\mathbf{N}(s_0,t) + o((s-s_0)^2)\end{aligned} \quad (9.3.1)$$

where $\boldsymbol{\tau}$ and $\mathbf{N}$ are unit tangents and normal vectors and κ is the curvature. These are defined by the formulae

$$\boldsymbol{\tau} = \mathbf{x}_s \quad , \quad \mathbf{N} = \frac{1}{\kappa}\mathbf{x}_{ss} \quad (9.3.2)$$

As explained in section (3.3), we introduce a ray coordinate system (ξ, t) such that the $t =$ constant curves representing successive positions of Ω_t and $\xi =$ constant curves that are suitably defined rays. The curves with the natural parameter $s =$ constant, in general, do not represent rays. Hence, we choose another variable $\xi = \xi(s,t)$ and define a quantity G by

$$G = \xi_s^{-1} \ , \ G^2 = x_\xi^2 + y_\xi^2 \quad (9.3.3)$$

so that

$$\boldsymbol{\tau} = \frac{1}{G}(x_\xi, y_\xi) \quad \text{and} \quad \mathbf{N} = \frac{1}{G}(y_\xi, -x_\xi) \quad (9.3.4)$$

From (9.3.2)

$$\kappa\mathbf{N} = \frac{1}{G^2}(x_{\xi\xi}, y_{\xi\xi}) + \xi_{ss}(x_\xi, y_\xi) \quad (9.3.5)$$

Taking its scalar product first by $\boldsymbol{\tau}$ and then by $\mathbf{N}$ we get

$$\xi_{ss} = -(x_\xi x_{\xi\xi} + y_\xi y_{\xi\xi})/G^4 \quad (9.3.6)$$

and

$$\kappa = (x_{\xi\xi}y_\xi - x_\xi y_{\xi\xi})/G^3 = -\frac{1}{G}\Theta_\xi \quad (9.3.7)$$

respectively, where Θ is the angle which $\mathbf{N}$ makes with the x-axis.

The derivatives of $\mathbf{N}$ and $\boldsymbol{\tau}$ with respect to ξ are given by

$$\mathbf{N}_\xi = -\kappa G\boldsymbol{\tau}, \ \boldsymbol{\tau}_\xi = \kappa G\mathbf{N} \quad (9.3.8)$$

We write down some obvious identities in matrix notations, since we shall use them frequently in this form:

$$(N_1, N_2)\begin{pmatrix} N_1 \\ N_2 \end{pmatrix} = 1, \ (\tau_1, \tau_2)\begin{pmatrix} \tau_1 \\ \tau_2 \end{pmatrix} = 1, \ (N_1, N_2)\begin{pmatrix} \tau_1 \\ \tau_2 \end{pmatrix} = 0 \quad (9.3.9)$$

and

$$\begin{pmatrix}\tau_1\\ \tau_2\end{pmatrix}(\tau_1,\tau_2)+\begin{pmatrix}N_1\\ N_2\end{pmatrix}(N_1,N_2)=\begin{pmatrix}1&0\\0&1\end{pmatrix} \tag{9.3.10}$$

9.3.2 Geometrical compatibility conditions

Consider a function $z(x,y,t)$ which could be density or pressure or a component of a velocity vector. On either side of the moving curve Ω_t, z and its partial derivatives are assumed to be sufficiently smooth. As in the previous chapters, let $[z](\xi,t)=z_r|_{\Omega_t}-z_\ell|_{\Omega_t}$ represent the jump across Ω_t.

We introduce notations

$$\begin{aligned} A_0 &= [z],\ A_1=[z_x]N_1+[z_y]N_2 \\ A_2 &= [z_{xx}]N_1^2+2[z_{xy}]N_1N_2+[z_{yy}]N_2^2 \end{aligned} \tag{9.3.11}$$

The quantities A_0, A_1 and A_2 are functions of ξ and t only. Our aim in this section is to express $[z_x], [z_y], [z_{xx}], [z_{xy}]$ and $[z_{yy}]$ in terms of A_0, A_1, A_2, their derivatives, with respect to the surface coordinate ξ and geometrical characteristics of the curve Ω_t.

From Hadamard's lemma (see the result (3.3.39))

$$A_{0\xi}=(x_\xi,y_\xi)\begin{pmatrix}[z_x]\\ [z_y]\end{pmatrix} \tag{9.3.12}$$

Taking left multiplication of both sides by the column vector $\binom{x_\xi}{y_\xi}=G\binom{\tau_1}{\tau_2}$, we get

$$GA_{0\xi}\begin{pmatrix}\tau_1\\ \tau_2\end{pmatrix}=G^2\begin{pmatrix}\tau_1\\ \tau_2\end{pmatrix}(\tau_1,\tau_2)\begin{pmatrix}[z_x]\\ [z_y]\end{pmatrix}$$

Using the identity (9.3.10), we finally deduce

$$\begin{pmatrix}[z_x]\\ [z_y]\end{pmatrix}=A_1\begin{pmatrix}N_1\\ N_2\end{pmatrix}+\frac{1}{G}A_{0\xi}\begin{pmatrix}\tau_1\\ \tau_2\end{pmatrix} \tag{9.3.13}$$

which is called the *geometrical compatibility condition* of the first order.

Replacing z by z_x and again by z_y in (9.3.13) we get

$$\begin{pmatrix}[z_{xx}]\\ [x_{xy}]\end{pmatrix} = A_1^{(x)}\begin{pmatrix}N_1\\ N_2\end{pmatrix} + \frac{1}{G}[z_x]_\xi\begin{pmatrix}\tau_1\\ \tau_2\end{pmatrix} \tag{9.3.14}$$

$$\begin{pmatrix}[z_{yx}]\\ [z_{yy}]\end{pmatrix} = A_1^{(y)}\begin{pmatrix}N_1\\ N_2\end{pmatrix} + \frac{1}{G}[z_y]_\xi\begin{pmatrix}\tau_1\\ \tau_2\end{pmatrix} \tag{9.3.15}$$

where

$$A_1^{(x)} = [z_{xx}]N_1 + [z_{xy}]N_2,\ A_1^{(y)} = [z_{yx}]N_1 + [z_{yy}]N_2 \tag{9.3.16}$$

From the definitions of A_2, $A_1^{(x)}$ and $A_1^{(y)}$; and $[z_{xy}] = [z_{yx}]$, we deduce

$$A_2 = A_1^{(x)}N_1 + A_1^{(y)}N_2 \tag{9.3.17}$$

The relation (9.3.14-15) can be written in the matrix form

$$\begin{pmatrix}[z_{xx}] & [z_{xy}]\\ [z_{yx}] & [z_{yy}]\end{pmatrix} = \begin{pmatrix}A_1^{(x)}\\ A_1^{(y)}\end{pmatrix}(N_1, N_2) + \frac{1}{G}\begin{pmatrix}[z_x]\\ [z_y]\end{pmatrix}_\xi (\tau_1, \tau_2) \tag{9.3.18}$$

Taking the transpose of (9.3.18), we get

$$\begin{pmatrix}[z_{xx}] & [z_{yx}]\\ [z_{xy}] & [z_{yy}]\end{pmatrix} = \begin{pmatrix}N_1\\ N_2\end{pmatrix}(A_1^{(x)}, A_1^{(y)}) + \frac{1}{G}\begin{pmatrix}\tau_1\\ \tau_2\end{pmatrix}([z_x], [z_y])_\xi \tag{9.3.19}$$

Premultiplying (9.3.18-19) by (N_1, N_2) and equating the results on the right hand sides, we get

$$(A_1^{(x)}, A_1^{(y)}) = A_2(N_1, N_2) + \frac{1}{G}(N_1, N_2)\begin{pmatrix}[z_x]\\ [z_y]\end{pmatrix}_\xi (\tau_1, \tau_2)$$

which, with the help of (9.3.13), leads to

$$(A_1^{(x)}, A_1^{(y)}) = A_2(N_1, N_2) + \frac{1}{G}(A_{1\xi} + \kappa A_{0\xi})(\tau_1, \tau_2) \tag{9.3.20}$$

Thus, the first term in (9.3.18) has been expressed in terms of A_2, $A_{1\xi}$ and $A_{0\xi}$. To evaluate the second term, we differentiate (9.3.13) with respect to ξ, use (9.3.8) and substitute in (9.3.18) to get

$$\begin{pmatrix}[z_{xx}] & [z_{xy}]\\ [z_{yx}] & [z_{yy}]\end{pmatrix} = A_2\begin{pmatrix}N_1\\ N_2\end{pmatrix}(N_1, N_2) + \frac{1}{G}(A_{1\xi} + \kappa A_{0\xi})$$

$$\left\{\begin{pmatrix}\tau_1\\ \tau_2\end{pmatrix}(N_1,N_2)+\begin{pmatrix}N_1\\ N_2\end{pmatrix}(\tau_1,\tau_2)\right\}$$
$$+\ \left\{G^{-2}(A_{0\xi\xi}-\Gamma A_{0\xi})-\kappa A_1\right\}\begin{pmatrix}\tau_1\\ \tau_2\end{pmatrix}(\tau_1,\tau_2) \quad (9.3.21)$$

where

$$\Gamma=\frac{1}{G}G_\xi \;=\; \frac{x_\xi x_{\xi\xi}+y_\xi y_{\xi\xi}}{G} \qquad (9.3.22)$$

The relation (9.3.21) is the geometrical compatibility condition of the second order.

9.3.3 Some results in a ray coordinate system

In general, a ray, i.e., the ξ = constant curve of a given moving front Ω_t, and the front Ω_t at time t, do not meet orthogonally. However, when we consider a gas dynamic shock running into a gas at rest, the shock rays are orthogonal to the shock fronts. From now onward, *we shall assume that the successive positions of the surface Ω_t and the rays form a pair of orthogonal families of curves in the (x, y)-plane.* Since the ray velocity (x_t, y_t) is in the direction of (N_1, N_1) and $x_tN_1 + y_tN_2 = C$, it follows that the rays are given by

$$\begin{pmatrix}x_t(\xi,t)\\ y_t(\xi,t)\end{pmatrix}=C(\xi,t)\begin{pmatrix}N_1\\ N_2\end{pmatrix} \qquad (9.3.23)$$

From (3.3.15) with $T = 0$ and (9.3.7) we get

$$G_t=-\kappa CG \qquad (9.3.24)$$

From the definition

$$\begin{aligned}\mathbf{N} &= G^{-1}(y_\xi,-x_\xi)\\ \mathbf{N}_t &= -G_t\mathbf{N}/G+G^{-1}((y_t)_\xi,-(x_t)_\xi)\end{aligned}$$

Using (9.3.23 - 24), we get

$$\mathbf{N}_t=-\frac{1}{G}C_\xi\boldsymbol{\tau} \qquad (9.3.25)$$

Similarly, we can deduce

$$\boldsymbol{\tau}_t-\frac{1}{G}C_\xi\mathbf{N} \qquad (9.3.26)$$

9.3.4 Kinematical compatibility conditions

Let (ξ, t) be a shock ray coordinate system. In this section, we shall find expressions for $[z_t], [z_{xt}]$ and $[z_{yt}]$ in terms of A_0, A_1, A_2, their derivatives with respect to t and ξ and geometrical-kinematical characteristics of Ω_t. Using (9.3.23), we have

$$\begin{aligned} A_{0t} &= [z(x(\xi,t), y(\xi,t), t)]_t \\ &= C([z_x],[z_y])\begin{pmatrix} N_1 \\ N_2 \end{pmatrix} + [z_t] = A_1 C + [z_t] \end{aligned}$$

Hence, we get

$$[z_t] = A_{0t} - CA_1 \tag{9.3.27}$$

Now consider $[z_x]_t$ and $[z_y]_t$

$$\begin{pmatrix} [z_x] \\ [z_y] \end{pmatrix}_t = \begin{pmatrix} [z_{xt}] \\ [z_{yt}] \end{pmatrix} + \begin{pmatrix} [z_{xx}] & [z_{xy}] \\ [z_{yx}] & [z_{yy}] \end{pmatrix}\begin{pmatrix} x_t \\ y_t \end{pmatrix} \tag{9.3.28}$$

The left hand side of (9.3.28) is (see (9.3.13))

$$\left\{ A_1 \begin{pmatrix} N_1 \\ N_2 \end{pmatrix} + \frac{1}{G} A_{0\xi} \begin{pmatrix} \tau_1 \\ \tau_2 \end{pmatrix} \right\}_t$$

Using the results of (9.3.21, 23, 25 and 26) in (9.3.28) we get

$$\begin{pmatrix} [z_{xt}] \\ [z_{yt}] \end{pmatrix} = (A_{1t} - CA_2 + \frac{1}{G} C_\xi A_{0\xi})\begin{pmatrix} N_1 \\ N_2 \end{pmatrix} + \frac{1}{\sqrt{G}}(A_{0t} - CA_1)_\xi \begin{pmatrix} \tau_1 \\ \tau_2 \end{pmatrix} \tag{9.3.29}$$

The relations (5.3.27 and 29) are called kinematical compatibility conditions of the first and second order.

9.4 Dynamical compatibility conditions

The kinematics of the shock front briefly mentioned in sections 9.1 and 9.2 assumes a known solution on both sides of the shock and then describes how the shock surface propagates or fits itself into the two known states. The theory is of little practical use since, in all practical problems, the state behind the shock is influenced by the shock and is to be determined as a part of the problem along with the shock position except for the very special case of a single conservation law or weak shock theory (see section 9.2). However, we have shown in the Chapters 7 and 8 that for one-dimensional problems, an infinite system of compatibility conditions does lead to a practical method of solving the whole problem. We called this method a new theory of shock dynamics (NTSD).

The derivation of the compatibility conditions on a curved shock for gas dynamics equations, called *dynamic compatibility conditions*, has been quite challenging. Maslov (1980), developed a general theory for a system of first order conservation laws using the theory of distribution and gave a very elegant mathematical method to derive the infinite system of compatibility conditions for a curved shock in the form of transport equations. Srinivasan and Prasad (1985) used Maslov's method to derive the first compatibility condition for the complete set of gas dynamics equations, including the energy equation. It was found later that Maslov's uniqueness lemma was not correct and this affected the higher order compatibility conditions. Ravindran and Prasad (1993) worked out the correct form of the second compatibility condition. This, as in the case of Maslov's method, involved quite complex algebraic operations and it was not easy to derive the third compatibility condition. Finally, explicit forms of the first three compatibility conditions were derived for gas dynamic equations by Lazarev, Ravindran and Prasad (1998) following the method of Grinfel'd (1978), who had used tensor analysis to work out the conditions for the equations of nonlinear elastic materials.

Meanwhile, Anile and Russo (1986, 1988) presented a general procedure for the derivation of the infinite system of compatibility conditions not just for gas dynamic or elasticity equations but for an arbitrary hyperbolic system of quasilinear partial differential equations. Anile and Russo's method, for a general system, is very important from the point of view of a theoretical development. However, one cannot use this to derive explicit forms of the compatibility conditions in a particular system such as the gas dynamic equations since the algebraic operations become prohibitively complex even for the second compatibility condition. It is in this context that Grinfel'd's procedure is more powerful since by its use we can deduce the higher order compatibility conditions for a particular system in a simpler and more systematic way. Grinfel'd's method uses a heavy dose of tensor analysis and an average researcher in shock propagation may find it difficult to master the technique in a short period. Therefore, in this section, we present a simpler version (due to Lazarev) of Grinfel'd's procedure using only matrix notation for the gas dynamic equations in two-space-dimensions.

9.4.1 The first set of dynamical compatibility conditions

We consider the propagation of a shock front Ω_t in a polytropic gas with γ as the constant ratio of specific heats. Assume that the velocity components u, v, pressure p and density ρ are $C^\infty(\mathbf{R}^3)$ except for a discontinuity of the first kind on Ω_t and assume that the shock front propagates into a gas in uniform state at rest, so that

$$\rho_r = \rho_0 \ , \ p_r = p_0 \ , \ u_r = 0 = v_r \tag{9.4.1}$$

Then the ray coordinate system satisfies (9.3.23).

We write the equations of two-dimensional motion in the form

$$\rho_t + (u, v)\begin{pmatrix} \rho_x \\ \rho_y \end{pmatrix} + \rho(u_x + v_y) = 0, \tag{9.4.2}$$

$$\begin{pmatrix} u_t \\ v_t \end{pmatrix} + \begin{pmatrix} u_x & u_y \\ v_x & v_y \end{pmatrix}\begin{pmatrix} u \\ v \end{pmatrix} + \frac{1}{\rho}\begin{pmatrix} p_x \\ p_y \end{pmatrix} = 0 \tag{9.4.3}$$

$$p_t + (u, v)\begin{pmatrix} p_x \\ p_y \end{pmatrix} + \gamma p(u_x + v_y) = 0 \tag{9.4.4}$$

We denote the jumps in density, velocity and pressure p by

$$D_0 = [\rho], \ u_0 = [u], \ v_0 = [v], \ S_0 = [p]$$

and the jump in the normal component of the velocity by

$$H_0 = [N_1 u + N_2 v] = N_1 u_0 + N_2 v_0$$

These notations may be compared with those in section 8.2. The Rankine–Hugoniot conditions on the shock Ω_t give

$$\left.\begin{aligned} &u_0 = H_0 N_1, \ v_0 = H_0 N_2, \qquad S_0 = \rho_0 C H_0, \ H_0 = \tfrac{D_0 C}{\rho_0 - D_0}, \\ &C^2 = a_0^2 \frac{2(\rho_0 - D_0)}{2\rho_0 + (\gamma - 1)D_0}, \ a_\ell^2 = \frac{a_0^2 \rho_0 (2\rho_0 - D_0(\gamma + 1))}{(\rho_0 - D_0)(2\rho_0 + (\gamma - 1)D_0)} \end{aligned}\right\} \tag{9.4.5}$$

where $a_0^2 = \gamma p_0/\rho_0$. The only unknown, which appears on the right hand side of (9.4.5) is D_0.

We introduce a number of symbols

$$\left.\begin{aligned} u_1 &= (N_1, N_2)\begin{pmatrix} [u_x] \\ [u_y] \end{pmatrix}, \quad v_1 = (N_1, N_2)\begin{pmatrix} [v_x] \\ [v_y] \end{pmatrix} \\ H_1 &= (N_1, N_2)\begin{pmatrix} u_1 \\ v_1 \end{pmatrix}, \quad D_1 = (N_1, N_2)\begin{pmatrix} [\rho_x] \\ [\rho_y] \end{pmatrix} \\ S_1 &= (N_1, N_2)\begin{pmatrix} [p_x] \\ [p_y] \end{pmatrix} \end{aligned}\right\} \tag{9.4.6}$$

$$\left.\begin{aligned} u_2 &= (N_1, N_2)\begin{pmatrix} [u_{xx}] & [u_{xy}] \\ [u_{yx}] & [u_{yy}] \end{pmatrix}\begin{pmatrix} N_1 \\ N_2 \end{pmatrix} \\ v_2 &= (N_1, N_2)\begin{pmatrix} [v_{xx}] & [v_{xy}] \\ [v_{yx}] & [v_{yy}] \end{pmatrix}\begin{pmatrix} N_1 \\ N_2 \end{pmatrix} \\ H_2 &= (N_1, N_2)\begin{pmatrix} u_2 \\ v_2 \end{pmatrix}, \\ D_2 &= (N_1, N_2)\begin{pmatrix} [\rho_{xx}] & [\rho_{xy}] \\ [\rho_{yx}] & [\rho_{yy}] \end{pmatrix}\begin{pmatrix} N_1 \\ N_2 \end{pmatrix} \\ S_2 &= (N_1, N_2)\begin{pmatrix} [p_{xx}] & [p_{xy}] \\ [p_{yx}] & [p_{yy}] \end{pmatrix}\begin{pmatrix} N_1 \\ N_2 \end{pmatrix} \end{aligned}\right\} \tag{9.4.7}$$

Taking the jump of the equation of continuity (9.4.2) across Ω_t we get

$$[\rho_t] - ([u], [v])\begin{pmatrix} [\rho_x] \\ [\rho_y] \end{pmatrix} + (\rho_0 - D_0)([u_x] + [v_x]) = 0$$

which, after using (9.3.13) and (9.3.27), becomes

$$\begin{aligned} D_{0t} - CD_1 - (u_0, v_0)&\left\{D_1\begin{pmatrix} N_1 \\ N_2 \end{pmatrix} + D_{0\xi}\frac{1}{G}\begin{pmatrix} \tau_1 \\ \tau_2 \end{pmatrix}\right\} \\ + \ (\rho_0 - D_0)&\left\{(u_1, v_1)\begin{pmatrix} N_1 \\ N_2 \end{pmatrix} + \frac{1}{G}(u_0, v_0)_\xi\begin{pmatrix} \tau_1 \\ \tau_2 \end{pmatrix}\right\} = 0 \end{aligned}$$

Using (9.4.5) and (9.3.8) for the evaluation of the term $(u_0, v_0)_\xi$, we finally get the first dynamical compatibility condition from the equation of continuity

$$D_{0t} - (C + H_0)D_1 + (\rho_0 - D_0)H_1 = \kappa H_0(\rho_0 - D_0) \tag{9.4.8}$$

The jump of the equation of momentum (9.4.3) gives

$$\begin{pmatrix}[u_t]\\ [v_t]\end{pmatrix} - \begin{pmatrix}[u_x] & [u_y]\\ [v_x] & [v_y]\end{pmatrix}\begin{pmatrix}[u]\\ [v]\end{pmatrix} + \frac{1}{\rho_0 - D_0}\begin{pmatrix}[p_x]\\ [p_y]\end{pmatrix} = 0$$

which leads to

$$\begin{pmatrix}u_0\\ v_0\end{pmatrix}_t - (C + H_0)\begin{pmatrix}u_1\\ v_1\end{pmatrix} + \frac{1}{\rho_0 - D_0}\left\{S_1\begin{pmatrix}N_1\\ N_2\end{pmatrix} + \frac{1}{G}S_{0\xi}\begin{pmatrix}\tau_1\\ \tau_2\end{pmatrix}\right\} = 0$$

Multiplying on the left respectively by (N_1, N_2) and (τ_1, τ_2) we get the first normal and tangential components of the compatibility condition derived from the equation of momentum

$$H_{0t} - (C + H_0)H_1 + (\rho_0 - D_0)^{-1}S_1 = 0 \tag{9.4.9}$$

and

$$h_0 = G^{-1}(C + H_0)^{-1}(S_{0\xi}(\rho_0 - D_0)^{-1} - H_0C_\xi) \tag{9.4.10}$$

where $h_0 = G^{-1}(u_1, v_1)\binom{\tau_1}{\tau_2}$.

Following a similar procedure, from (9.4.4) we deduce

$$[p_t] - ([u], [v])\begin{pmatrix}[p_x]\\ [p_y]\end{pmatrix} + \gamma(p_0 - S_0)([u_x] + [v_y]) = 0$$

or

$$S_{0t} - CS_1 - H_0S_1 + \gamma(p_0 - S_0)\left\{(u_1, v_1)\begin{pmatrix}N_1\\ N_2\end{pmatrix} + G^{-1}(u_0, v_0)\begin{pmatrix}\tau_1\\ \tau_2\end{pmatrix}\right\} = 0$$

Hence we get the first dynamical compatibility condition from the energy equation in the form

$$S_{0t} - (C + H_0)S_1 + \gamma(p_0 - S_0)(H_1 - \kappa H_0) = 0 \tag{9.4.11}$$

It is convenient to combine (9.4.8 - 9 and 11) into one equation in a matrix form

$$\begin{pmatrix}D_0\\ H_0\\ S_0\end{pmatrix}_t + \begin{pmatrix}-(C + H_0) & \rho_0 - D_0 & 0\\ 0 & -(C + H_0) & (\rho_0 - D_0)^{-1}\\ 0 & \gamma(p_0 - S_0) & -(C + H_0)\end{pmatrix}\begin{pmatrix}D_1\\ H_1\\ S_1\end{pmatrix}$$

$$= \kappa H_0\begin{pmatrix}\rho_0 - D_0\\ 0\\ \gamma(p_0 - S_0)\end{pmatrix} \tag{9.4.12}$$

This contains all important members of the first set of compatibility conditions. Note that the condition (9.4.10) has not been included in this set. This expression for h_0 will be useful in the derivation of the next set of compatibility conditions.

9.4.2 The second set of dynamical compatibility conditions

To obtain the second set of compatibility conditions, we use the following procedure. We differentiate each of the equations (9.4.2 - 4) with respect to x and y, respectively, take the jump of the results across Ω_t and finally take the scalar product of the two jump equations with the unit normal vector. Moreover, after using this procedure on the two components of the momentum equations, we take once again the inner product with the unit normal vector (this gives the normal component of the vector form of the second compatibility condition from the momentum equation). Now we shall carry out the procedure in detail.

Differentiating (9.4.2) with respect to x and y, we get

$$\begin{pmatrix}\rho_{tx}\\ \rho_{ty}\end{pmatrix} + \begin{pmatrix}u_x & v_x\\ u_y & v_y\end{pmatrix}\begin{pmatrix}\rho_x\\ \rho_y\end{pmatrix} + \begin{pmatrix}\rho_{xx} & \rho_{xy}\\ \rho_{yx} & \rho_{yy}\end{pmatrix}\begin{pmatrix}u\\ v\end{pmatrix} + (u_x + v_y)\begin{pmatrix}\rho_x\\ \rho_y\end{pmatrix} + \rho\begin{pmatrix}u_{xx} + v_{xy}\\ u_{xy} + v_{yy}\end{pmatrix} = 0$$

Taking the jump across Ω_t, we get

$$\begin{pmatrix}[\rho_{tx}]\\ [\rho_{ty}]\end{pmatrix} - \begin{pmatrix}[u_x] & [v_x]\\ [u_y] & [v_y]\end{pmatrix}\begin{pmatrix}[\rho_x]\\ [\rho_y]\end{pmatrix} - \begin{pmatrix}[\rho_{xx}] & [\rho_{xy}]\\ [\rho_{yx}] & [\rho_{yy}]\end{pmatrix}\begin{pmatrix}[u]\\ [v]\end{pmatrix} - ([u_x] + [u_y])\begin{pmatrix}[\rho_x]\\ [\rho_y]\end{pmatrix} + (\rho_0 - D)\begin{pmatrix}[u_{xx}] + [v_{xy}]\\ [u_{xy}] + [v_{yy}]\end{pmatrix} = 0 \quad (9.4.13)$$

After we multiply this equation on the left by (N_1, N_2), we need to do the following calculations

$$(N_1, N_2)\begin{pmatrix}[\rho_{tx}]\\ [\rho_{ty}]\end{pmatrix} = D_{1t} - CD_2 + \frac{1}{G^2}D_{0\xi}C_\xi$$

$$-(N_1, N_2)\begin{pmatrix}[u_x] & [v_x]\\ [u_y] & [v_y]\end{pmatrix}\begin{pmatrix}[\rho_x]\\ [\rho_y]\end{pmatrix} = -\left((N_1, N_2)\begin{pmatrix}[u_x]\\ [u_y]\end{pmatrix}, (N_1, N_2)\begin{pmatrix}[v_x]\\ [v_y]\end{pmatrix}\right)\begin{pmatrix}[\rho_x]\\ [\rho_y]\end{pmatrix}$$

$$
\begin{aligned}
&= -(u_1, v_1)\left\{D_1\begin{pmatrix}N_1\\N_2\end{pmatrix} + \frac{1}{G}D_{0\xi}\begin{pmatrix}\tau_1\\\tau_2\end{pmatrix}\right\}\\
&= -D_1H_1 - \frac{1}{G}D_{0\xi}(u_1, v_1)\begin{pmatrix}\tau_1\\\tau_2\end{pmatrix}\\
&= -D_1H_1 - D_{0\xi}h_0 \quad \text{(see (9.4.10) for } h_0\text{)}
\end{aligned}
$$

$$
\begin{aligned}
-(N_1, N_2)\begin{pmatrix}[\rho_{xx}] & [\rho_{xy}]\\ [\rho_{yx}] & [\rho_{yy}]\end{pmatrix}\begin{pmatrix}[u]\\ [v]\end{pmatrix} &= -(N_1, N_2)\begin{pmatrix}[\rho_{xx}] & [\rho_{xy}]\\ [\rho_{yx}] & [\rho_{yy}]\end{pmatrix}\begin{pmatrix}N_1\\N_2\end{pmatrix}H_0\\
&= -D_2H_0
\end{aligned}
$$

$$
\begin{aligned}
-(N_1, N_2)\begin{pmatrix}[\rho_x]\\ [\rho_y]\end{pmatrix}([u_x] + [v_y]) &= -D_1([u_x] + [v_y])\\
&= -D_1\left((u_1N_1 + \frac{1}{G}u_{0\xi}\tau_1) + (v_1N_2 + \frac{1}{G}v_{0\xi}\tau_2)\right)\\
&= -D_1\left(H_1 + \frac{1}{G}(u_{0\xi}, v_{0\xi})\begin{pmatrix}\tau_1\\\tau_2\end{pmatrix}\right)\\
&= -D_1\left\{H_1 + \frac{1}{G}((u_0, v_0)\begin{pmatrix}\tau_1\\\tau_2\end{pmatrix})_\xi - \frac{1}{G}(u_0, v_0)\begin{pmatrix}\tau_1\\\tau_2\end{pmatrix}_\xi\right\}\\
&= -D_1(H_1 - \kappa H_0)
\end{aligned}
$$

$$
\begin{aligned}
&(\rho_0 - D)(N_1, N_2)\begin{pmatrix}[u_{xx}] + [v_{xy}]\\ [u_{xy}] + [v_{yy}]\end{pmatrix}\\
&\quad = (\rho_0 - D_0)\left\{(N_1, N_2)\begin{pmatrix}[u_{xx}]\\ [u_{xy}]\end{pmatrix} + N_1, N_2\begin{pmatrix}[v_{xy}]\\ [v_{yy}]\end{pmatrix}\right\}
\end{aligned}
$$

and now using the result (9.3.21) columnwise

$$
\begin{aligned}
&= (\rho_0 - D_0)\left\{u_2N_1 + \frac{1}{G}(u_{1\xi} + \kappa u_{0\xi})\tau_1 + v_2N_2 + \frac{1}{G}(v_{1\xi} + \kappa v_{0\xi})\tau_2\right\}\\
&= (\rho_0 - D_0)\left\{(u_2, v_2)\begin{pmatrix}N_1\\N_2\end{pmatrix} + \frac{1}{G}(u_1, v_1)_\xi\begin{pmatrix}\tau_1\\\tau_2\end{pmatrix} + \frac{\kappa}{G}(u_0, v_0)_\xi\begin{pmatrix}\tau_1\\\tau_2\end{pmatrix}\right\}
\end{aligned}
$$

$$= (\rho_0 - D_0)\left\{H_2 + \frac{1}{G}((u_1, v_1)\begin{pmatrix}\tau_1\\ \tau_2\end{pmatrix})_\xi - \frac{1}{G}(u_1, v_1)\begin{pmatrix}\tau_1\\ \tau_2\end{pmatrix}_\xi\right.$$

$$\left. + \frac{\kappa}{G}((u_0, v_0)\begin{pmatrix}\tau_1\\ \tau_2\end{pmatrix})_\xi - \frac{\kappa}{G}(u_0, v_0)\begin{pmatrix}\tau_1\\ \tau_2\end{pmatrix}_\xi\right\}$$

$$= (\rho_0 - D_0)\left\{H_2 + \frac{1}{G}(Gh_0)_\xi - \kappa H_1 - \kappa^2 H_0\right\}$$

The equation (9.4.13) finally gives the second dynamical compatibility condition from the equation of continuity

$$\begin{aligned} D_{1t} - (C + H_0)D_2 + (\rho_0 - D_0)H_2 & \\ = D_{0\xi}(h_0 - \frac{1}{G}C_\xi) + D_1(2H_1 - \kappa H_0) & \\ + (\rho_0 - D_0)\{\kappa^2 H_0 + \kappa H_1 - \frac{1}{G}(Gh_0)_\xi\} & \quad (9.4.14) \end{aligned}$$

Now we consider the momentum equation (9.4.3). Differentiation of the first component once with respect to x and then with respect to y, gives

$$\begin{pmatrix}u_{tx}\\ u_{ty}\end{pmatrix} + \begin{pmatrix}u_x & v_x\\ u_y & v_y\end{pmatrix}\begin{pmatrix}u_x\\ u_y\end{pmatrix} + \begin{pmatrix}u_{xx} & u_{yx}\\ u_{xy} & u_{yy}\end{pmatrix}\begin{pmatrix}u\\ v\end{pmatrix} + \frac{1}{\rho}\begin{pmatrix}p_{xx}\\ p_{xy}\end{pmatrix} - \frac{1}{\rho^2}p_x\begin{pmatrix}\rho_x\\ \rho_y\end{pmatrix} = 0$$

Taking the jump of this equation, we get

$$\begin{aligned} \begin{pmatrix}[u_{tx}]\\ [u_{ty}]\end{pmatrix} - \begin{pmatrix}[u_x] & [v_x]\\ [u_y] & [v_y]\end{pmatrix} + \begin{pmatrix}[u_x]\\ [u_y]\end{pmatrix} - \begin{pmatrix}[u_{xx}] & [u_{yx}]\\ [u_{xy}] & [u_{yy}]\end{pmatrix}\begin{pmatrix}[u]\\ [v]\end{pmatrix} & \\ + \frac{1}{\rho_0 - D_0}\begin{pmatrix}[p_{xx}]\\ [p_{xy}]\end{pmatrix} + \frac{1}{(\rho_0 - D_0)^2}[p_x]\begin{pmatrix}[\rho_x]\\ [\rho_y]\end{pmatrix} = 0 & \quad (9.4.15) \end{aligned}$$

Since

$$(N_1, N_2)\begin{pmatrix}[u_{tx}]\\ [u_{ty}]\end{pmatrix} = u_{1t} - Cu_2 + \frac{1}{G^2}u_{0\xi}C_\xi$$

$$-(N_1, N_2)\begin{pmatrix}[u_x] & [v_x]\\ [u_y] & [v_y]\end{pmatrix} = -(u_1, v_1)\left\{u_1\begin{pmatrix}N_1\\ N_2\end{pmatrix} + \frac{1}{G}u_{0\xi}\begin{pmatrix}\tau_1\\ \tau_2\end{pmatrix}\right\}$$

$$= - \quad u_1 H_1 - u_{0\xi} h_0$$

$$-(N_1, N_2)\begin{pmatrix} [u_{xx}] & [u_{yx}] \\ [u_{xy}] & [u_{yy}] \end{pmatrix}\begin{pmatrix} [u] \\ [v] \end{pmatrix} = -H_0 u_2,$$

$$\frac{1}{\rho_0 - D_0}(N_1, N_2)\begin{pmatrix} [p_{xx}] \\ [p_{xy}] \end{pmatrix} = \frac{1}{\rho_0 - D_0}\left\{ S_2 N_1 + \frac{1}{G}(S_{1\xi} + \kappa S_{0\xi})\tau_1 \right\}$$

and

$$\frac{1}{(\rho_0 - D_0)^2}[p_x](N_1, N_2)\begin{pmatrix} [\rho_x] \\ [\rho_y] \end{pmatrix} = \frac{1}{(\rho_0 - D_0)^2} D_1 (S_1 N_1 + \frac{1}{G} S_{0\xi}\tau_1)$$

Left multiplication of (9.4.15) by (N_1, N_2) gives

$$\begin{aligned} u_{1t} - (C + H_0)u_2 + \frac{1}{G^2}u_{0\xi}C_\xi - u_1 H_1 - u_{0\xi}h_0 \\ + \frac{1}{\rho_0 - D_0}\{S_2 N_1 + \frac{1}{G}(S_{1\xi} + \kappa S_{0\xi})\tau_1\} \\ + \frac{1}{(\rho_0 - D_0)^2} D_1 (S_1 N_1 + \frac{1}{G} S_{0\xi}\tau_1) = 0 \end{aligned} \qquad (9.4.16)$$

Similarly, the second component of the momentum equation gives

$$\begin{aligned} v_{1t} - (C + H_0)v_2 + \frac{1}{G^2}v_{0\xi}C_\xi - v_1 H_1 - v_{0\xi}h_0 \\ + \frac{1}{(\rho_0 - D_0)}\{S_2 N_2 + \frac{1}{G}(S_{1\xi} + \kappa S_{0\xi})\tau_2\} \\ + \frac{1}{(\rho_0 - D_0)^2} D_1 (S_1 N_2 + \frac{1}{G} S_{0\xi}\tau_2) = 0 \end{aligned} \qquad (9.4.17)$$

We combine the two equations (9.4.16 - 17) in one equation in the matrix form

$$\begin{pmatrix} u_1 \\ v_1 \end{pmatrix}_t - (C + H_0)\begin{pmatrix} u_2 \\ v_2 \end{pmatrix} + \frac{1}{\rho_0 - D_0} S_2 \begin{pmatrix} N_1 \\ N_2 \end{pmatrix}$$

$$+\left(\frac{1}{G^2}C_\xi - h_0\right)\begin{pmatrix}u_0\\v_0\end{pmatrix}_\xi - H_1\begin{pmatrix}u_1\\v_1\end{pmatrix} + \frac{D_1S_1}{(\rho_0-D_0)^2}\begin{pmatrix}N_1\\N_2\end{pmatrix}$$
$$+\left\{\frac{D_1}{(\rho_0-D_0)^2}S_{0\xi} + \frac{1}{\rho_0-D_0}(S_{1\xi}+\kappa S_{0\xi})\right\}\frac{1}{G}\begin{pmatrix}\tau_1\\\tau_2\end{pmatrix} = 0 \quad (9.4.18)$$

Premultiplying this by (N_1, N_2) and using the formulae for $\mathbf{N}_t$ and $\mathbf{N}_\xi$, we get the final form of the normal component of the second compatibility condition from the momentum equation

$$H_{1t} - (C+H_0)H_2 + \frac{1}{\rho_0-D_0}S_2 = (h_0 - \frac{1}{G^2}C_\xi)H_{0\xi} + H_1^2$$
$$-\frac{D_1S_1}{(\rho_0-D_0)^2} - h_0C_\xi \quad (9.4.19)$$

Now consider the energy equation (9.4.4). Differentiation with respect to x and y gives

$$\begin{pmatrix}p_{xt}\\p_{yt}\end{pmatrix} + \begin{pmatrix}u_x & v_x\\u_y & v_y\end{pmatrix}\begin{pmatrix}p_x\\p_y\end{pmatrix} + \begin{pmatrix}p_{xx} & p_{xy}\\p_{yx} & p_{yy}\end{pmatrix}\begin{pmatrix}u\\v\end{pmatrix} + \gamma(u_x+v_y)\begin{pmatrix}p_x\\p_y\end{pmatrix}$$
$$+\gamma p\left\{\begin{pmatrix}u_{xx}\\u_{xy}\end{pmatrix} + \begin{pmatrix}u_{xy}\\v_{yy}\end{pmatrix}\right\} = 0$$

Taking the jump across Ω_t, we get

$$\begin{pmatrix}[p_{xt}]\\[p_{yt}]\end{pmatrix} - \begin{pmatrix}[u_x] & [v_x]\\[u_y] & [v_y]\end{pmatrix}\begin{pmatrix}[p_x]\\[p_y]\end{pmatrix} - \begin{pmatrix}[p_{xx}] & [p_{xy}]\\[p_{yx}] & [p_{yy}]\end{pmatrix}\begin{pmatrix}u_0\\v_0\end{pmatrix}$$
$$-\gamma\left([u_x]+[v_y]\right)\begin{pmatrix}[p_x]\\[p_y]\end{pmatrix} + \gamma(p_0 - S_0)\left\{\begin{pmatrix}[u_{xx}]\\[u_{xy}]\end{pmatrix} + \begin{pmatrix}v_{xy}\\v_{yy}\end{pmatrix}\right\} = 0 \quad (9.4.20)$$

Since

$$(N_1,N_2)\begin{pmatrix}[p_{xt}]\\[p_{yt}]\end{pmatrix} = S_{1t} - CS_2 + \frac{1}{G^2}S_{0\xi}C_\xi,$$
$$-(N_1,N_2)\begin{pmatrix}[u_x] & [v_x]\\[u_y] & [v_y]\end{pmatrix}\begin{pmatrix}[p_x]\\[p_y]\end{pmatrix} = -([p_x]u_1 + [p_y]v_1) = -S_1H_0 - S_{0\xi}h_0,$$
$$-(N_1,N_2)\begin{pmatrix}[p_{xx}] & [p_{xy}]\\[p_{yx}] & [p_{yy}]\end{pmatrix}\begin{pmatrix}u_0\\v_0\end{pmatrix} = -S_2H_0,$$
$$-\gamma([u_x]+[v_y])(N_1,N_2)\begin{pmatrix}[p_x]\\[p_y]\end{pmatrix} = -\gamma([u_x]+[v_y])S_1$$

$$= -\gamma(H_1 + \frac{1}{G}(u_{0\xi}\tau_1 + v_{0\xi}\tau_2))S_1$$

$$= -\gamma S_1 \left\{ H_1 + \frac{1}{G}((u_0, v_0)\begin{pmatrix}\tau_1\\ \tau_2\end{pmatrix})_\xi - \frac{1}{G}(u_0, v_0)\begin{pmatrix}\tau_1\\ \tau_2\end{pmatrix}_\xi \right\}$$

$$= -\gamma S_1(H_1 - \kappa H_0)$$

$$\gamma\ (p_0 - S_0)(N_1, N_2)\left\{ \begin{pmatrix}[u_{xx}]\\ [u_{xy}]\end{pmatrix} + \begin{pmatrix}[u_{xy}]\\ [v_{yy}]\end{pmatrix} \right\} \gamma(p_0 - S_0)\{u_2 N_1$$

$$= +v_2 N_2 + \frac{1}{G}((u_{1\xi} + \kappa u_{0\xi})\tau_1 + (v_{1\xi} + \kappa v_{0\xi})\tau_2)\Big\}$$

$$= \gamma(p_0 - S_0)\left\{ H_2 + \frac{1}{G}(u_1, v_1)_\xi \begin{pmatrix}\tau_1\\ \tau_2\end{pmatrix} + \frac{1}{G}\kappa(u_0, v_0)_\xi \begin{pmatrix}\tau_1\\ \tau_2\end{pmatrix} \right\}$$

$$= \gamma(p_0 - S_0)\left\{ H_2 + \frac{1}{G}(gh_0)_\xi - \kappa H_1 - \kappa^2 H_0 \right\}$$

Left multiplication of (9.4.20) by (N_1, N_2) gives

$$S_{1t} - (C + H_0)S_2 + \gamma(p_0 - S_0)H_2 = S_{0\xi}(h_0 - \frac{1}{G}C_\xi)$$

$$+(\gamma + 1)S_1 H_1 - \gamma\kappa S_1 H_0$$

$$+\gamma(p_0 - S_0)\{\kappa^2 H_0 + \kappa H_1 - (h_{0\xi} + \Gamma h_0)\} \qquad (9.4.21)$$

where Γ is given by (9.3.22).

We finally write the second set of compatibility conditions in the matrix form

$$\begin{pmatrix} D_1 \\ H_1 \\ S_1 \end{pmatrix}_t + \begin{pmatrix} -(C + H_0) & (\rho_0 - D_0) & 0 \\ 0 & (C + H_0) & (\rho_0 - D_0)^{-1} \\ 0 & \gamma(p_0 - S_0) & -(C + H_0) \end{pmatrix} \begin{pmatrix} D_2 \\ H_2 \\ S_2 \end{pmatrix}$$

$$= \begin{pmatrix} D_{0\xi}(h_0 - \frac{1}{G^2}C_\xi) + D_1(2H_1 - \kappa H_0) + M_1 \\ H_{0\xi}(h_0 - \frac{1}{G^2}C_\xi) + H_1^2 - \frac{D_1 S_1}{(\rho_0 - D_0)^2} - h_0 C_\xi \\ S_{0\xi}(h_0 - \frac{1}{G^2}C_\xi) + S_1((\gamma + 1)H_1 - \gamma\kappa H_0) \\ +\gamma(p_0 - S_0)M_1 \end{pmatrix} \qquad (9.4.22)$$

where

$$M_1 = \frac{1}{\rho_0}(\kappa^2 H_0 + \kappa H_1 - (h_{0\xi} + \Gamma h_0)) \tag{9.4.23}$$

We note that the left hand side of (9.4.12 and 22) are exactly the same as the compatibility conditions (8.2.6 and 10) for a plane shock propagation in one-space dimension. For a plane shock, the curvature $\kappa = 0$ and all derivatives with respect to ξ vanish. Hence, all terms on the right hand side of (9.4.12) vanish. All terms on the right hand side of (9.4.22) except those in f_1 given by (8.2.10) also vanish for a plane shock.

9.4.3 First and second set of equations in the shock ray theory

We collect here all equations required for the calculation of successive positions of a two-dimensional shock front and distribution of shock strength, say D_0.

The shock position is given by (9.3.23)

$$\frac{\partial x}{\partial t} = N_1 C \tag{9.4.24}$$

$$\frac{\partial y}{\partial t} = N_2 C \tag{9.4.25}$$

where $\frac{\partial}{\partial t}$ represents a partial differentiation keeping ξ fixed, i.e., it represents time-rate of change along a shock ray. Let Θ represent the angle which the normal to the shock front Ω_t at time t, makes with the x-direction. Then

$$N_1 = \cos\Theta \ , \qquad \mathbf{N}_2 = \sin\Theta \tag{9.4.26}$$

Substituting (9.4.26) in any one of the four equations in (9.3.25 - 26), we get

$$\frac{\partial \Theta}{\partial t} = -\frac{1}{G} C_\xi \tag{9.4.27}$$

We note that $\tan\Theta = -\tau_1/\tau_2 = -x_\xi/y_\xi$. Curvature κ and Θ are related by

$$\kappa = -\frac{1}{G}\Theta_\xi \tag{9.4.28}$$

The metric g is given by (9.3.3) but it evolved according to the law (9.3.24), i.e.,

$$G_t = -\kappa C G \tag{9.4.29}$$

The set of five equations (9.4.24 - 25, 27 - 29) for x, y, Θ, κ and G is not closed since the shock velocity C depends on the shock intensity D_0 through the equation (9.4.5). The evolution of D_0 and the other two quantities H_0 and S_0 is given by the first set of compatibility conditions, which is

$$\begin{pmatrix} D_0 \\ H_0 \\ S_0 \end{pmatrix}_t + P \begin{pmatrix} D_1 \\ H_1 \\ S_1 \end{pmatrix} = \kappa H_0 \begin{pmatrix} \rho_0 - D_0 \\ 0 \\ \gamma(\rho_0 - S_0) \end{pmatrix} \tag{9.4.30}$$

where the matrix P is given by the expression below (8.2.6).

The set of equations with addition of (9.4.30) is again not closed due to the presence of new quantities D_1, H_1, S_1. Their evolution along a shock ray is given by the second set of compatibility conditions

$$\begin{pmatrix} D_1 \\ H_1 \\ S_1 \end{pmatrix}_t + P \begin{pmatrix} D_2 \\ H_2 \\ S_2 \end{pmatrix} = \mathbf{F}_1 \tag{9.4.31}$$

where the column vector $\mathbf{F}_1$ is given by the expression on the right hand side of (9.4.22). All quantities in $\mathbf{F}_1$ depend on $G_0, D_0, H_0, S_0, D_1, H_1, S_1$ and tangential derivatives of D_0, H_0 and S_0. Further compatibility conditions can be determined.

The shock ray theory consists of solving the infinite system of compatibility conditions along with equations (9.4.24 - 29). However, for each $i (i = 0, 1, 2, \ldots)$ only one of D_i, H_i and S_i is independent. For $i = 0$, this is the well-known result (9.4.5) as a consequence of the Rankine–Hugoniot conditions.

For $i = 1, 2, 3, \ldots$, we introduced $\pi_1, \pi_2, \pi_3, \ldots$ in section 8.2 and expressed D_i, H_i and S_i in terms of π_i for each i. We can follow exactly the same procedure and reduce the ith set of the compatibility condition to a scalar equation giving evolution of π_i. This gives us an infinite set of compatibility conditions, each one of these being in a scalar form. Let us assume that the initial position of the shock Ω_0 (which determines $x(\xi, 0), y(\xi, 0)$ and $\Theta(\xi, 0)$), the distribution of the shock strength $D_0(\xi)$ (or $\pi_0(\xi)$) on Ω_0 and the partial derivatives of

ρ or u and v or p on Ω_0 (so that initial values of D_i or H_i or S_i which would determine $\pi_i(\xi, 0), i = 1, 2, 3, \ldots$) are known. Then the shock ray equations (9.4.24 - 29) along with the infinite system of scalar equations can be solved to give the successive positions of the shock front and the distribution of the shock strength, etc., at later times. This can be treated as *an extension of the Huyghen's construction of the wavefront to a shock front.* However, it is important to note that a shock front is *not self-propagating*, since in the initial data we need also the values of the normal derivatives, i.e., information not just on the shock front but in a neighbourhood of Ω_0.

It is now clear how to develop the new theory of shock dynamics, which would require setting one of the three variables D_n, H_n, S_n or a linear combination π_n in the nth compatibility condition (in scalar form) equal to zero. As shown in the case of a plane shock propagation in section 8.4, NTSD would be quite an efficient procedure for solving many practical problems, at least in the intermediate time range. The theory has been applied to two-dimensional problems by Singh and Singh (1999) but their initial conditions for π_0 and π_1 need improvement.

9.5 A weak shock ray theory

We use an explicit form of the first two scalar compatibility conditions for a shock of arbitrary strength derived by Ravindran and Prasad (1993), which has the advantage of identifying each term clearly. They are expressed in terms of μ, μ_1 and μ_2 defined on the shock front

$$\left.\begin{aligned} \mu &= \frac{\rho_\ell - \rho_0}{\rho_0} = -\frac{D_0}{\rho_0}, \\ \frac{\partial \mu}{\partial \mathbf{N}} &= -\frac{D_1}{\rho_0} = N_1\frac{\partial \mu}{\partial x} + N_2\frac{\partial \mu}{\partial y} = \mu_1 \text{ say} \\ N_1^2\frac{\partial^2 \mu}{\partial x^2} + 2N_1N_2\frac{\partial^2 \mu}{\partial x \partial y} &+ N_2^2\frac{\partial^2 \mu}{\partial y^2} = -\frac{D_2}{\rho_0} = \mu_2, \text{ say} \end{aligned}\right\} \tag{9.5.1}$$

The expression of C in terms of μ is

$$C = a_0 \left\{\frac{2(1+\mu)}{2-(\gamma-1)\mu}\right\}^{\frac{1}{2}} \tag{9.5.2}$$

In order to give the explicit form, we denote a point on the shock front by (X, Y) and introduce the spatial shock ray derivative

$$\frac{d}{ds} = \frac{1}{a_0}\left(\frac{\partial}{\partial t} + N_1 C\frac{\partial}{\partial x} + N_2 C\frac{\partial}{\partial y}\right) \tag{9.5.3}$$

We note that $\frac{d}{ds}$ here is $\frac{1}{a_0}\frac{\partial}{\partial t}$ of (9.4.24 - 29) where it is a derivative in the (t, ξ)-plane. The shock rays (9.4.24 - 25 and 27) become

$$\frac{dX}{ds} = N_1 C/a_0, \quad \frac{dY}{ds} = N_2 C/a_0 \tag{9.5.4}$$

and

$$\frac{a_0}{C}\frac{d\Theta}{ds} = -\frac{\gamma+1}{2(1+\mu)Q}\frac{\partial \mu}{\partial T} \tag{9.5.5}$$

where

$$\frac{\partial}{\partial T} = \frac{1}{G}\frac{\partial}{\partial \xi} = N_1\frac{\partial}{\partial y} - N_2\frac{\partial}{\partial x} \tag{9.5.6}$$

and

$$Q = 2 - (\gamma - 1)\mu \tag{9.5.7}$$

The explicit form of the first compatibility condition is

$$\frac{a_0}{C}\frac{d\mu}{ds} = -\mu\frac{Q}{S}\left[2\frac{\partial\Theta}{\partial T} + \frac{\gamma+1}{1+\mu}\mu_1\right] \tag{9.5.8}$$

where

$$S = 8 + (5 - 3\gamma)\mu + (\gamma^2 - 1)\mu^2 \tag{9.5.9}$$

The second compatibility condition is

$$\begin{aligned}\frac{a_0}{C}\frac{d\mu_1}{ds} &= -\frac{1}{\zeta_1}\left[\frac{(1+\gamma)\mu}{2(1+\mu)^3}\mu_2 + \zeta_2\mu_1^2 - \zeta_3\mu_1\frac{\partial\Theta}{\partial T}\right.\\ &+ \left.\zeta_4\left(\frac{\partial\Theta}{\partial T}\right)^2 + \zeta_5\frac{\partial^2\mu}{\partial T^2} + \zeta_6\left(\frac{\partial\mu}{\partial T}\right)^2\right]\end{aligned} \tag{9.5.10}$$

where $\zeta_1, \zeta_2, \zeta_3, \zeta_4, \zeta_5, \zeta_6$ are quite complicated functions of μ (see Ravindran and Prasad (1993) where $\frac{\partial}{\partial T}$ is the negative of $\frac{\partial}{\partial T}$ used here).

A NTSD can be formulated by dropping the term μ_2 in (9.5.10). Computation with this formulation in the one-dimensional piston problem did not give very good results when the shock strength became large – probably due to the vanishing of a denominator in the complicated expressions of the coefficients. However, as we saw in chapter 8, the numerical results are very good in the formulation of the previous section.

The shock ray equations take an extremely simple and useful form for a weak shock. In this case, we assume

$$\mu = 0(\epsilon),\ \frac{\partial \mu}{\partial T} = 0(\epsilon),\ \mu_1 = 0(1),\ \mu_2 = O\left(\frac{1}{\epsilon}\right) \tag{9.5.11}$$

where ϵ is a small quantity. If we retain only the terms of order ϵ, many terms in (9.5.10) drop out and finally we get

$$\frac{dX}{ds} = (1 + \frac{\gamma+1}{4}\mu)\cos\Theta,\ \frac{dY}{ds} = (1 + \frac{\gamma+1}{4}\mu)\sin\Theta \tag{9.5.12}$$

$$\frac{d\Theta}{ds} = -\frac{\gamma+1}{4}\frac{\partial \mu}{\partial T} \tag{9.5.13}$$

$$\frac{d\mu}{ds} = -\frac{\mu}{2}\frac{\partial \Theta}{\partial T} - \frac{(\gamma+1)\mu}{4}\mu_1 \tag{9.5.14}$$

and

$$\frac{d\mu_1}{ds} = -\frac{1}{2}\mu_1\frac{\partial \Theta}{\partial T} - \frac{\gamma+1}{2}\mu_1^2 - \frac{(\gamma+1)\mu}{4}\mu_2 \tag{9.5.15}$$

Thus the shock ray theory for a weak shock provides a set of 5 equations. This set of equations can be closed in a NTSD by setting $\mu_2 = 0$ in (9.5.15). It is interesting to note that the equation for μ_1 is very simple; the first term on the right hand side represents an increase or decrease in the magnitude of μ_1 due to geometric convergence or divergence of the rays and the second term is due to the usual genuine nonlinearity observed in chapter 7 for the single conservation law. These equations have been used to solve interesting problems by Kevlahan (1996).

The assumption $\mu_2 = O\left(\frac{1}{\epsilon}\right)$ in (9.5.11) is mathematically justified. If $\mu = O(\epsilon)$ and the normal derivative $\langle \mathbf{N}, \nabla \rangle$ for quantities of order 1 satisfying short wave approximation near the wavefront is $O\left(\frac{1}{\epsilon}\right)$, $\mu_2 = O\left(\frac{1}{\epsilon}\right)$. It is mathematically satisfying and physically very important that the weak shock ray theory can be derived from WNLRT. We shall do this in the next chapter. The derivation there is far easier because it avoids working with the complicated expressions which we had to deal with in this chapter for a shock of arbitrary strength and a weak shock limit.

Chapter 10

Propagation of a curved weak shock

In this chapter, we present a derivation of the equations of a weak shock ray theory from those of WNLRT, a physically realistic conservation form of NTSD and finally, some results of numerical solutions of these equations for converging shock fronts starting from various kinds of initial geometry. Distributions of the shock strength μ and the normal derivative μ_1, defined in (9.5.1), have been varied in order to bring out some interesting results. The effects of changing the initial strength of the shock or that of the normal derivative, and also the effect of initial curvature on the formation, propagation and separation of kinks have been studied. Then we have studied the ultimate shape and decay of shocks with initially periodic shapes and plane shocks with a dent and bulge, and interpreted these results as corrugational stability of a shock front. Finally, we have presented a comparison of these results with those obtained from other theories.

10.1 Governing equations of the NTSD

A system of equations of shock ray theory consists of the ray equations derived from a shock manifold partial differential equation and an infinite system of compatibility conditions along a shock ray. We emphasize again that unlike the well known geometrical optics theory for the propagation of a one parameter family of wavefronts across which wave amplitude is continuous, the shock ray theory with infinite system of compatibility conditions is exact. This is because geometrical optics requires high frequency approximation which is satisfied exactly for a shock front. Even under the formulation of the NTSD with just two compatibility conditions, the equations for a curved shock propagation are very complex and so far, only one attempt (Singh and Singh (1999)) has been made to use them to compute shock geometry and amplitude distribution on it at a later time for a shock of arbitrary strength. However, much more work, especially in the formulation of conservation laws from the two compatibility needs to be done. This problem did not arise in the case of one-dimensional piston problem dealt in chapter 8 where the NTSD was found to be very successful even in dealing with a strong shock produced by an accelerating piston. We have seen at in the end of the last chapter that, under suitable assumptions, these equations for a weak shock reduce to a rather simpler set of equations (9.5.12 - 15) which we shall use in this chapter.

The derivation of the equations (9.5.12 - 15) of the NTSD for a weak shock (we simply call it NTSD omitting "for a weak shock" henceforth in this chapter) is quite clear but requires extremely complex algebraic calculations. Hence, we first present one more derivation of these equations (Prasad (2000)) from the equations of WNLRT, derived under short wave or high frequency assumption. Consider a weak shock front propagating into a polytropic gas at rest ahead of it. Then the shock will be followed by a one parameter family of nonlinear waves belonging to the same characteristic field (or mode). Each one of these wavefronts will catch up with the shock, interact with it and then disappear. A nonlinear wave, while interacting with the shock will be instantaneously coincident with it in the short wave assumption. The ray equations (6.1.1 - 2) of the WNLRT in three-space-dimensions for a particular nonlinear wavefront in notations of section 4.4 where we have used $\epsilon a_0 \tilde{w}$ for w, are

$$\frac{d\mathbf{x}}{dt} = \mathbf{n}a_0\left(1+\epsilon\frac{\gamma+1}{2}\tilde{w}\right) \tag{10.1.1}$$

and

$$\frac{d\mathbf{n}}{dt} = -\epsilon\frac{\gamma+1}{2}a_0\mathbf{L}\tilde{w} \tag{10.1.2}$$

where $\epsilon\tilde{w}$ represents the amplitude of the wave in terms of which perturbations due to the waves in the density ρ, fluid velocity $\mathbf{q}$ and pressure p are given by

$$\rho - \rho_0 = \epsilon\rho_0\tilde{w}\ , \quad \mathbf{q} = \epsilon\mathbf{n}a_0\tilde{w}\ , \quad p - p_0 = \epsilon\rho_0 a_0^2\tilde{w} \tag{10.1.3}$$

The transport equation (6.1.3) for the amplitude $\tilde{w}$ on the nonlinear wavefront is

$$\frac{d\tilde{w}}{dt}\left\{\frac{\partial}{\partial t} + a_0\left(1+\epsilon\frac{\gamma+1}{2}\right)\tilde{w}\langle\mathbf{n},\nabla\rangle\right\}\tilde{w} = -\frac{1}{2}a_0\langle\nabla,\mathbf{n}\rangle\tilde{w} \tag{10.1.4}$$

Now we use the theorem 9.2.1.

We denote the unit normal to the shock front by $\mathbf{N}$. For the linear wavefront just ahead of the shock and instantaneously coincident with it (this is actually a linear wavefront moving with the ray velocity $\mathbf{N}$ multiplied by the local sound velocity a_0), $\tilde{w}$=0 and the bicharacteristic velocity is $\mathbf{N}a_0$. For the nonlinear wavefront just behind the shock and instantaneously coincident with it, we denote the amplitude $\tilde{w}$ by μ. Then μ is the shock amplitude of the weak shock under consideration. Using the theorem and the results (10.1.1-2) with $\mathbf{n} = \mathbf{N}$, we get for a point $\mathbf{X}$ on the shock ray

$$\frac{d\mathbf{X}}{dT} = \frac{1}{2}\left\{a_0\mathbf{N} + \mathbf{N}a_0\left(1+\epsilon\frac{\gamma+1}{2}\mu\right)\right\} = \mathbf{N}a_0\left(1+\epsilon\frac{\gamma+1}{4}\mu\right) \tag{10.1.5}$$

$$\frac{d\mathbf{N}}{dT} = -\frac{1}{2}\left\{0+\epsilon\frac{\gamma+1}{2}a_0\mathbf{L}\mu\right\} = -\epsilon\frac{\gamma+1}{4}a_0\mathbf{L}\mu \tag{10.1.6}$$

where T is the time measured while moving along a shock ray. We take $\tilde{w} = \mu$ and $\mathbf{n} = \mathbf{N}$ in (10.1.4) and write it as

$$\begin{aligned}\frac{d\mu}{dT} &\equiv \left\{\frac{\partial}{\partial t} + a_0\left(1+\epsilon\frac{\gamma+1}{4}\mu\right)\langle\mathbf{N},\nabla\rangle\right\}\mu \\ &= -\frac{1}{2}a_0\langle\nabla,\mathbf{N}\rangle\mu - \epsilon\frac{\gamma+1}{4}\mu\langle\mathbf{N},\nabla\rangle\tilde{w}\end{aligned} \tag{10.1.7}$$

where we note that since μ is defined only on the shock front (and also on the instantaneously coincident nonlinear wavefront behind it but not on the other members of the one parameter family of wavefronts following it), the normal derivative $\langle\mathbf{N},\nabla\rangle\mu$ does not make sense mathematically. We introduce a new variable, defined on the shock front (μ, μ_1 and μ_2 here are not the same as those in (9.5.11))

$$\mu_1 = \epsilon\left\{\langle\mathbf{N},\nabla\rangle\tilde{w}\right\}|_{\text{shock front}} \tag{10.1.8}$$

where ϵ appears to make $\mu_1 = O(1)$ since we wish to consider a variation of $\tilde{w}$ on a length scale over which the fast variable θ (introduced in the derivation of (10.1.4), see expression (4.4.4)) varies. This variable is $\theta = \phi/\epsilon$, where ϕ satisfies the characteristic partial differential equation $\phi_t + \langle q, \nabla\phi\rangle + a|\nabla\phi| = 0$.

The equation (10.1.7) leads to the first compatibility condition along a shock ray

$$\frac{d\mu}{dT} = \Omega_s\mu - \frac{\gamma+1}{4}\mu\mu_1 \tag{10.1.9}$$

where

$$\bar{\Omega}_s = -\frac{1}{2}a_0\langle\nabla, \mathbf{N}\rangle \tag{10.1.10}$$

is the value of the mean curvature of the nonlinear wavefront instantaneously coincident with the shock from behind.

To find the second compatibility condition along a shock, we differentiate (10.1.4) in the direction of $\mathbf{n}$ but on the length scale over which θ varies. On this length scale, $\mathbf{n}$ is constant and we get, after rearranging some terms,

$$\begin{aligned}\left\{\frac{\partial}{\partial t} + a_0\left(1 + \epsilon\frac{\gamma+1}{4}\tilde{w}\right)\langle\mathbf{n}, \nabla\rangle\right\}\langle\mathbf{n}, \nabla\rangle\tilde{w} &= -\tfrac{1}{2}\langle\nabla, \mathbf{n}\rangle\langle\mathbf{n}, \nabla\rangle\tilde{w} \\ &\quad -\epsilon\tfrac{\gamma+1}{4}\tilde{w}\langle\mathbf{n}, \nabla\rangle^2\tilde{w} - \epsilon\tfrac{\gamma+1}{4}\left\{\langle\mathbf{n}, \nabla\rangle\tilde{w}\right\}^2\end{aligned} \tag{10.1.11}$$

Writing this equation for the wavefront instantaneously coincident with the shock, multiplying it by ϵ and introducing a variable μ_2 by

$$\mu_2 = \epsilon^2\left\{\langle\mathbf{n}, \nabla\rangle^2\tilde{w}\right\}|_{\text{shock front}} \tag{10.1.12}$$

we get

$$\frac{d\mu_1}{dT} = \bar{\Omega}_s\mu_1 - \frac{\gamma+1}{4}\mu_1^2 - \frac{\gamma+1}{4}\mu\mu_2 \tag{10.1.13}$$

which is the second compatibility condition along shock rays given by (10.1.5 - 6). Note that in μ, μ_1 and μ_2 in (9.5.11) are the same as $\epsilon\mu, \mu_1$ and $\frac{1}{\epsilon}\mu_2$, respectively in this section. Hence the equations (9.5.12 - 15) are the same equations as the equations (10.1.5 - 6, 9 and 13) derived in this section.

Similarly, higher order compatibility conditions can be derived. Thus, for the Euler's equations, we have derived the infinite system of compatibility conditions for a weak shock just from the dominant terms of WNLRT. Since the shock ray theory (SRT) can be derived from the WNLRT, the latter is more general than SRT. However, the results obtained by SRT are quantitatively different from those by WNLRT (see section 10.4.3).

As we have already mentioned, the shock ray theory is an exact theory (weak shock assumption is another independent assumption) but since there are infinite number of compatibility conditions on it, it is impossible to use it for computing shock propagation. We now use the new theory of shock dynamics (NTSD) according to which the system of equations (10.1.5 - 6, 9 and 13) can be closed by dropping the term containing μ_2 in the equation (10.1.13).

This step is justified in the case $\mu_1 > 0$, which occurs very frequently in applications such as a blast wave. For a single conservation law, the equation (7.2.3) shows that when $\phi'(\xi) > 0$ *i.e.*, when $\mu_1 > 0$, the second derivative u_{xx} at the shock, namely, μ_2 monotonically decreases as t increases and for large t, $\mu_2 \sim \frac{1}{t^3}$. The numerical results presented in Tables 7.4.1 - 2 clearly justify use of a theory with $\mu\mu_2$ neglected.

When we consider propagation of even stronger shocks in gas dynamics, the results of Chapter 8 show that neglecting the term $\mu\mu_2$ in the second compatibility condition gives good results not only when $\mu_1 > 0$ but also when μ_1 is less than zero (as exemplified by the accelerating piston problem after the initial push of the piston). Thus NTSD is quite a robust method for gas dynamics.

The final equations of the NTSD, which we use in this paper, for the propagation of a weak shock propagation are (10.1.5 - 6, 9 and 13) with the third term on the right hand side of (10.1.13) omitted.

10.2 Conservation form of the equations for a two-dimensional shock propagation

We first consider a non-dimensional co-ordinate system which has been non-dimensionalised with respect to the sound speed a_0 of the polytropic gas at rest ahead of the shock and a suitable length scale L in the problem and use the same symbols x, y and t for the non-dimensional co-ordinates. Then we introduce a shock ray co-ordinate system (ξ, t) such that ξ = constant represents a shock ray (which for a shock moving into a polytropic gas at rest is orthogonal to the successive positions of the shock) and t= constant represents the shock front. We denote by M the Mach number of the shock i.e., the shock velocity divided by the constant speed of sound in the gas ahead of it, i.e.,

$$M = 1 + \epsilon \frac{\gamma + 1}{4} \mu \tag{10.2.1}$$

then Mdt represents an element of length along a ray. Let G be the metric corresponding to the variable ξ i.e., $G\,d\xi$ is an element of length, say l along the shock front at a time t. We can measure l from a suitable point on the shock front, say the point where the ray $\xi = 0$ meets the shock front. Then

$$l = \int_0^{\xi} G d\xi \tag{10.2.2}$$

We choose the origin of the (x, y)-plane to be a suitable point on the initial shock front and ray $\xi = 0$ to be the one which starts from $(0, 0)$ at $t = 0$. We also define a quantity N by

$$N = \frac{\gamma + 1}{4} \mu_1 \tag{10.2.3}$$

Let Θ be the angle that a shock ray (i.e., the normal to the shock front at time t) makes with the x–axis. In the ray co-ordinate system, the subscript t denotes the non-dimensional time derivative along the shock ray i.e., ∂_t stands for $\frac{d}{dT}$ in the expression (10.1.7) with T now representing non-dimensional time. Then the equations (10.1.5 - 6, 9 and 13) giving the successive positions $(X(t,\xi), Y(t,\xi))$ of a weak shock and distributions of M and N on it, take the form

$$X_t = M \cos \Theta, \qquad Y_t = M \sin \Theta \tag{10.2.4}$$

$$\Theta_t + \frac{1}{G} M_\xi = 0 \tag{10.2.5}$$

$$M_t + \frac{1}{2}\frac{1}{G}(M - 1)\Theta_\xi + (M - 1)N = 0 \tag{10.2.6}$$

$$N_t + \frac{1}{2}\frac{1}{G} N \Theta_\xi + 2N^2 = 0 \tag{10.2.7}$$

where we have omitted the last term on the right hand side of (10.1.13). To make this a complete set, we need to add an equation for G. This equation is (9.3.24) which in the notations of the present section is

$$G_t - M\Theta_\xi = 0 \tag{10.2.8}$$

The equations (10.2.4) define rays. Equations (10.2.5 and 8) are geometric conditions. The other two, namely (10.2.6 - 7), are dynamic compatibility conditions along a shock ray.

We eliminate Θ_ξ from (10.2.6 - 8) to get two relations from which we derive a homogeneous relation between M_t, N_t and G_t in the form

$$2NM_t + \frac{(M-1)N}{2GM}G_t - (M-1)N_t = 0 \qquad (10.2.9)$$

For a weak shock under consideration, $0 < M-1 << 1$, we can replace $\frac{M-1}{M}$ by $(M-1)$ and get an integral of (10.2.9) in the form $G(M-1)^4N^{-2} = h(\xi)$, where $h(\xi)$ can be obtained from the distribution of G, M and N on the initial position of the shock front. initially, ξ is chosen to be the arc-length along the shock front at $t = 0$. Therefore, initially, $G = 1$. By changing ξ to a function of ξ, (say ξ'), it is possible to choose $h(\xi) = 1$. This gives

$$G = (M-1)^{-4}N^2 \qquad (10.2.10)$$

We denote ξ' by ξ itself.

The system of equations (10.2.4 - 7) where G is given by (10.2.8) forms a complete set of equations to give successive positions of a shock front and distribution of M and N. The system (10.2.5 - 8) decouples from the equations (10.2.4) and forms a hyperbolic system with eigenvalues $0, \pm\sqrt{\frac{M-1}{2G^2}}$ in the (ξ, t)-plane.

The characteristic fields of the non-zero eigenvalues are genuinely nonlinear and in the solutions of (10.2.5 - 7), shocks appear in the (ξ, t)-plane. Study of these shocks which map into kinks on the shock front in the (x, y)-plane, would require conservation form of the equations (10.2.5 - 7).

Two physically realistic conservation laws follow from the equations (3.3.13 - 14)

$$(G\sin\Theta)_t + (M\cos\Theta)_\xi = 0, \ (G\cos\Theta)_t - (M\sin\Theta)_\xi = 0 \qquad (10.2.11)$$

As argued at the end of section 3.3.3, these are the only two conservation laws which can conserve distance in two independent directions in (x, y)-plane.

The third conservation form is obtained from the consideration of the flow of energy along a ray tube. Eliminating Θ_ξ from (10.2.6

and 8), we get

$$M_t + \frac{M-1}{2GM} G_t + (M-1)N = 0 \tag{10.2.12}$$

Since the second term has $M-1$ in the numerator, we can set $M = 1$ in the denominator as explained earlier. This leads to a form

$$\left(G(M-1)^2\right)_t + 2G(M-1)^2 N = 0 \tag{10.2.13}$$

In a small amplitude theory, the quantity $G(M-1)^2$ is interpreted as the energy density along a ray tube (per unit length in ξ variable). This quantity remains constant as we move with the wave along the ray in the linear theory i.e., $(G(M-1)^2)_t = 0$. In a nonlinear theory, there is an additional term due to dissipation through the shock and we take (10.2.13) as the third conservation law. There appears to be no other physically realistic guiding principle from which another conservation law can be deduced. Since (10.2.10) has been obtained as an integral (though an approximate one) of the governing equations (2.5.6 - 8), we treat it as a relation valid along a shock ray even when it crosses a kink. This is an assumption, the function $h(\xi)$ mentioned following (10.2.9) need not be the same on the two sides of the kink. A detailed discussion of this point is available in Whitham (1974). Thus, we take the fourth conservation law as

$$\left(G(M-1)^4 N^{-2}\right)_t = 0 \tag{10.2.14}$$

The equations (10.2.11, 13 and 14) form the system of conservation laws which with appropriate initial values, can be solved numerically in (ξ, t)-plane. By numerically integrating the two equations (10.2.4), the solution can be mapped onto the (x, y)-plane giving the successive positions of the shock front and distribution of amplitude on them.

Experimental results of Sturtevant and Kulkarni (1976) show that in the linear case a complex wavefield develops near the focus of the wavefront. If the shock wave is very weak, the wavefront emerges from focus crossed and folded (in accordance with the predictions of geometric acoustics theory), whereas, in the stronger shock case, the caustic of the linear theory is resolved and the shock front beyond the focus is uncrossed, as predicted by the theory of shock dynamics. In both cases, the behaviour at the focus is nonlinear. Explanation or determination of the transition from the linear to fully nonlinear behaviour (of uncrossed geometry with kinks) is a very important and challenging mathematical problem. It appears that the present mathematical formulation may not be able to describe this transition because as $M \to 1$, G tends to infinity according to (10.2.10) or equivalently, (10.2.14).

We present the results of extensive numerical solutions of the conservative form of four equations for a quite general form of initial data for M and N and also X and Y giving initial position of shock. We are able to numerically simulate the experimentally observed behaviour when the caustic is resolved. A total variation bounded (TVB) scheme based on the Lax-Friedrichs flux (Shu (1987), Cockburn, Lin and Shu (1989)) has been used to solve this system of conservation laws. The source term in the equation was handled through Strong splitting (LeVeque and Yee (1990)). Effect of varying the initial curvature and also the effects of varying the initial Mach strength M and normal derivative N on the formation and propagation of kinks have been studied.

10.3 Initial conditions, results and discussion

We consider an initially concave shock front moving from left to right (i.e., in the positive direction of x–axis into a gas at rest).

Let

$$X(\xi,0) = X_0(\xi), \qquad Y(\xi,0) = Y_0(\xi) \tag{10.3.1}$$

denote the initial shock front with an initial distribution of amplitude M and N given by

$$M(\xi,0) = M_0(\xi), \;\; N(\xi,0) = N_0(\xi) \tag{10.3.2}$$

The initial value of Θ is obtained from (10.3.1) and that of G is obtained from (10.2.10)

$$\Theta(\xi,0) = \Theta_0(\xi), \;\; G(\xi,0) = G(M_0(\xi), N_0(\xi)) = G_0(\xi) \tag{10.3.3}$$

The system (10.2.11 and 13 - 14) subject to the initial conditions (10.3.1 - 3) is solved using the finite difference scheme. For details, see Monica and Prasad (2001).

After solving the system of equations (10.2.11 and 13 - 14) with initial conditions (10.3.1 - 3), we get the shock rays and the shock position following the method described in 6.3. We present below some main results extracted from extensive numerical computation.

10.3.1 Propagation of a shock front initially parabolic in shape

The initial shock front is taken as

$$y^2 = bx, \qquad 1 \leq b \leq 8, \quad |y| \leq z \tag{10.3.4}$$

extended on either side by the tangents for $|y| > z$. M_0 is prescribed on the initial front as a symmetric function of Θ_0. We take

$$M_0 = \alpha e^{-\beta \Theta_0^2} \tag{10.3.5}$$

where the parameter α is a measure of the strength of the initial shock front and β is a measure of the rate of change of M_0 along the shock front. The distribution of amplitude on the shock front can be varied by varying α and β.

We also prescribe the initial value of N_0. For most cases, N_0 is taken to be a constant and has been varied from 0.1 to 1.0.

(a) Effect of varying the initial amplitude distribution

The results of computations are presented for $b = 2, 4, 8$. For each value of b, the initial distribution of α was varied between 1.1 and 1.2. We present below some of the representative cases. We start with a general qualitative behaviour from one case.

Figs. 10.3.1 with $b = 8$, $z = 1$ are typical of the results of computation, the first figure showing the successive positions of the shock front and other figures showing distributions of M, Θ, N and G on the shock at different times marked by the values of t from $t = 0$ to 10 at time interval two. The shock rays, except the central one, initially tend to converge but later deviate from a straight path due to non-uniform distribution of the amplitude and become parallel to the x-axis. Thus, the formation of the caustic (which happens in the linear theory) is avoided and the propagating shock front becomes plane in the central region. A pair of kinks, shown in Fig.10.3.1[a] by dots, appear on the shock. The position of the kink on the shock is accurately determined by the sharp gradient from the M–l graph in Fig. 10.3.1[b]. The ultimate shape of the shock front as t tends to infinity, consists of a plane central disc separated on the two sides from straight wings by the two kinks. In this and all the cases considered in numerical computation, it is found that the kinks move away from one another leaving an ever increasing shock disc. There is, though, a definite condition (3.3.37) involving the values of M on the two sides of the kink which, if satisfied, would cause the shock disc to shrink. As we shall see at the end of section 10.4.3, both $M-1$ and N tend to zero on the wings as well as on the disc as $t \to \infty$.

The strength M_0 on the shock at $t = 0$ is prescribed to be constant in Fig. 10.3.1. Due to the convergence of rays, the shock strength in the central part increases initially but it decreases on those rays which start from the straight part (i.e., $|y| > 1$) of the initial shock front. At a critical time t_c, depending on the values of α and β, the shock strength M attains a maximum value on the central ray and after that, it continuously decreases along all rays. The Θ–l graph in Fig. 10.3.1[c] shows that till time $t = 10$, Θ is not constant on the disc implying that the central disc is not plane, but it tends to be plane as time increases. The value of N, the gradient of the gas density (or velocity) just behind the shock has also been prescribed to be a constant equal to 0.15. Since the nonlinear waves from behind catch up with the shock and then disappear from the flow field, N decreases continuously but with different rates along different rays. The N–l graph in Fig. 10.3.1[d] shows that from $N = 0.15$ at $t = 0$, the value of N has decreased to less than 0.06 at $t = 10$. It is important to note that the shape of the shock front is qualitatively the same as that of the nonlinear wavefront obtained in section 6.3. Fig. 10.3.1[e] shows that the metric G, which was constant initially with a value of about 14, decreases significantly in the central part of the shock front. For an efficient numerical computation, the lower bound of G should be strictly greater than zero.

A number of computations changing the value of M_0 all over the initial front by changing the value of α (say 1.15, 1.12, 1.10) and keeping the same all the other parameters in the initial data of the case represented in Fig. 10.3.1. have been done. It has been found that as α increases

- N decreases faster along each ray

- the shock disc becomes larger and
- the shock front moves faster - an obvious result

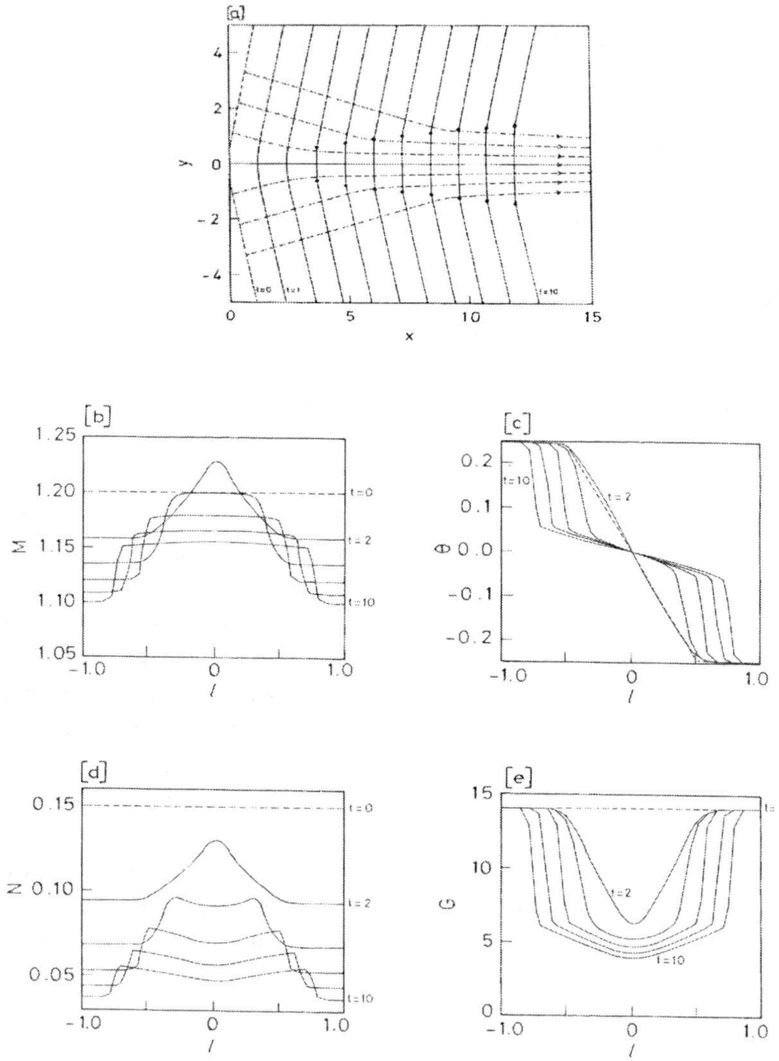

Fig. 10.3.1: Propagation of a weak nonlinear shock front starting from an initial shock front wherein the central part is a parabola $y^2 = 8x$ for $|y| < 1$. The initial amplitude distribution is given by $M_0 = 1.2$ and $N_0 = 0.15$. The dotted curves in [a] are the rays.

It was also found that the maximum value M, which is attained at $t = t_c$ on the central ray, always appears before the kinks appear. Next, the initial value of M_0 is so prescribed that it is at maximum (minimum) at the center and decreases (increases) monotonically to a common value on the two sides. These values of M_0 are chosen in such a way that they coincide on the straight line part, but are different on the parabolic part of the initial front. In Fig. 10.3.2, an initial shock front has been chosen with values $b = 8$, $z = 1$ and taken $\alpha = 1.2$, $\beta = 0.5$ for M_0 to be maximum at the center (say Case I) and $\alpha = 1.1301$, $\beta = -0.5$ for M_0 minimum at the center (say Case II). In both the cases, the M_0 on the straight line part is 1.1645. Figs. 10.3.2[c] - [d] show the comparison of the graph of M in the two cases. The solid lines correspond to Case I and the dotted lines correspond to Case II. Although the maximum value of M at a given time in Case I is more than the corresponding value in Case II before the critical time (i.e., the time at which the shock front develops kinks), the behaviour is reversed after the critical time. We note that the smaller value of M_0 in the central region leads to a delay in the formation of kinks and to a slight lagging behind of the central disc initially. However, as time increases, the constant value on the wings ultimately determines the geometry of the shock front and the amplitude distribution. This is expected as the value of M_0 is changed only on a small bounded interval near the origin on the ξ–axis leaving the value of M_0 unchanged outside this interval. The distribution of M_0 on this outside infinite part decides the ultimate shape and the shock strength of the shock front.

An interesting result to note from Figs. 10.3.2[c] - [d] is that as M_0 is decreased, the central part of the shock front remains more concave for a longer time leading to the convergence of the rays for a longer time which ultimately pushes the value of M to a larger value. This unexpected result was also observed in the WNLRT (Prasad and Sangeeta (1999)).

(b) Effect of varying the initial N

The value of N represents the gradient in the normal direction of the gas density just behind the shock and hence is a measure of the rate of interaction of the nonlinear waves with the shock. As we have noted earlier, and as will be seen later, this interaction for N >0, leads to the ultimate decay of the shock strength to zero. Larger

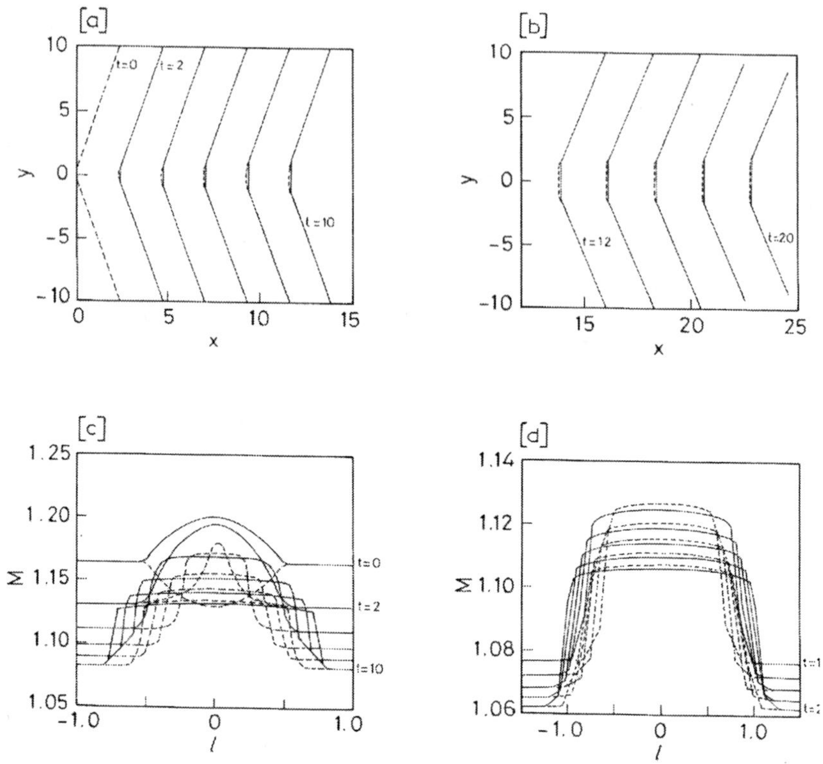

Fig. 10.3.2: Comparison of the results when the amplitude distribution is decreased on the parabolic part of the initial shock front (b=8, z=1). [d] shows $M - l$ graph for $t = 12, 14, \ldots, 20$.
solid lines: results for $\alpha = 1.2$, $\beta = 0.5$
dotted lines: results for $\alpha = 1.1301$, $\beta = -0.5$

value of N means higher rate of interaction leading to faster decay of the shock. All the calculations presented in this chapter correspond to $N > 0$ i.e., expansion waves behind the shock. Taking $N<0$ initially may lead to very large and even infinite value of the shock strength $M-1$ showing that the small amplitude assumption breaks down and this theory is no longer valid. Fig. 10.3.3 presents results of computation with $N_0 = 0.15$ and 0.5, keeping all the other parameters the same. We note that when N_0 is increased everywhere on the initial shock front (and not just on the curved part)

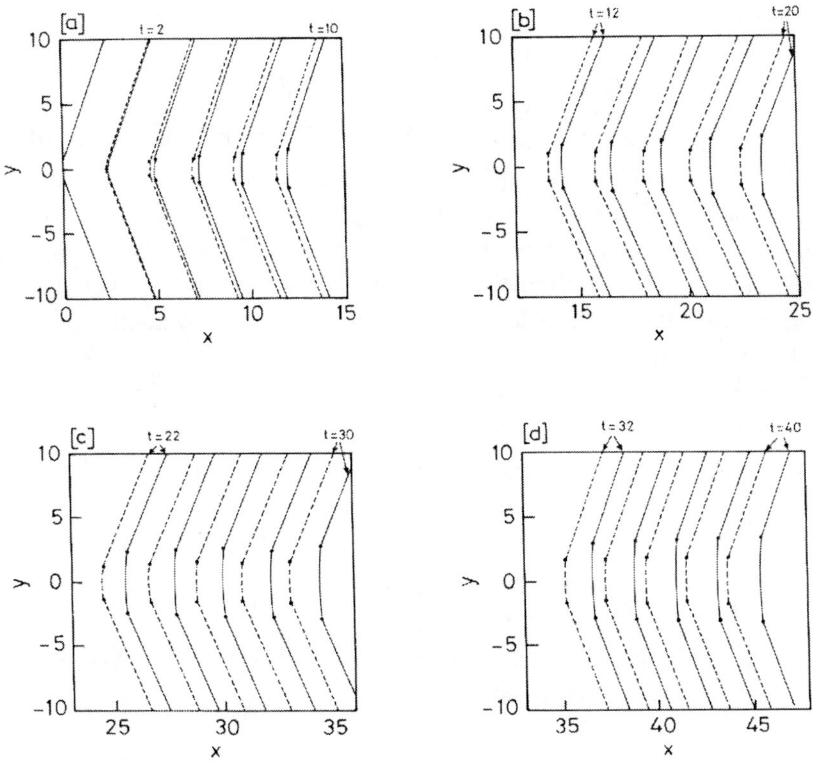

Fig. 10.3.3: It shows the effect of increase in the value of of N_0, on the shock position. Initial shock front is for $b = 8$, $z = 1$ with $\alpha = 1.2, \beta = 0.0$
solid lines: $N_0 = 0.15$
dotted lines: $N_0 = 0.5$

- shock strength decreases everywhere on the shock front faster as t increases (see Monica (1999), for figures showing this result))
- the shock position lags behind
- the shock disc becomes smaller
- the kinks are formed later
- $M_{max}(t) - M_{min}(t)$ initially increases and then decreases with time

where we define

$$M_{max}(t) = \max_{\xi \in \mathbb{R}} M(\xi, t), \qquad M_{min}(t) = \min_{\xi \in \mathbb{R}} M(\xi, t)$$

(c) Effect of varying geometry of the initial shock front

We first consider the effect of changing z in (10.3.4), so that the inclination to the x-axis of the wings of the initial shock front changes without changing the curvature at the center. When z in (10.3.4) is increased from $z = 1$ to $z = 2$, the central curved part of the front is increased and the inclination of the wings with the x-axis is decreased but the curvature at the center of the front does not change with z. A comparison of the results shows that when z is increased

- the rays from the wings converge more strongly leading to a stronger shock strength in the center,
- the shock disc between the two kinks becomes smaller
- the value of N decreases more slowly showing that the interaction of the waves from behind is slower

Next, when curvature of the initial shock front at the center is increased by increasing the value of b. This leads to

- a higher value of the shock strength $M - 1$ and the normal derivative N near the center
- a smaller shock disc

However, as time increases, $M-1$ and N on the disc seem to approach the same value when z is kept fixed. This is to be expected as the long time behaviour of the solution is determined by the initial value of the shock strength on the wings.

(d) Effect of increasing the shock strength on the initial shock front

An increase of α leads to an obvious result of the shock moving faster. It also results in an increase in the distance separating the two kinks (see also Prasad and Sangeeta (1999), Fig. 9).

10.3.2 Propagation of a shock front with initially sinusoidal shape and periodic amplitude distribution

We consider now the initial shock front to be of a periodic sinusoidal shape

$$x = 0.2 - 0.2A\cos(\frac{\pi y}{B}), \quad \text{where } A = 1, B = 2 \tag{10.3.6}$$

where M_0 is prescribed as in (10.3.5) with $\alpha = 1.2$ and $\beta = 0.0$ and $N_0 = 0.1$. Figs. 10.3.4 show the shock fronts at successive times and the associated rays by broken lines. The convergence and divergence of the rays is seen quite clearly in Fig. 10.3.4[a] i.e., during the time interval (0,10). With increasing time, the rays become straight lines parallel to the x-axis and the shock front tends to become planar.

A pair of kinks appear in each period of the shock front at $t \approx 2$. Two kinks suitably paired first move closer to one another, interact, producing a new pair of kinks which move apart and the process continues. After a long period of time, the amplitude $M-1$ decays to zero along each ray. The total variation of M in each period keeps on decreasing with increasing time and $M-1 \to 0$ as $t \to \infty$. After a long time, even when the shock front has become almost a straight line, it is not difficult to locate the position of the kinks. At any time t, the kinks are easily located from the (M, l)-graph as shocks (see Fig. 10.3.5).

As before, we define $M_{max}(t)$ ($M_{min}(t)$) as the maximum (minimum) Mach number attained on the front at a given time t. $M_{max}(t)$ attains a maximum value greater than $M_{max}(0)$ at $t = 2$, which is approximately the time when the kinks first appear. Similarly, $M_{min}(t)$ attains its minimum value at $t = 6$, the time when the kinks first interact. The Figs. 10.3.6 show that $M_{max}(t)$ and $M_{min}(t)$ both tend to 1 and $M_{max}(t) - M_{min}(t) \to 0$ as t $\to \infty$. We note that when the kinks interact, the values $M_{max}(t)$ and $M_{min}(t)$ suddenly jump from smaller values to larger values – the amount of the jump tends to zero as $t \to \infty$.

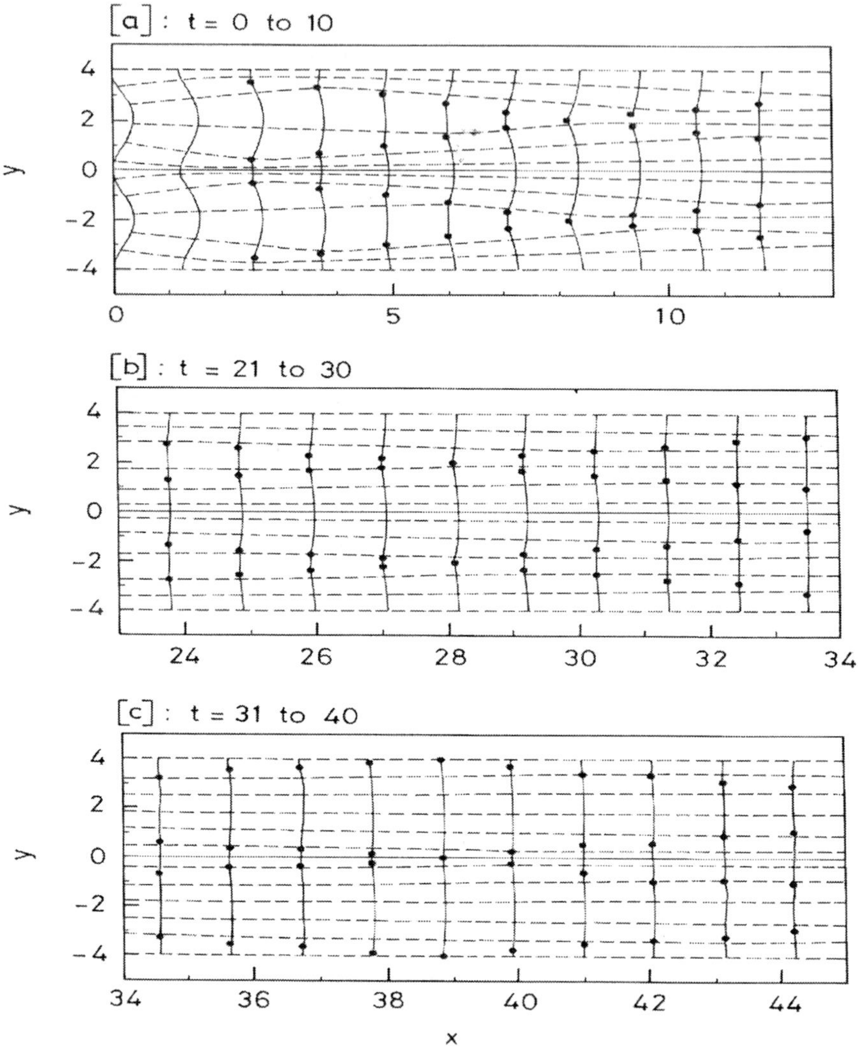

Fig. 10.3.4: Successive positions of an initially sinusoidal shock front and rays plotted at $t = 0, 1, 2, 3, \cdots, 40$. Initial shock front is $x = 0.2 - 0.2\cos(\frac{\pi y}{2})$ with M_0=1.2 and $N_0 = 0.1$.

The two kinks which approach each other and interact producing a new pair correspond to shocks belonging to two different characteristic fields with characteristic velocities $\pm\sqrt{\frac{M-1}{2G^2}}$. The shock speed in (ξ, t)-plane is given by $\pm\,\{[M^2]/[G^2]\}^{1/2}$. When these shocks interact, a new pair of shocks is produced, which leaves behind it a much higher value of M as seen in Fig. 10.3.5. The corresponding results for interaction of kinks on a nonlinear wavefront is given by the relation (6.2.18).

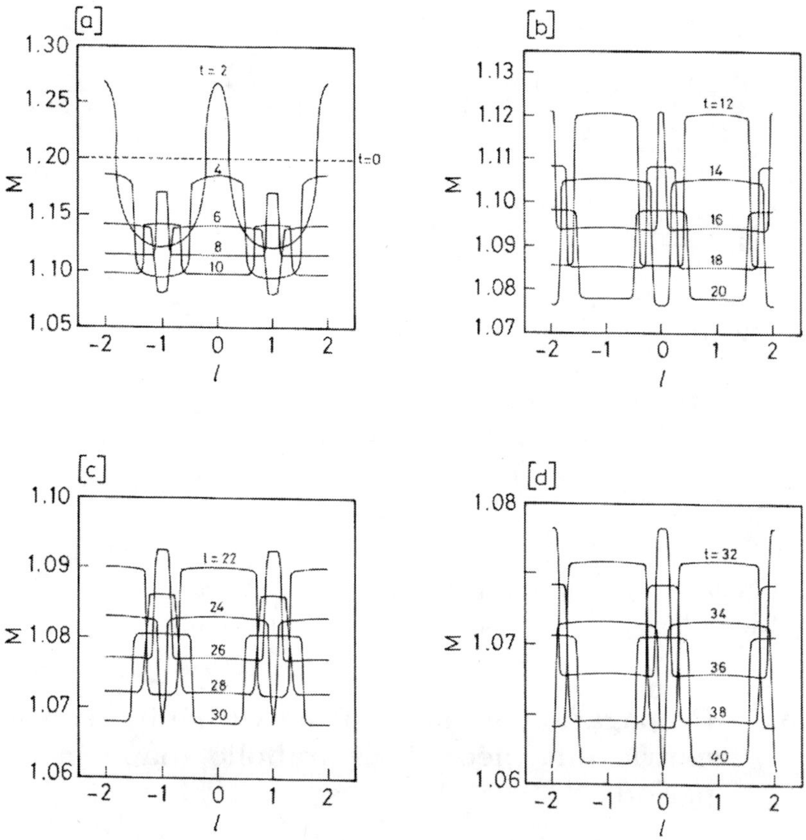

Fig. 10.3.5: Variation of the Mach number with respect to l in the case of an initially sinusoidal shock (it should be noted that graphs in figures [a] - [d] have different scales of M). The value of M changes suddenly after the kinks interact.

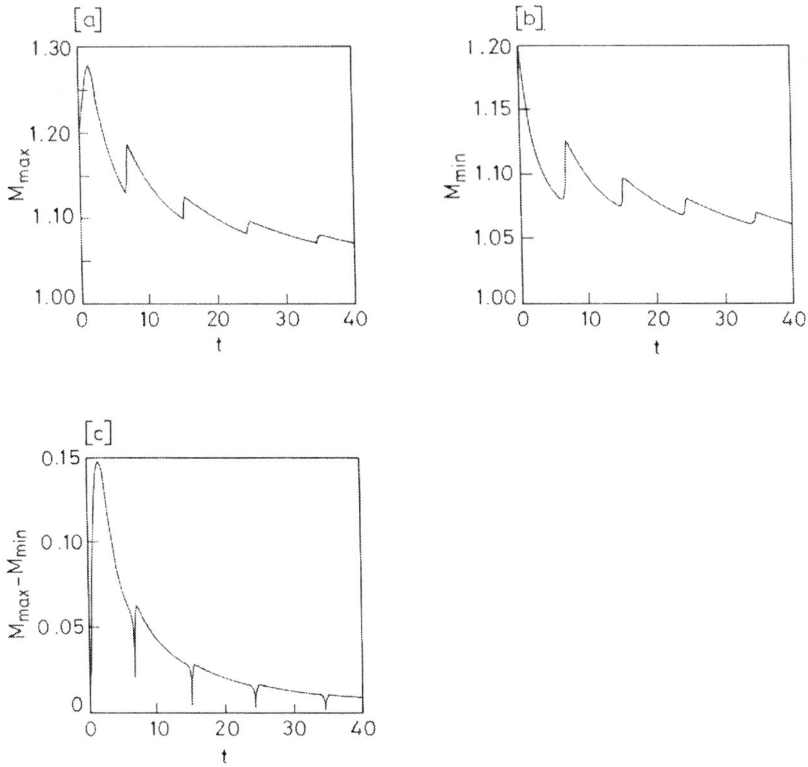

Fig. 10.3.6: Graphs of M_{max}, M_{min}, $(M_{max}-M_{min})$ with time when the initial shock front is sinusoidal.

10.3.3 Propagation of a shock front with initially asymmetric but piecewise parabolic shape in each period

The initial shock front is taken as

$$x = \begin{cases} 0.6 - \dfrac{(y+6.4)^2}{8}, & -6.4 < y < -6.0 \\ -\dfrac{y+0.2}{10}, & -6.0 < y < -0.4 \\ \dfrac{y^2}{8}, & -0.4 < y < 1.0 \\ \dfrac{y-0.5}{4}, & 1.0 < y < 2.4 \\ 0.6 - \dfrac{(y-3.4)^2}{8}, & 2.4 < y < 3.4 \end{cases}$$

and this configuration is repeated periodically in y. M_0 has been prescribed as 1.1.

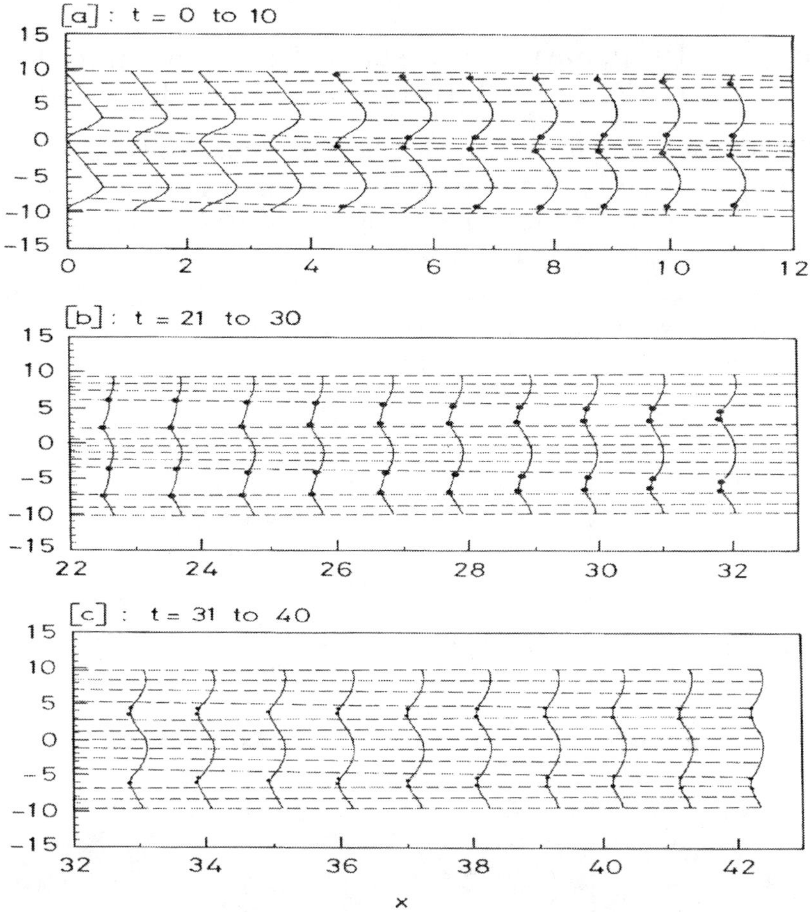

Fig. 10.3.7: Successive positions of the shock and associated rays starting from a periodic shock, asymmetric and with piecewise parabolic shape in each period.

Here too, the behaviour of the shock front is similar to the sinusoidal case except that the shape remains asymmetric in each period with increasing time, the rays become parallel to the x–axis and the shock front tends to become planar. This case is shown in Fig. 10.3.7. We note that the shock front in the previous case has become almost plane at $t = 30$ but this is not so in this case. It is not clear whether it is due to the fact that the initial shock front has a bigger oscillation in its shape or due to the lack of symmetry in its shape in each period.

10.3.4 Propagation of a shock front with initially periodic but arbitrary shape in each period

An arbitrary periodic shape is not possible numerically but we choose here a shape so far removed from the symmetric shape that it may be considered practically a representative of any unsymmetric shape. The initial shock front is formed in each period by joining a series of parts of parabolas and straight lines, the slope being continuous at the points where they are joined. This curve is given by

$$x = \begin{cases} \dfrac{(y+4.4)^2}{4}, & -5.0 < y < -3.4 \\ \dfrac{y+3.9}{2}, & -3.4 < y < -3.2 \\ -\dfrac{(y+2.2)^2}{4} + 0.6, & -3.2 < y < -1.2 \\ -\dfrac{y+0.5}{2}, & -1.2 < y < -1.0 \\ \dfrac{y^2}{4}, & -1.0 < y < 0.4 \\ \dfrac{y-0.2}{5}, & 0.4 < y < 2.0 \\ -\dfrac{(y-2.4)^2}{4} + 0.4, & 2.0 < y < 2.8 \\ -\dfrac{y-4.6}{5}, & 2.8 < y < 4.4 \\ \dfrac{(y-4.8)^2}{4}, & 4.4 < y < 5.0 \end{cases}$$

in $|y| \leq 5$ and extended for all y periodically. The results of computation for this case are shown in Fig. 10.3.8.

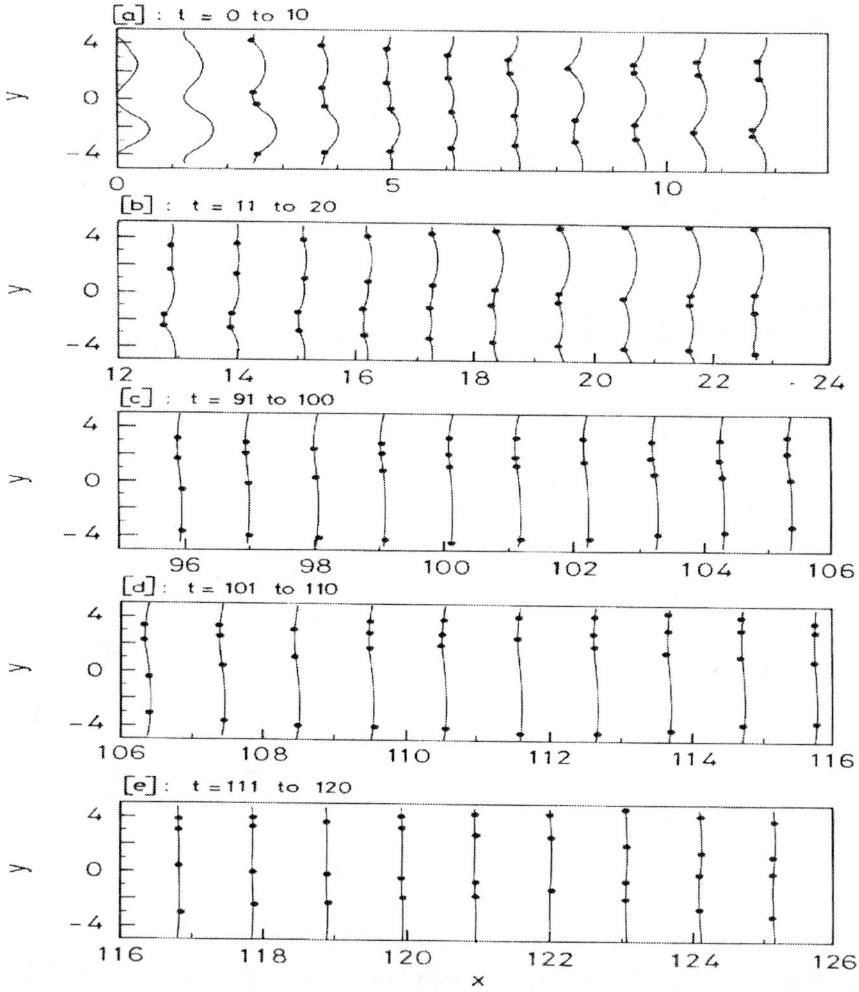

Fig. 10.3.8: Long term behaviour of a shock of periodic shape but far removed from a symmetric shape in each period.

10.3.5 When the initial shock front has a single smooth dent or bulge

We take the initial shock front to be given by

$$x = \pm e^{-(\frac{y^2}{2})} \tag{10.3.7}$$

where '+' represents a bulge and '–' represents a dent. The initial Mach number is prescribed as

$$M_0 = \alpha e^{-\beta x^2} \tag{10.3.8}$$

Note that as $y \to \pm\infty$, $x \to 0$ and $M_0 \to \alpha$. The value of N is taken to be a constant equal to 0.1.

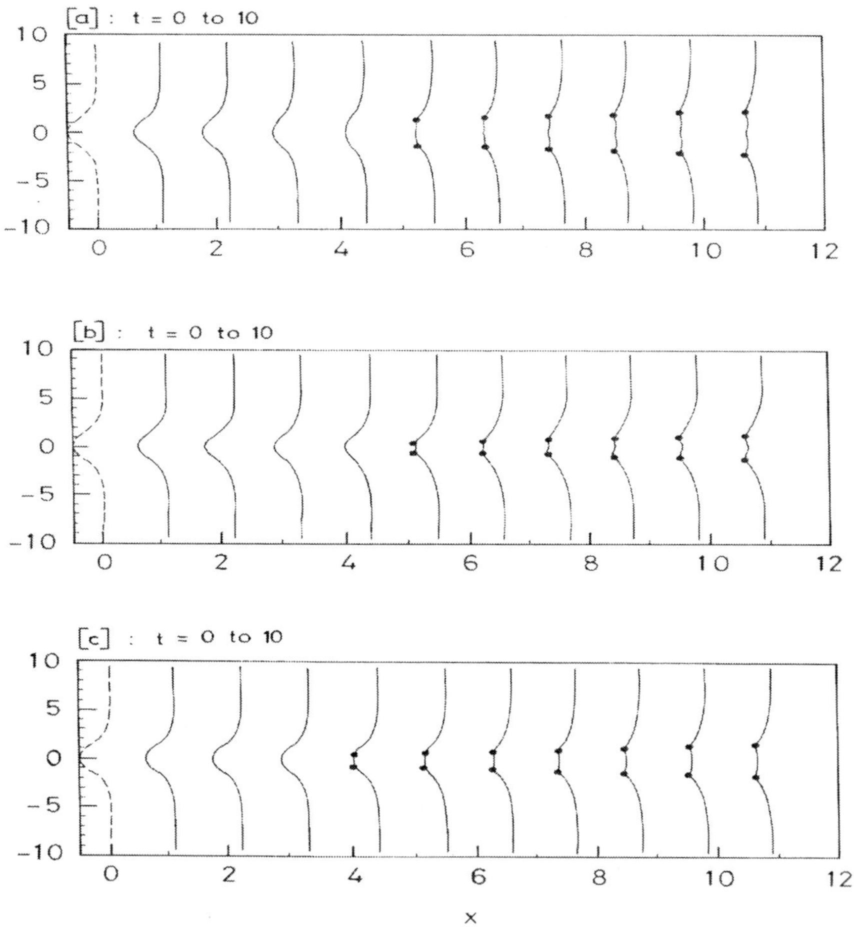

Fig. 10.3.9: Successive positions of a shock front which initially has a smooth dent: $x = -\frac{1}{2}\exp(-y^2/4)$, $M_0 = 1.12\exp(-\beta x^2)$, N_0=0.1
(a) $\beta = -0.05$, (b) $\beta = 0.02$, (c) $\beta = 0.0$.

We first discuss the results when there is a dent. Figs. 10.3.9 and 10.3.10 give the successive positions of the shock front for some representative cases. In Fig 10.3.9[a], initially the shock Mach number at the center is minimum, in Fig. 10.3.9[b] it is maximum and Fig. 10.3.9[c] it has the same value everywhere on the front. The graphs representing the distribution of M, N and Θ with l at various times can be found in (Monica (1999)). The amplitude first increases in the initially dented region, the increase being maximum at the center. Elsewhere, it continues to decrease but it decreases more rapidly near the outer edges of the dent (i.e., near $l = \pm 1.25$). A pair of kinks is formed which moves away from the center. The dent tends to become plane but in this process, the center of the dent moves faster in such a way that the central portion becomes convex with the two kinks on its two sides contained in two dents as seen in Fig. 10.3.9. It has also been found that as the amplitude of the

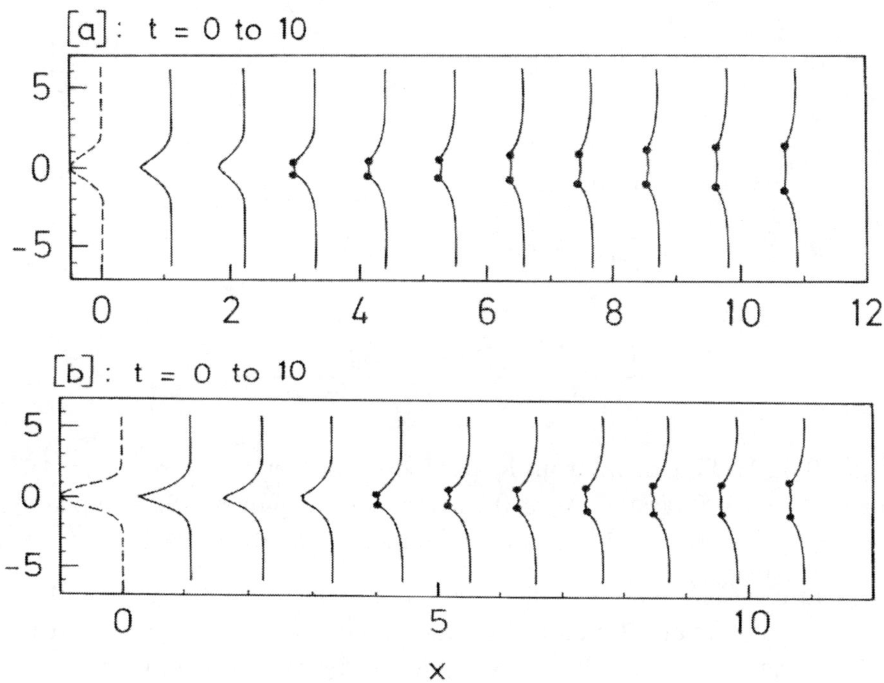

Fig. 10.3.10: Successive positions of a shock front which initially has a smooth deep dent:
(a) $x = -\frac{1}{2}\exp(-y^2), M_0 = 1.12\exp(-\beta x^2), \beta = 0.1, N_0 = 0.1$
(b) $x = -\exp(-y^2), M_0 = 1.12\exp(-\beta x^2), \beta = 0.02, N_0 = 0.1.$

shock in the dent increases (i.e., β decreases), the shock strength M as well as the gradient of the density N behind the shock increases (all measured at a fixed time, say t=1).

When the dent is quite deep (Fig. 10.3.10[b]), it appears that a kink has developed at the center but a detailed study of (M, l)-graph at $t = 1$ shows that it is a pair of kinks which appear very close to the center. At this time, the shock strength M rises very rapidly at the center to a value about 1.45, elsewhere, it remains small and close to the initial value near 1.1. Soon the two kinks move away, the shock develops a convex part at the center separated by kinks from a concave part on either side (see also the result in Fig. 10.3.9 [c]).

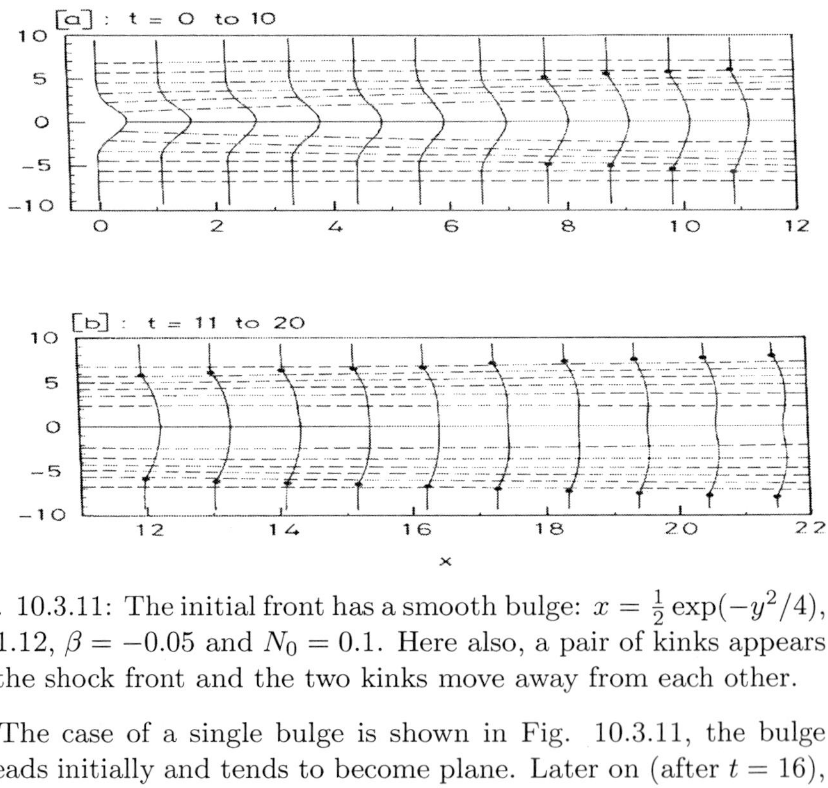

Fig. 10.3.11: The initial front has a smooth bulge: $x = \frac{1}{2}\exp(-y^2/4)$, α=1.12, $\beta = -0.05$ and $N_0 = 0.1$. Here also, a pair of kinks appears on the shock front and the two kinks move away from each other.

The case of a single bulge is shown in Fig. 10.3.11, the bulge spreads initially and tends to become plane. Later on (after $t = 16$), it becomes concave at the center followed by two convex portions – one on either side. A pair of kinks (one on either side) is formed quite early and they move away from one another.

10.4 Comparison with other theories

10.4.1 Qualitative verification of the shape of the front obtained by DNS to support the kink theory

As already mentioned in the introduction, numerical results of Kevlahan (1996) provide enough justification of the validity of the present theory. We present here just one result of numerical solution of full gas dynamic equations.

The most important aspect of the results obtained by us is the resolution of the linear caustic and appearance of kinks on the shock front (also on nonlinear wavefronts in chapter 6). Numerical solution of Euler's equations of gas dynamics also leads to the same result: resolution of a linear caustic and appearance of kinks. This has been shown by computation for a curved piston problem with piston shape given by

$$y^2 = x/2 \quad \text{for} \quad |y| < 0.2, \;\; y = \pm(x/0.8 + 0.1) \text{ for } |y| > 2 \quad (10.4.1)$$

and initial piston velocity equal to 1, piston acceleration also equal to 1. The result of numerical computation has been shown in Fig. 10.4.1. The numbers on the contours represent the values of the particle velocity in x-direction on the contour. We first notice that the figure shows a leading shock front with a well-defined kink across which the direction of the shock changes abruptly; on one side is the central disc and the other side the wing. This clearly agrees with the qualitative picture of the shape of the shock front obtained by NTH. In addition, there is a rapid compression region starting from the kink on the shock. The structure of the kink as a shock front with continuously turning direction of the tangent, can be found on a length scale (along the shock) much smaller than the radius of the curvature of the piston and has been discussed below. A numerical solution of the shock front using the shock ray theory of this chapter takes very little computer time, say only 20% of the time of the results shown in Fig. 10.4.1.

10.4.2 Comparison with earlier theories

The first and simplest theory of the curved weak shock propagation to calculate the shock position is the CPW theory and consists in fitting a weak shock in a one parameter family of nonlinear waves

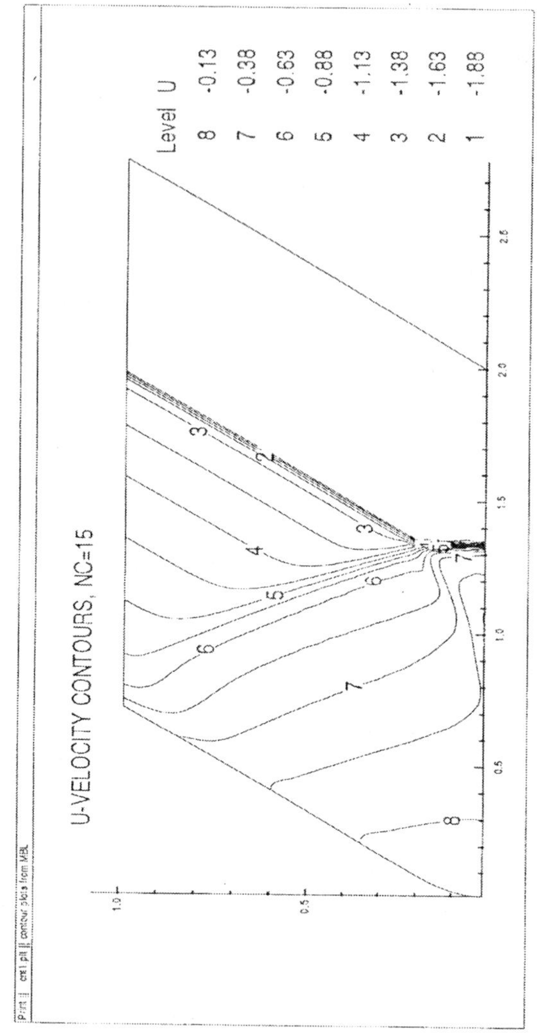

Fig. 10.4.1: A shock front with a kink has been captured in the numerical solution of an accelerating curved piston problem using full gas dynamic equations in conservation form. The figure represents particle velocity contours (in the direction of the piston motion).

moving along linear rays. We deduced it in section 4.2. In this theory, a caustic appears whenever the initial wavefront is concave and amplitude becomes infinite showing that the small amplitude theory is no longer valid. A shock front fitted in the solution obtained by the above method can not be continued in the caustic region where diffraction becomes important.

The diffraction effect, in the caustic region of the linear theory, takes place on a length scale of the order of $\epsilon^{1/2}$ in a direction transverse to the direction of the rays. To capture the diffraction phenomenon in a nonlinear wave, Hunter (1997) expanded the solution $\mathbf{u}$ of the system of conservation laws for $\mathbf{u} : I\!R^{m+1} \to I\!R^n$ in the form

$$(\mathbf{f}^{(0)}(\mathbf{u}))_t + \sum_{\alpha=1}^{m} (\mathbf{f}^{(\alpha)}(\mathbf{u}))_{x_\alpha} = 0 \tag{10.4.2}$$

$$\mathbf{u} = \epsilon \mathbf{u}_0(t, \mathbf{x}, \theta, \eta) + \epsilon^{3/2}\mathbf{u}_1(t, \mathbf{x}, \theta, \eta) + \epsilon^2 \mathbf{u}_2(t, \mathbf{x}, \theta, \eta) + O(\epsilon^{5/2}) \tag{10.4.3}$$

where θ is related to the phase function ϕ and η is a variable related to a new function $\psi(t, \mathbf{x})$ by

$$\theta = \phi/\epsilon\,, \quad \eta = \psi(t, \mathbf{x})/\epsilon^{3/2} \tag{10.4.4}$$

It turns out that ϕ is the linear phase function and ψ is constant along the linear rays, confirming that the diffraction effects are captured on a length scale of order $\epsilon^{1/2}$ in the transverse direction. Expressing the amplitude of $\mathbf{u}_0$ in terms of a scalar $\tilde{w}$ and that of $\mathbf{u}_1$ in terms of v and taking ϕ and ψ to be linear functions of $t, x_1, \cdots, x_n$; it is possible to derive the 2-dimensional Burgers' equation

$$\tilde{w}_t + (\frac{1}{2}\tilde{w}^2)_\theta + v_\eta = 0 \tag{10.4.5}$$

$$\tilde{w}_\eta - v_\theta = 0 \tag{10.4.6}$$

The numerical solution of (10.4.5 - 6) by Hunter and Brio (2000) does show both the fish tail pattern of the very weak shock (see Hunter (1997) figures on pp. 242-3) and the nonlinear Mach stem pattern for a moderately weak shock (Hunter (1997), Figures on pp. 240-1) as observed by Sturtevant and Kulkarni (1976). These are important and local solutions but it will be difficult to derive these local results as a part of the global solution starting from a concave shock front, as done in this chapter. What is important to note is that if we look at the shock front in (x, y)-plane on a length scale of the order of $1/\epsilon$, then the finer structure containing the continuously turning tangent of the figures on pages 237, 239 and 240 will reduce to a point – the leading shock will become a shock with a kink. Our theory does not give the flow behind the shock and hence the Mach stem or rapid compression region behind the two branches of the main shock can not be obtained by our theory. Thus, we suggest that the Mach stem pattern solution of the two-dimensional Burgers' equation gives the structure of a kink on a shock front. Of course, the two-dimensional Burgers' equation has a limitation on its own validity for kinks which join the shocks on its two sides at angles which are not small. The two-dimensional Burgers' equation (10.4.5 - 6) has been derived assuming that ϕ=0 represents a plane shock. It will be interesting to study the original two-dimensional Burgers equation without an assumption that ϕ is a linear function of x_1 and x_2.

10.4.3 Comparison with weakly nonlinear ray theory

Now we discuss a comparison of the results of Prasad and Sangeeta (1999) using the WNLRT and the shock ray theory presented in the previous sections. Unlike the equation (10.2.6), the transport equation for the WNLRT is homogeneous (i.e., without a term like $(M-1)N$) and the second transport equation (10.2.7) is absent. The shapes of the converging nonlinear wavefronts and initially sinusoidal or periodic wavefronts are almost the same as those obtained in this paper. Looking at these shapes alone, (and also those obtained by Kevlahan (1996)), it is not possible to say more about the asymptotic results as $t \to \infty$. The computation with WNLRT also shows that in the periodic case, not only does $\lim_{t\to\infty}\left(m_{\substack{\max\\ \xi\in I\!R}}(t) - m_{\substack{\min\\ \xi\in I\!R}}(t)\right)$ tend to zero, as seen here, but also that $m_{max}(t)$ and $m_{min}(t)$ both approach constant values greater than 1. Thus for a periodic wavefront, m approaches a constant value uniformly as $t \to \infty$. The work of Glimm and Lax (1970) shows that this asymptotic value is reached as $1/t$. The numerical results in this paper show that the Mach number M always decays to 1 (for N >1) and Θ tends to a constant as $t \to \infty$. For an initially periodic shock, Θ tends to *zero* (i.e., the shock front tends to become plane) and in the case of the initially concave shock front, Θ tends to *zero* on the central disc and a constant value on the wings. This implies that on a fixed ray (ξ = constant), for large t, we may set $\Theta_\xi=0$ in (10.2.6) and (10.2.7). This gives

$$M_t + N(M-1) = 0 \tag{10.4.7}$$

and

$$N_t + 2N^2 = 0 \tag{10.4.8}$$

Solving these two equations, we get the usual law of decay of shocks in one-dimensional Burgers' equation

$$M - 1 = \frac{M^* - 1}{(2N^*(t-t^*)+1)^{1/2}}, \quad N = \frac{N^*}{2N^*(t-t^*)+1} \tag{10.4.9}$$

where M^* and N^* are the values of M and N at $t = t^*$ and it is assumed that this ray does not pass through a kink during the time interval (t, t^*). This shows that the shock strength $M-1$ tends to *zero* as $O(\frac{1}{t^{1/2}})$ and N tends to zero as $O(\frac{1}{t})$. This is a major difference from the results of WNLRT. A comparison of the results obtained from NTSD and WNLRT is presented in Figs. 10.4.2 and 10.4.3.

10.4.4 Comparison with Whitham's theory

Whitham's shock dynamics (1957, 1959) is very simple from the point of view of applications. Even though the theory does not properly take into account the interaction of the shock with the flow behind it, it has provided good results for the successive positions of a shock and the kinks in some cases (Henshaw, Smyth and Schwendeman (1986), Schwendeman (1988)) especially for strong shocks. For a weak shock, the equations of Whitham's shock dynamics expressed in terms of the Mach number of the shock, are exactly the same as the differential form of the equations of nonlinear ray theory, but the relation of the Mach number M of a shock front and m of a nonlinear wavefront to the perturbation $\tilde{w}$ of WNLRT (or the corresponding value μ in this chapter) are different. We first note that μ and the non-dimensional $\tilde{w}$ are related by

$$\tilde{w}|_{shock} = \mu \tag{10.4.10}$$

The relations between M, m and $\tilde{w}$ (or μ) are

$$M = 1 + \epsilon\frac{\gamma+1}{4}\mu, \qquad m = 1 + \epsilon\frac{\gamma+1}{2}\tilde{w} \tag{10.4.11}$$

For comparison of our results in this chapter with Whitham's shock dynamics and WNLRT, we have taken a few cases of a converging shock and prescribed the same value of w (or μ) on the initial front in all the three cases namely, the NTSD used in this chapter, Whitham's theory and WNLRT. In addition, we shall have to prescribe N for the NTSD. The use of our conservation form of the equations makes the calculation by Whitham's theory very simple.

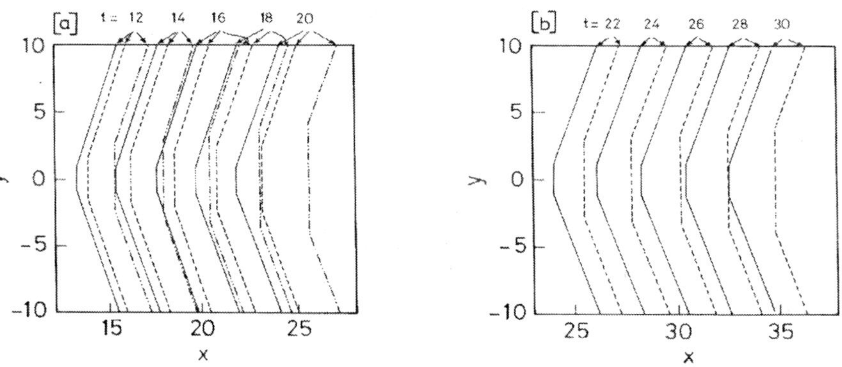

Fig. 10.4.2: Comparison of the NTSD and Whitham's theory
[a] This figure compares the successive positions of the shock fronts and nonlinear wavefronts starting with the same initial front: $y^2 = 8x$, for $|y| < 1$, and the same amplitude distribution on it $(\mu = 1/6) : \alpha = 1.1, \beta = 0.0$.
Continuous curves: shock front by NTSD (N_0=0.15),
$$M_0 = 1 + 0.6\mu$$
Dotted curves: shock front by Whitham's shock dynamics
$$M_0 = 1 + 0.6\mu$$
Broken curves: nonlinear wavefronts (Prasad and Sangeeta (1999))
$$M_0 = 1 + 1.2\mu$$
[b] The comparison of the case in Fig. 6.2[a] continues for times that are greater. The size of the central disc varies considerably and is much larger in the case of Whitham's front than that given by NTSD.

The results show that in early stages, the nonlinear wavefront by WNLRT is a little ahead of the shock front by NTSD and Whitham's theory, but all three differ considerably as t increases. The nonlinear wavefront has moved so far away from the other two that it has not been shown after $t = 20$ in Fig. 10.4.2. It is true that qualitatively, the geometry of the fronts is similar but Whitham's theory gives a bigger shock disc. However, the most striking result seen from Fig.10.4.3 is that the shock strength by Whitham's theory attains a constant value on the disc and the wings, whereas that obtained by the NTSD decreases with time. This causes the former to move ahead of the latter; the separation between the kinks keeps on increasing with time. This, of course, is expected as Whitham's theory does not properly take into account the interaction of the shock with the nonlinear waves which catch up with it from behind.

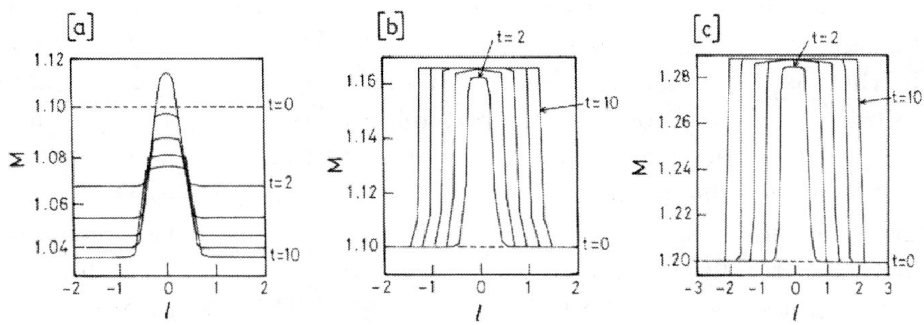

Fig. 10.4.3: Comparison of the Mach number distribution on the fronts (at times $0, 2, 4, \ldots, 10$) with the same amplitude (excess density μ) distribution on the initial front for the case represented in Fig. 10.4.2 with
(a) NTSD (b) Whitham's theory (c) WNLRT
Note: The Mach number at the center of the fronts rises considerably in all three cases but it becomes constant on the central disc for the Whitham's shock front and nonlinear wavefront. As seen in all previous cases with $N_0 > 0$, the Mach number (and hence the wave amplitude) decreases with time on all parts of the shock front by NTSD.

Fig. 10.4.2 shows that at time $t = 30$ the central disc of Whitham's shock has moved about 10% ahead of the distance travelled by the shock calculated by NTSD. Fig. 10.4.3 shows that the Mach number at the center of the fronts rises considerably in all three cases but it becomes constant on the central disc of the Whitham's shock front and another much higher constant on the nonlinear wavefront. As seen in all the previous cases with $N_0 > 0$, the Mach number (and hence the wave amplitude) decreases with time on all shock rays of NTSD. At $t = 20$ the wave amplitude in terms of $M - 1$ on the shock disc by NTSD has become less than half of the maximum value 1.166 attained at the center of the shock by Whitham's theory. The shock disc in Whitham's shock increases in size much faster.

10.5 Corrugational stability and persistence of a kink

Corrugational stability, by which we mean that a plane shock is stable, seems to have been discussed first by Gardner and Kruskal (1964) in the case of a magneto-hydrodynamic shock. Though Whitham (1974) uses his theory of shock dynamics to discuss this problem qualitatively, he deduces important results that *(i)* a perturbation like a single bulge or dent will decay as $t^{-1/2}$ and *(ii)* a perturbation like sinusoidal shape will decay as t^{-1}. Anile and Russo obtained an exact stability criterion in 1986 (see Anile et al., (1993)) for corrugational stability. The WNLRT and the NTSD are ideally suited to discuss the corrugational stability of a nonlinear wavefront and a shock front respectively. In the case of a perturbation of a plane front into a periodic front (like the one discussed in section 10.3), we can use results like those available in Glimm and Lax (1970) to the WNLRT to prove rigorously that the front tends to be plane as t^{-1}. However, the following the method of section 6.2.3 we can show that Glimm–Lax assumption "interaction of two shocks of the same family always produces a shock of the same family plus a rarefaction wave of the opposite family" is not satisfied even in the simpler system of the pair of conservation laws of the WNLRT. In spite of this, extensive numerical computation by Prasad and Sangeeta (1999) with WNLRT shows that a plane nonlinear wavefront is stable. Computations in this chapter indicate corrugational stability of a plane shock front. Here we have some additional results with dents and bulges on plane shock fronts – all these results point to the corrugational stability.

The corrugational stability is a result of genuine nonlinearity in the characteristic fields with characteristic velocities $\pm\left(\frac{M-1}{2G^2}\right)^{1/2}$ of the system (10.2.5 - 7). The shocks in (ξ, t)-plane, which are mapped into kinks, seem to cause dissipation of the *kinetic energy*. Further mathematical investigation in this dissipation is required. The equations of WNLRT are homogeneous but one of the equations of the NTSD – namely the equation (10.2.13) has a source term. For $N > 0$, as seen from (10.4.9), this additional term causes the shock amplitude $M-1$ to tend to zero. This is typical of a plane shock in gas dynamics (as can be seen from the results on model equation $u_t + \left(\frac{1}{2}u^2\right)_x = 0$ in chapter 7). Combining both these features, we interpret our results as *the genuine nonlinearities in the two characteristic fields cause decay of the shock amplitude and perturbation in its geometrical shape leading to corrugational stability of a shock but, in addition to this, the non-homogeneous term in the equation* (10.2.12) *causes the shock strength to tend to zero when* $N > 0$. When both these effects are included, the law of decay as pointed out by Whitham may have to be modified. We need further investigation on this.

The persistence of a kink is an interesting phenomenon, which we observe from the long time computation. As we have seen in all the cases considered by us, a kink may appear on an initially smooth shock front, but once it is formed it persists for all the time until it meets another kink. The persistence of a kink follows from the similar property of a shock in a genuinely nonlinear characteristic field: *a shock, once formed can not terminate at a finite distance in* (ξ, t)*-plane.* A proof of this statement is available in section 1.7. We further notice that the interaction of a pair of shocks (whether the shocks belong to the same characteristic field or different characteristic fields) always produces another pair of shocks (Baskar and Prasad, - in preparation). Thus, the number of kinks on the shock front always remains the same after some time has elapsed and when no new kinks are being produced.

References

1. Airy, G B (1845), Tides and waves, article 192, Encyc. Metrop. (Lamb, H (1932) *Hydrodynamics*, p 259-62 Cambridge University Press, Cambridge).

2. Arnold, V I (1993), *Singularities of caustics and wavefronts*, Kluwer Academic Publishers, Dordrecht.

3. Anile, M A, Hunter, J K, Pantano, P and Russo, G (1993), *Ray Methods for Nonlinear Waves in Fluids and Plasmas*, Longman, Essex.

4. Anile, M A and Russo, G (1986), Generalised wavefront expansion I: higher order corrections for the propagation of a weak shock wave, *Wave Motion*, **8**, 243-58.

5. Anile, M A and Russo, G (1988), Generalised wavefront expansion II: the propagation of step shocks, *Wave Motion*, **10**, 3-18.

6. Baskar, S, Potdar, N and Szeftel, J (1999), Riemann problem for the equations of a weakly nonlinear ray thoery, private communication.

7. Baskar, S and Prasad, P, Riemann problem for kinematical conservation laws and geometrical features of nonlinear wavefronts, in preparation.

8. Bhatnagar, P L and Prasad, P (1970), Study of self-similar and steady flows near singularities, Part I, *Proc. Roy. Soc.*, London, **A315**, 569-84.

9. Bhatnagar, P L and Prasad, P (1971), Study of the self-similar and steady flows near singularities, Part II: a case of multiple characteristic velocity, *Proc. Roy. Soc.*, London, **A 322**, 45-62.

10. Bhatnagar, P L (1979), *Nonlinear Waves in One-dimensional Dispersive Systems*, Oxford Mathematical Monograph Series, Oxford University Press, Oxford.

11. Brio, M and Hunter, J K (1922), Mach reflection for two-dimensional Burgers' equation, *Physica D*, **60**, 194-207.

12. Buchal, R N and Keller, J B (1960), Boundary layer problems in diffraction theory, *Comm. Pure Appl. Math.*, **13**, 85-144.

13. Burton, C V (1893), On plane and spherical sound waves of finite amplitude, *Philosophical Magazine*, **35**, 317-333.

14. Choquet-Bruhat, Y (1969), Ondes asymptotiques et approchees pour de systems d'equations aux derive′es partialles nonline′aires *J. Math. Pures et Appl.*, **48**, 117-58.

15. Cockburn, B, San, Y-L and Shu, C W (1989), TVB Runge-Kutta local projection discontinuous Galerkin finite element method for conservation laws III: one-dimensional systems, *J. Compu. Phys.*, **84**(1), 90-113.

16. Courant, R and Friedrichs, K O (1948), *Supersonic Flow and Shock Saves,* Interscience Publishers, New York.

17. Courant, R and Hilbert, D (1953), *Methods of Mathematical Physics*, Vol. I, John Wiley & Sons, New York.

18. Courant, R and Hilbert, D (1962), *Methods of mathematical physics*, Vol II : Partial Differential Equations, John Wiley & Sons, New York.

19. Duff, G F D (1960), The Cauchy problem for elastic waves in an anisotropic medium, *Phil. Trans. Roy. Soc.* London, A**252**, 249-73.

20. Epstein, M and Sniatycki, J (1992), Fermat's principle in elastodynamics, *J. of Elasticity*, **27**, 45-56.

21. Evans, L C (1998), *Partial Differential Equations*, American Mathematical Society, Providence, RI.

22. Gȧrding, L (1980), *Singularities in Linear Wave Propagation*, Springer-Verlag, New York.

23. Gardner, C S and Kruskal, M D (1964), Stability of plane magnetohydrodynamic shocks, *Phys. Fluids* **7**, 700-6.

24. Gardner, C S and Morikawa, G K (1960), *Similarity in the asymptotic behaviour of collision-free hydromagnetic waves and water waves*, Courant Inst. Math. Sci. Rep., N. 90-9082, 1.

25. Gel'fand, I M (1962), Some problems in the theory of quasi-linear equations, *Amer. Math. Soc. Translations*, Ser. **2**, 29, 295-381.

26. Gel'fand, I M and Shilov, G E (1964), *Generalized Functions*, Vol.I: *Properties and Operations*, Academic Press, New York.

27. Giles, M, Prasad, P and Ravindran, R (1995), *Conservation form of equations of three dimensional front propagation*, Technical Report, Department of Mathematics, Indian Institute of Science, Bangalore.

28. Glimm, H and Lax, P D (1970), *Decay of solutions of systems of nonlinear hyperbolic conservation laws*, Amer. Math. Soc. Memoir, No. 101, American Mathematical Society, Providence, RI.

29. Grinfel'd, M A (1978), Ray method for calculating the wave-front intensity in nonlinear elastic material, *PMM J. Appl. Math. and Mech.*, **42**, 958-77.

30. Gubkin, K E (1958), Propagation of discontinuities in sound waves, *PMM J. Appl. Math. and Mech.*, **22**, 787-93.

31. Guderley, G (1942), Strake Kugelige und zylindrische Verdichtungs stosse inder Nahe des Kugelmittelpunktes bzw. der Zylinderachse, *Luftfahrtforschung* 19, No. 9.

32. Goursat, E (1917), *A Course in Mathematical Analysis*, Vol II, Part two, Differential Equations, republished in 1959 by Dover, New York.

33. Harten, A (1983), High resolution schemes for hyperbolic conservation laws, *J. Comput. Phys.* **49**, 357-93.

34. Henshaw, W D, Smyth, N F and Schwendeman, D W (1986), Numerical shock propagation using geometrical shock dynamics, *J. Fluid Mech.*, **171**, 519-45.

35. Hunter, C (1963), Similarity solutions for the flow into a cavity, *J. Fluid Mech.* **15**, 289-305.

36. Hunter, J K (1995), *Asymptotic equations for nonlinear hyperbolic waves*, Surveys in Applied Mathematics, Vol II, Keller, J.B., McLaughlin, W and Papanicolaou, C., Eds., Plenum Press, New York.

37. Hunter, J K (1997), Nonlinear wave diffraction, *Geometrical Optics and Related Topics*, Colombini, F. and Lerner, N., Eds., Berkhäuser, 221-43, Boston.

38. Hunter, J K and Brio, M (2000), Weak shock reflection, *J. Fluid Mech.*, **410**, 235-61.

39. Jeffrey, A (1976), *Quasilinear Hyperbolic Systems and Waves*, Pitman Publishing, London.

40. John, F (1982), *Partial Differential Equations*, 4th Edit. Springer-Verlag, New York.

41. Kantrowitz, A R (1947), The formation and stability of normal shock waves in channel flow, National Advisory Committee for Aeronautics, USA, Technical Note Number 1225.

42. Keller, J B (1954), Geometrical acoustics I: the theory of weak shock waves, *J. Appl. Phys.*, **25**, 938-47.

43. Kevlahan, N K R (1994), *Structure and shocks in turbulence.* Ph.D. Thesis, University of Cambridge, U.K.

44. Kevlahan, N K R (1996), The propagation of weak shocks in non-uniform flow, *J. Fluid Mech.* **327**, 167-97.

45. Kluwick, A von (1971), Zur Ausbreitung schwacher stoesse in dreidimensionalen instationaeren stroemungen, *Z Angew. Math. Mech.*, **51**, 225-32.

46. Kulikovskii, A G and Slobodkina, F A (1967), Equilibrium of arbitrary steady flows at transonic points, *PMM J. Appl. Math. Mech.*, **21**, 623-30.

47. Kuo, Y H (1951), On the stability of two dimensional smooth transonic flows, *J. Aero. Sci.*, **18**, 1-6, 54.

48. Kovner, I (1990), Fermat principle in arbitrary gravitational fields, *Astrophysical J.*, **351** 114-20.

49. Lang, H de (1992), Christiaann Huygens, originator of wave optics, in *Huygens' Principle 1690-1990: Theory and Applications*, Block, H, Ferwerda, H A and Kuinken, H K Eds., North-Holland, Amsterdam.

50. Lax, P D (1957), Hyperbolic system of conservation laws II, *Comm. Pure Appl. Math.*, **10**, 537-66.

51. Lax, P D (1963), *Lectures on hyperbolic partial differential equations*, Stanford University, Stanford.

52. Lazarev, M P, Prasad, P and Singh, S K (1995), Approximate solution of the one-dimensional piston problem, *Z. Angew. Math. Phys.*, **46**, 752-71.

53. Lazarev, M P, Ravindran, R and Prasad, P (1998), Shock propagation in gas dynamics: explicit form of higher order compatability conditions, *Acta Mechanica*, **126**, 139-55.

54. LeVeque, R J and Yee, H C (1990), A study of numerical methods for hyperbolic conservation laws with stiff source term, *J. Compu. Phys.*, **86**, 187-210.

55. Lighthill, M J (1949), A technique for rendering approximate solutions to physical problems uniformly valid, *Phil. Mag.*, **44**, 1179-1201.

56. Lighthill, M J (1954), Viscosity effects in sound waves of finite amplitude, *Surveys in Mechanics*, Batchelor, G K and Davis, R M Eds., Cambridge University Press, Cambridge, 250-351.

57. Ludford, G S S (1951), The classification of one-dimensional flows and the general shcok problem of a compressible, viscous, heat-conducting fluid, *J. Aeronaut. Sci.*, **18**, 830-34.

58. Ludwig, D (1966), Uniform asymptotic expansions at a caustic, *Commun. Pure Appl. Math.*, **19**, 215-50.

59. Luckacova-Medvidova, M, Morton K W and Warnecke G (2000), Evolution Galerkin methods for hyperbolic systems in two space dimensions, *Mathematics of Computation*, **69**, 1355-1384.

60. Maslov, V P (1980), Propagation of shock waves in an isentropic non-viscous gas, *J. Sov. Math.*, **13**, 119-163 (Russian publication 1978).

61. Monica, A (1999), *Propagation of a Curved Weak Shock Front*, Ph.D. Thesis, Indian Institute of Science, Bangalore.

62. Monica, A and Prasad, P (2001) Propagation of a curved weak shock front, revised version under consideration by *J. Fluid Mech.*, **434**, 119-151.

63. Morawetz, C S (1964), Non-existence of transonic flow past a profile, *Comm. Pure. Appl. Math.*, **17**, 357-67.

64. Morton, K W, Prasad, P and Ravindran, R (1992), *Conservation form of nonlinear ray equations*, Technical report 2. Department of Mathematics, Indian Institute of Science, Bangalore.

65. Nieuwland, G Y (1966), *Theoretical design of shock free transonic flow around aerofoil sections*, Aerosp. Proc, 5B Congress Int. Counc. Aerosp. Sci.

66. Nieuwland, G Y and Spee, B M (1968), *Transonic shock free flow, fact or fiction?* AGARD Specialists' Meeting on Transonic Flows (Paris), NLR MP 68004.

67. Nityananda, R and Samuel, J (1992), Fermat's principle in general relativity, *Phys. Review D.*, **45**, 3862-64.

68. Pack, D C (1960), A note on the breakdown of continuity in the motion of a compressible fluid, *J. Fluid Mech.*, **8**, 103-8.

69. Parker, D F (1969), Nonlinearity, relaxation and diffusion in acoustic and ultrasonics, *J. Fluid Mech.*, **39**, 793-815.

70. Parker, D F (1971), An asymptotic theory for oscillatory nonlinear signals, *J. Inst. Math. Appl.*, **7**, 92-110.

71. Petrovsky, I G (1954), *Lectures on Partial Differential Equations*, Wiley-Interscience, New York.

72. Prasad, P (1969), A class of one dimensional steady state flows and general shock structure problem in radiation-gas-dynamics, *Quart. J. Mech. Appl. Math.*, **22**, 333-53.

73. Prasad, P (1973), Nonlinear wave propagation on a steady transonic flow, *J. Fluid Mech.*, **57**, 721-37.

74. Prasad, P (1975), Approximation of perturbation equations in a quasilinear hyperbolic system in the neighbourhood of a bicharacteristic, *J. Math. Anal. Appl.*, **50**, 470-82.

75. Prasad, P (1977), *Introduction to hyperbolic partial differential equations and nonlinear waves,* Lecture Note, Department of Applied Mathematics, Indian Institute of Science, Bangalore.

76. Prasad, P (1982), Kinematics of a multi-dimensional shock of arbitrary strength in an ideal gas, *Acta Mechanica*, **45**, 163-76.

77. Prasad, P (1987), Extension of Huyghen's construction of a wavefront to a nonlinear wavefront and a shock front, *Curr. Sci.*, **56**, 50-54.

78. Prasad, P (1990), On shock dynamics, *Proc. Indian Acad. Sci.*, Mathematical Sci., **100**, 87-92.

79. Prasad, P (1993), *Propagation of a Curved Shock and Nonlinear Ray Theory*, Pitman Researches in Mathematics Series, No. 292, Longman, Essex.

80. Prasad, P (1994), A nonlinear ray theory, *Wave Motion*, **20**, 21-31.

81. Prasad, P (1995), Formation and propagation of singularities on a nonlinear wavefront and a shockfront, *J. of Indian Inst. of Sci.*, **75**, 517-35.

82. Prasad, P (1997), Nonlinearity, Conservation Law and Shocks, Part I: Genuine Nonlinearity and Discontinuous Solutions, *RESONANCE- Journal of Science Education*, Indian Academy of Sciences, Bangalore, **2**, 2, 8-18.
Nonlinearity, Conservation Law and Shocks, Part II: Stability

Consideration and Examples, *RESONANCE- Journal of Science Education*, Indian Academy of Sciences, Bangalore, 1997, **2**, No.7, 8-19.

83. Prasad, P (2000), An asymptotic derivation of weakly nonlinear ray theory, *Proc. Indian Acad. Sci. Math. Sci.*, **110**, 4, 431-47.

84. Prasad, P and Krishnan, E V (1977), Nonlinear wave propagation in a two-dimensional steady transonic flow, *J. Fluid Mech.*, **82**, 17-28.

85. Prasad, P and Ravindran, R (1977), A theory of nonlinear waves in multi-dimensions with special reference to surface-water-waves, *J. Inst. Math. Appl.*, **20**, 9-20.

86. Prasad, P and Ravindran, R (1985), *Partial Differential Equations*, Wiley Eastern, Delhi and John Wiley & Sons, New York.

87. Prasad, P and Ravindran, R (1990), A new theory of shock dynamics, Part - II numerical solution, *Appl. Math. Lett.*, **3**, 107-9.

88. Prasad, P, Ravindran, R and Sau, A (1991), On the characteristic rule for shocks *Appl. Math. Lett.*, **4**, 5-8.

89. Prasad, P and Sangeeta, K (1999), Numerical simulation of converging nonlinear wavefronts, *J. Fluid Mech.*, **385**, 1-20.

90. Ramanathan, T M, Prasad, P and Ravindran, R (1984), On the propagation of a weak shock front: theory and application, *Acta Mechanica*, **51**, 167-77.

91. Ramanathan, T M (1985), *Huygens' Method of Construction of Weakly Nonlinear Wavefronts and Shock Fronts*, Ph.D. Thesis, Indian Institute of Science, Bangalore.

92. Rao, P S (1956), Supersonic bangs, I & II, *Aeron. Quart.*, **7**, 21-45; 134-55.

93. Ravindran, R (1979), Trapped waves in the neighbourhood of a sonic type of singularity, *J. Fluid Mech.*, **95**, 465-75.

94. Ravindran, R and Prasad, P (1985), Kinematics of a shock front and resolution of a hyperbolic caustic, *Advances in Nonlinear Waves*, Vol II, 77-99, Pitman Research Notes in Mathematics Series, No **111**, Pitman, London.

95. Ravindran, R and Prasad, P (1990), A new theory of shock dynamics, Part-I: analytic considerations, *Appl. Math. Lett.*, **3**, 77-81.

96. Ravindran, R and Prasad, P (1993), *On infinite system of compatibility conditions along a shock ray*, Quart. J. Appl. Math. Mech. **46,** 131-40.

97. Ravindran, R, Sunder, S and Prasad, P (1994), Long time behaviour of the solution of system of equations from new theory of shock dynamics, *Computers Math. Appl.*, **27**, 91-104.

98. Reddy, A S, Tikekar, V G and Prasad, P (1982), Numerical solution of hyperbolic equations by method of bicharacteristic, *J. Math. Phy. Sci.*, **16**, 575-603.

99. Roy, R and Ravindran, R (1988), A note on the equivalence of the shock manifold equations, *Acta Mechanica*, **73**, 239-44.

100. Sangeeta, K (1996), *Numerical Simulation of Converging Nonlinear Wavefronts*, Ph.D. Thesis, Indian Institute of Science, Bangalore.

101. Schwendeman, D W (1988), Numerical shock propagation in non-uniform media, *J. Fluid Mech.*, **188**, 383-410.

102. Scott-Russell (1844) Report on waves, *Proc. Roy. Soc.* Edinburg, **20**, 319-20.

103. Sedov, L I (1959), *Similarity and Dimensional Methods in Mechanics*, Academic Press, New York.

104. Shu, C W (1987), TVB uniformly higher order schemes for conservation laws, *Math. of Computation*, **49**, 105-21.

105. Singh, S K and Singh, V P (1999), An approximate solution of two-dimensional convex piston problem, *Z. Angew. Math. Phys.*, **50**, 206-21.

106. Smoller, J (1983), *Shock Waves and Reaction-diffusion Equations*, Springer-Verlag, New York.

107. Sommerfeld, A and Runge, J (1911), Application of Vector Calculus to the Fundamentals of Geometrical Optics, *Ann. Phys.*, **35**, 277-98.

108. Spee, B M (1971), Nationaal Lucht-en Ruimtevaartlabotorium, NLRTR 69122 U.

109. Srinivasan, R and Prasad, P (1985), On the propagation of a multi-dimensional shock of arbitrary strength, *Proc. Indian Acad. Sci., Math. Sci.*, **49**, 27-42.

110. Srinivasan, R (1987), *A Mathematical Theory on the Propagation of Multi-dimensional Shock and Nonlinear Waves*, Ph.D. Thesis, Indian Institute of Science, Bangalore.

111. Stanyukovich, K P (1960), *Unsteady Motion of Continuous Media*, Pergamon Press, London.

112. Sturtevant, B and Kulkarni, V A (1976), The focusing of weak shock waves, *J. Fluid Mech.*, **73,** 651-71.

113. Sunder, S, Prasad, P and Ravindran, R (1992), A case study with new theory of shock dynamics, *Appl. Math Lett.*, **5**, 89-92.

114. Tabak, E and Rosales, R R (1994), Weak shock focusing and von-Neumann paradox of oblique shock reflection, *Phys. Fluids* **6**, 1874-92.

115. Tagare, S G and Prasad, P (1970), Instability of the self-similar flow into a cavity, *Proc. Indian Acad. Sci.*, **A71**, 219-24.

116. Tanuity, T (1974), Reductive perturbation methods and far field wave equations, *Prog. Theor. Phys. Suppl.*, **35**.

117. Tanuity, T and Wei, C C (1968), Reductive perturbation method in nonlinear wave propagation I., *J. Phys. Soc. Japan*, **24**, 941-66.

118. Thomas, T Y (1961), *Plastic Flow and Fracture in Solids*, Academic Press, New York.

119. Truesdell, C A and Toupin, R A (1960), *The Classical Field Theories*, Handbuch der Physik, Vol III/1, Springer-Verlag, Berlin.

120. Varley, E and Cumberbatch, E (1965), Nonlinear theory of wave front propagation, *J. Inst. Maths. Applcs*, **1**, 101-12.

121. Vincenti, V G and Traugott, S C (1971), The coupling of radiative transfer and gas motion, *Annu. Rev. Fluid Mech.*, **3**, 89-117.

122. von-Mises, R (1950), On the thickness of a steady shock wave, *J. Aeronaut. Sci.*, **17**, 551-54.

123. Whitham, G B (1956), On the propagation of weak shock waves, *J. Fluid Mech.*, **1**, 290-318.

124. Whitham, G B (1957), A new approach to problems of shock dynamics, Part I, Two dimensional problems, *J. Fluid Mech.*, **2**, 146-71.

125. Whitham, G B (1959), A new approach to shock dyanmics, Part II, Three-dimensional problems, *J. Fluid Mech.*, **5**, 369-86.

126. Whitham, G B (1959a) Some comments on wave propagation and shock wave structure with application to magnetohydrodynamics. *Comm. Pure Appl. Math.*, **12**, 113-58.

127. Whitham, G B (1974), *Linear and Nonlinear Waves*, John Wiley & Sons, New York.

128. Yee, H C, Warming, R F and Harten, A (1985), Implicit total variation diminishing (TVD) schemes for steady state calculations, *J. Comput. Phys.*, **57**, 327-60.

129. Zel'dovich, Ya B and Raizer Yu P (1967), *Physics of Shock Waves and High Temprature Hydrodynamic Phenomena*, Vol. II, Academic Press, New York.

Index

arête 113
bicharacteristic
  curves 56, 69
  lemma 70
Burgers' equation 2, 82, 180
  two-dimensional 317
BKPS theory 177, 180
Cauchy problem 5
caustic 112, 122
characteristic
  equation 53, 63, 69
  conoid 54, 63
  variable 52, 82, 86
conservation law 7, 49
  kinematical 119, 213, 295
corrugational stability 322
CPW theory 145, 146
discontinuity of first kind 127
eikonal equation 145, 147, 158
elementary wave solutions 214
elementary shape 215
  interaction 220
energy
  specific internal 50
  total 50
entropy
  condition 14, 15, 32
  function 29
  of gas 69
equal area rule 41, 45
Euler's equations 49, 68
  conservation laws 50, 68
  Riemann invariants 89, 92
  simple wave 91
Fermat's principle 104, 105, 109
genuine nonlinearity 2, 7, 87
Hadamard's lemma 128
high frequency approximation 97
Hopf's result 38
Huygens' wavefront construction 57, 98, 207, 286
hyperbolic system
  in two independent variables 47
  in multi-dimensions 64, 67
  canonical form 51, 64, 67
jump conditions 9, 137, 142
kinematical conservation laws 119, 213, 295
kink 112, 116, 123
Lax-Oleinik formula 41
Lorenz transformation 56
de Laval nozzle 175
Mean curvature 60
N wave 21
normal conoid 63
NTSD (new theory of shock dynamics) 236, 239, 286
  for a weak shock 288, 290, 293
propagation of discontinuities 60, 74
Rankine-Hugoniot condition 10, 142
ray 56, 69, 102
  coordinate system 118
  tube 59, 111
reducible system 77
Riemann invariant 82, 86
Riemann problem 217
self-propagating - *see* wavefront
self-similar solution 188
  stability 191
shock 4, 15
  compatibility conditions 229, 249, 268, 274
  corrugational stability 322
  dynamics 144, 245, 319
  entropy condition 14, 15, 32
  front 37
  persistence 33, 35
  ray theory 284
  rays 265
  stability condition 14
  structure 26, 27

weak shock ray theory 286
simple wave 77, 80, 84, 86, 94, 211
centered 12, 83, 88
SME (shock manifold equation) 265
for a weak shock 266
singularity
of steady equations 181
sonic type 175, 181, 188
sonic point 174
space-like surface 54, 64
spherical wavefront 99
stability
condition for shocks 14
steady flow near singularity 187
accelerating transonic flow 187
self-similar solutions 188
time-like direction 55, 65
transonic flow 173, 175, 187
controversy 174
wave equation 53
wavefront 97
local determinacy 116
nonlinear 37
self-propagation 37, 57, 115, 286
weak solution 8
WNLRT (weakly nonlinear ray theory) 112, 143, 149, 155, 158, 205
polytropic gas 205
validity 226